Wood Modification

Chemical, Thermal and Other Processes

Wiley Series in Renewable Resources

Series Editor

Christian V. Stevens, Department of Organic Chemistry, Ghent University, Belgium

Titles in the Series

Wood Modification: Chemical, Thermal and Other Processes
Callum A.S. Hill

Forthcoming Titles

Renewables-based Technology – Sustainability Assessment
J. Dewulf

Starch Biology, Structure and Functionality
A. Huber & W. Praznik

Wood Modification
Chemical, Thermal and Other Processes

CALLUM A.S. HILL

School of Agricultural and Forest Sciences, University of Wales, Bangor

John Wiley & Sons, Ltd

Telephone (+44) 1243 779777

Email (for orders and customer service enquiries): cs-books@wiley.co.uk
Visit our Home Page on www.wiley.com

Reprinted with corrections July 2006

Other Wiley Editorial Offices

John Wiley & Sons Inc., 111 River Street, Hoboken, NJ 07030, USA

Jossey-Bass, 989 Market Street, San Francisco, CA 94103-1741, USA

Wiley-VCH Verlag GmbH, Boschstr. 12, D-69469 Weinheim, Germany

John Wiley & Sons Australia Ltd, 42 McDougall Street, Queensland 4064, Australia

John Wiley & Sons (Asia) Pte Ltd, 2 Clementi Loop #02-01, Jin Xing Distripark, Singapore 129809

John Wiley & Sons Canada Ltd, 22 Worcester Road, Etobicoke, Ontario, Canada M9W 1L1

Wiley also publishes its books in a variety of electronic formats. Some content that appears in print may not be available in electronic books.

Library of Congress Cataloging in Publication Data

Hill, Callum A.S.
Wood modification : chemical, thermal and other processes / Callum A.S. Hill.
p. cm. — (Wiley series in renewable resources)
Includes bibliographical references and index.
ISBN-13: 978-0-470-02172-9 (cloth : alk. paper)
ISBN-10: 0-470-02172-1 (cloth : alk. paper)
1. Wood—Preservation. I. Title. II. Series.
TA424.H55 2006
620.1′2—dc22

2005021404

British Library Cataloguing in Publication Data

A catalogue record for this book is available from the British Library

ISBN-10 0-470-02172-1
ISBN-13 978-0-470-02172-9

Typeset in 10/12pt Times by Integra Software Services Pvt. Ltd, Pondicherry, India
Printed and bound in Great Britain by CPI Antony Rowe, Chippenham and Eastbourne
This book is printed on acid-free paper responsibly manufactured from sustainable forestry in which at least two trees are planted for each one used for paper production.

I should like to dedicate this book to my wife Joanne and our children for their support and understanding; and to my late father Alan Bernard Hill and my good friend Cameron Settle, who both made beautiful things from wood.

Contents

Foreword

In this book of the series, a 'more traditional' area of research in the field of renewable resources is discussed. It is indeed a long-standing topic to utilize wood for a vast number of applications; however, it is a very active and changing field of research. The modification of wood needs to be compatible with the increasing demands at the environmental level, the social role of forests and woods' and the technical characteristics needed to create certain materials. This book certainly paves the way to give the professional reader a taste of this area and offers an up-to-date overview of the different kinds of modification and the impact on the durability of the materials.

The success of a book strongly depends on the author, on his approach and his expertise. I was particular happy that Callum Hill took on this project in view of his extensive knowledge in the field. He has succeeded in presenting an overview of the modification of wood in sufficient detail and with great attention to the literature. Thanks' Callum.

This book will be very helpful to scientists in the field of modification of wood, but hopefully also to a variety of people in other disciplines either in the academic world, in industry or in the area of policy.

Christian V. Stevens
Faculty of Bioscience Engineering
Ghent University, Belgium
Series Editor, '*Renewable Resources*'
June 2005

Series Preface

Renewable resources, their use and modification are involved in a multitude of important processes with a major influence on our everyday lives. Applications can be found in the energy sector, chemistry, pharmacy, the textile industry, paints and coatings, to name but a few.

The area interconnects several scientific disciplines (agriculture, biochemistry, chemistry, technology, environmental sciences, forestry, etc.), which makes it very difficult to have an expert view on the complicated interaction. Therefore, the idea of creating a series of scientific books, focusing on specific topics concerning renewable resources, has been very opportune and can help to clarify some of the underlying connections in this area.

In a very fast changing world, trends are not only characteristic for fashion and political standpoints; also, science is not free from hypes and buzzwords. The use of renewable resources is again more important nowadays; however, it is not part of a hype or a fashion. As the lively discussions among scientists continue about for how many years we will still be able to use fossil fuels, opinions ranging from 50 to 500 years, they do agree that the reserve is limited and that it is essential not only to search for new energy carriers but also for new material sources.

In this respect, renewable resources are a crucial area in the search for alternatives for fossil-based raw materials and energy. In the field of energy supply, biomass and renewable-based resources will be part of the solution, alongside other alternatives such as solar energy, wind energy, hydraulic power, hydrogen technology and nuclear energy.

In the field of material sciences, the impact of renewable resources will probably be even bigger. Integral utilization of crops and the use of waste streams in certain industries will grow in importance, leading to a more sustainable way of producing materials.

Although our society was much more (almost exclusively) based on renewable resources centuries ago, this disappeared in the Western world in the 19th century. Now it is time to focus again on this field of research. However, this should not mean a 'retour à la nature', but it should be a multidisciplinary effort on a highly technological level to perform research towards new opportunities, to develop new crops and products from renewable resources. This will be essential to guarantee a level of comfort for a growing number of people living on our planet. It is 'the' challenge for the coming generations of scientists to develop more sustainable ways to create prosperity and to fight poverty and hunger in the world. A global approach is certainly favoured.

This challenge can only be dealt with if scientists are attracted to this area and are recognized for their efforts in this interdisciplinary field. It is therefore also essential that consumers recognize the fate of renewable resources in a number of products.

Furthermore, scientists do need to communicate and discuss the relevance of their work. The use and modification of renewable resources may not follow the path of the genetic engineering concept in view of consumer acceptance in Europe. Related to this aspect, the series will certainly help to increase the visibility of the importance of renewable resources.

Being convinced of the value of the renewables approach for the industrial world, as well as for developing countries, I was myself delighted to collaborate on this series of books focusing on different aspects of renewable resources. I hope that readers become aware of the complexity, the interaction and interconnections, and the challenges of this field and that they will help to communicate on the importance of renewable resources.

I certainly want to thank the people at Wiley from the Chichester office, especially David Hughes, Jenny Cossham and Lyn Roberts, in seeing the need for such a series of books on renewable resources, for initiating and supporting it, and for helping to carry the project through to the end.

Last but not least, I want to thank my family, especially my wife Hilde and my children Paulien and Pieter-Jan, for their patience and for giving me the time to work on the series when other activities seemed to be more inviting.

Christian V. Stevens
Faculty of Bioscience Engineering
Ghent University, Belgium
Series Editor, '*Renewable Resources*'
June 2005

Preface

The idea for writing this book came about through discussions with Chris Stevens at the University of Ghent in Belgium. It was considered that this was a good time to write such a book, in view of the rapid developments in the subject over the past decade and, in particular, because of the progress made in commercialization of technologies in this area. The task of writing this book has been immense, but very enjoyable. Although the subject matter of this book is concerned with wood modification, many of the processes described are generic and could be applied to other lignocellulosic materials. In writing this book, I have tried to do justice to the area and cover all the topics in depth. However, as with any textbook, it is necessary to work within editorial constraints. Hence, although the bulk of this subject is covered, it is inevitable that some areas are covered in greater detail than others. In deciding which areas to concentrate upon, I have taken account of other textbooks that have dealt with some of these topics in depth, and made reference to these where appropriate.

This area is undergoing huge developments at the present time, driven in part by environmental concerns regarding the use of wood treated with certain preservatives. There has been considerable commercial interest shown in wood modification over the past decade, with products based upon thermal modification and furfurylation now being actively marketed. The next few years will see the commercialization of acetylation and impregnation modification. This is a new industry, but one that has enormous potential.

The technologies that have been developed for wood modification have required the combined efforts of many hundreds of researchers worldwide. However, it would be impossible to study the literature in this area without encountering the pioneering work of Alfred J. Stamm of the Forest Products Laboratory in Madison, Wisconsin, USA. Nearly every area of wood modification can trace its origins back to the seminal work of Stamm. All those who work in wood modification owe him a great debt.

I hope that this book proves useful to all those with an interest in wood modification. Finally, I should like to say that although I have made every effort to ensure that this book is error-free, any mistakes that may have crept through the very thorough review procedure are entirely my responsibility.

Callum A.S. Hill

Preface

List of Abbreviations

ASE	Anti-shrink efficiency
ASTM	American Society for Testing and Materials
ATR	Attenuated total reflection infrared specroscopy
AWPA	American Wood Preservers Association
BC	Bulking coefficient
BET	Brunauer–Emmett–Teller
BPD	Biocidal Products Directive
BPO	Benzoyl peroxide
CCA	Copper chrome arsenic
CP-MAS	Cross-polarized magic angle spinning
CTMP	Chemithermomechanical pulp
DHBP	Dihydroxy benzophenone
DIPCI	Diisopropylcarbodiimide
DMDHEU	Dimethylol dihydroxy ethylene urea
DMF	Dimethyl formamide
DTA	Differential thermal analysis
DP	Degree of polymerization
E'	Dynamic Young's modulus
EDXA	Electron diffuse X-ray analysis
EELS	Electron energy loss spectroscopy
EETMS	β-(3,4, Epoxycyclohexyl)trimethoxysilane
EMC	Equilibrium moisture content
EMC_R	Reduced equilibrium moisture content
ESR	Electron spin resonance
EU	European Union
EW	Earlywood
FA	Furfuryl alcohol
FAO	Food and Agriculture Organisation of the United Nations
FTIR	Fourier transform infrared
FSP	Fibre saturation point
GAI	Gross annual increment
GC-MS	Gas chromatography mass spectrometry
GMA	Glycidyl methacrylate

GPa	Giga pascal
Gt	Giga tonne
ha	Hectare
HDI	1,6-Diisocyanatohexane
HEBP	2-Hydroxy-4-(2, 3-epoxypropoxy)benzophenone
HEMA	2-Hydroxymethylmethacrylate
H–H	Hailwood–Horrobin
HMDA	Hexamethylene diamine
HMDSO	Hexamethyldisiloxane
HPLC	High-pressure liquid chromatography
IBS	Internal bond strength
IR	Infrared
IPCC	Intergovernmental Panel on Climate Change
IPTES	3-Isocyanatopropyltriethoxysilane
IS	Isocyanate
ISS	Interfacial shear strength
JWPA	Japanese Wood Preserving Association
LCA	Life cycle analysis (assessment)
LMWDA	Low molecular weight diffusible agent
LOI	Limiting oxygen index
LVL	Laminated veneer lumber
LW	Latewood
MA	Maleic anhydride
MAPP	Maleic anhydride polypropylene
MC	Moisture content
MDF	Medium density fibreboard
MDI	Diphenylmethane diisocyanate
MEE	Moisture exclusion efficiency
MF	Melamine formaldehyde
MG	Maleic acid glycerol
ML	Middle lamella
MMA	Methylmethacrylate
MMF	Methylolated melamine formaldehyde
MOE	Modulus of elasticity
MOR	Modulus of rupture
MPa	Mega pascal
MTES	Methyltriethoxysilane
MTMS	Methyltrimethoxysilane
MUF	Melamine urea formaldehyde
MPF	Melamine phenol formaldehyde
MW	Molecular weight
NAI	Net annual increment
NMA	*N*-methylolacrylamide
NMR	Nuclear magnetic resonance
OD	Oven dry
OH	Hydroxyl

OSB	Oriented strand board
PA	Phthalic anhydride
PDMSO	Poly(dimethylsiloxane)
PEG	Poly(ethyleneglycol)
PF	Phenol formaldehyde
PGMA	Poly(glycerol methacrylate)
PhNCO	Phenyl isocyanate
PMPPIC	Poly(methylene) poly(phenylene) isocyanate
PP	Poly(propylene)
PRF	Phenol resorcinol formaldehyde
psi	Pounds per square inch
PTES	Propyltriethoxysilane
PTMS	*n*-Propyltrimethoxysilane
PTSC	*p*-Toluene sulphonyl chloride
PVAc	Poly(vinylacetate)
PVC	Poly(vinylchloride)
RH	Relative humidity
S %	Percentage swelling coefficient
SA	Succinic anhydride
SEM	Scanning electron microscopy
SEM-EDXA	Scanning electron microscopy electron diffuse X-ray analysis
SN_2	Bimolecular nucleophilic substitution
TBTO	Tributyl tin oxide
TDI	Toluene diisocyanate
TEA	Triethylamine
TEM	Transmission electron microscopy
TES	Tetraethoxysilane
TFA	Trifluoro acetic acid
TGA	Thermogravimetric analysis
TMS	Tetramethoxysilane
TMP	Thermomechanical pulp
TMPS	γ-Methacryloxypropyltrimethoxysilane
TPS	Tetrapropoxysilane
USGS	United States Geological Survey
UV	Ultraviolet
VC	Volume change
VOC	Volatile organic compound
VTMS	Vinyltrimethoxysilane
WPG	Weight percentage gain
WS	Water soak
XPS	X-ray photoelectron spectroscopy

1

The Use of Timber in the Twenty-first Century

1.1 Introduction

There is an increasing need to develop technologies in which renewable materials are used as direct replacements for nonrenewables. Our current rates of consumption of nonrenewables are high and in most cases increasing, but the reserves from which they are obtained are finite and exhaustible. Our present patterns of consumption are not sustainable in the long term. Although this problem appears to be unique to the 21st century, these concerns are not new. The idea that resource scarcity could act as a constraint upon economic development can be traced back to the writings of Thomas Malthus, who showed that expanding populations will outstrip their food supplies. The concepts that he discussed apply equally to all finite resources. In the event, advances in technology have tended to compensate for resource scarcity, but this process cannot continue indefinitely.

It is only comparatively recently that we have become acutely aware of the need to utilize resources in a sustainable manner. The concept of sustainability began to receive attention during the 1970s and was first formalized internationally in the World Conservation Strategy of 1980. The initial concepts were taken from the idea of *sustainable yield*, as applied in forestry and agriculture. This is defined as the amount of crop that can be harvested without compromising the capacity of future harvests to produce an equal crop. The level of consumption of a resource to support an activity should meet the needs of the present, whilst ensuring that sufficient resources are available to meet the needs of the future. The concepts of sustainability include social, economic and environmental factors. All three must be taken into account if the absolute sustainability of a process is to be determined.

As the study of the interactions between the environment and economic processes has developed, there has been increasing emphasis placed upon analyses of materials/energy

Wood Modification: Chemical, Thermal and Other Processes C. Hill

flows within economies. This has led to the development of the subject of biophysical economics, which views the economy in terms of flows of energy and materials within ecosystem processes. As a consequence, thermodynamic principles have become closely involved in the construction of biophysical economic models. The incorporation of environmental considerations into economics will be an important factor in improving the competitiveness of renewable materials.

1.2 Nonrenewables: a Finite and Exhaustible Resource

The consumption of nonrenewable materials tends to exhibit a classic relationship, which was first demonstrated by M. King Hubbert in an analysis of oil production from the 48 contiguous states of the USA. The shape of this 'Hubbert curve' is similar to, although not identical with, a normal distribution (Figure 1.1). With some variation, this pattern of production/consumption is exhibited for all nonrenewables. The finite life span of nonrenewable reserves has been commented upon by many workers in the area.

It is important to distinguish between the commonly used terms, resource, reserve base and reserve:

- A resource is the amount of material that is known to exist plus the quantity that is thought to exist.
- A reserve base is the amount of material in the resource that meets certain physical criteria for extraction.
- A reserve is the amount of material in a reserve base that can be economically extracted at the present time.

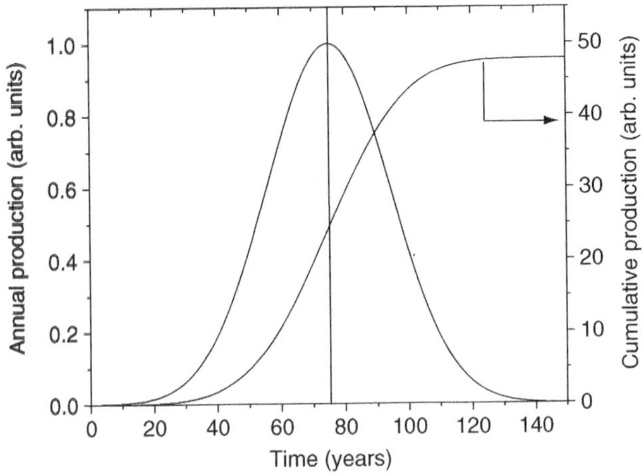

Figure 1.1 An example of a Hubbert curve, representing an idealized history of resource extraction. The amount of material extracted in a year is represented by the bell-shaped curve, whereas the cumulative amount of material extracted is given by the sigmoidal curve.

The extraction of nonrenewable materials follows a common pattern, in that the highest-quality reserves are the first to be extracted, and as these are worked out, progressively lower-quality reserves are then processed. As technology advances and the price of the commodity rises, the reserves increase.

However, although the level of reserves will rise as technology improves and prices increase, eventually there comes a point at which the effort (energy) expended in extracting the material is greater than the advantage gained by using that material. A good example would be the extraction of low-quality oil reserves, where a point is reached at which the energy expended in extracting the oil is greater than the energy obtained from the oil once it is extracted. Another example of this is illustrated by considering the amount of gold existing in seawater, with a total amount of 10 million tonnes being present in the world's oceans. However, this is at such a low concentration (10 parts per trillion) that extraction would be hopelessly uneconomic.

The US Geological Survey (USGS) produces annual mineral commodity summaries, from which it is possible to crudely estimate reserve lifetimes (Figure 1.2).

Calculations of this type are very approximate (obtained by dividing the current reserves by the annual production), but although the figures may be criticized in terms of absolute accuracy, the basic principle that the reserves will eventually be depleted cannot be denied.

The use of nonrenewable resources is characterized by linear mass/energy flows through economic systems. A resource is extracted, processed, utilized and ultimately discarded. Until recently, the final stage of this linear throughput has not been a factor when considering the economics of a process. Physical processes involving chemical transformations are subject to the laws of thermodynamics. An inevitable consequence associated with any chemical transformation is the unavoidable related increase in the entropy of the environment. This results in dissipation, the conversion of high-quality energy and materials into lower-quality forms (a decrease in organization at the atomic/

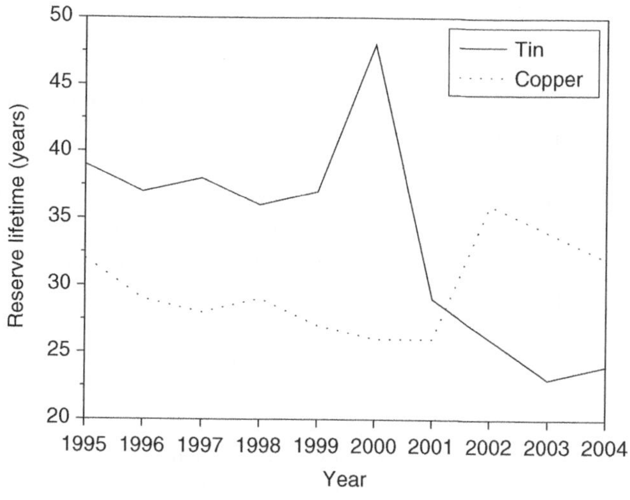

Figure 1.2 *The reserve lifetimes of copper and zinc ores, estimated from the annual data supplied by the US Geological Survey.*

molecular level). A classic example would be the combustion of fossil fuels, where the high-grade heat (that which can do useful work) is emitted into the environment as low-grade heat, and the atoms in the fuel are lost into the environment as high-entropy gases (such as carbon dioxide).

Natural processes are distinguished from the material and energy throughputs of economic processes in two main ways:

- Cyclic materials flows, where the wastes from a metabolic process become the feedstocks for other metabolic processes.
- Assimilative rather than dissipative processes, where atoms are taken from a dissipated state and formed into organized structures.

Of course, natural systems do not operate outside of the laws of thermodynamics. The process of assimilating atoms to form organized structures requires energy, and this energy is derived (with a few exceptions) from the sun, via photosynthetic processes.

The challenge for mankind in the 21st century is to design our industrial processes so that they become integrated with natural metabolic processes. This is why the study of renewable materials is becoming so important.

1.3 Renewable Materials

Mankind has used (and hopefully will continue to use) renewable materials for millennia, but the idea of using renewable materials as industrial feedstocks began to be taken seriously in the 1930s in the USA (Geiser, 2001; Finlay, 2004). George Washington Carver was an early pioneer of this idea, who developed many industrial products derived from peanuts. The idea of using agricultural crops for industrial feedstocks was also promoted by William J. Hale, an organic chemist who published a book called *The Farm Chemurgic* in 1934. This was the start of the chemurgy movement. Hale argued that it was important from a strategic point of view for the USA not to rely upon foreign imports in order to support its domestic chemicals industry. This is an idea that seems just as pertinent now as it did then.

Hale worked with Wheeler McMillen, Thomas Edison, Irenee DuPont and Henry Ford to promote the chemurgy principles. Thomas Edison was interested in using the wild flower goldenrod as a feedstock for rubber production. Henry Ford had a particular interest in producing industrial organic chemicals from soybeans. In 1938, Ford constructed the first of several industrial soy processing plants to make soy oil based enamels for car body paints, using the glycerol by-product in shock absorbers. He supported a great deal of research into crop-based products that could be used for producing car components.

The chemurgy movement grew rapidly during the time of the Great Depression, and the advent of the Second World War resulted in a huge amount of activity to derive as many products as possible from domestic sources, for strategic reasons. However, with the end of the war, the development of many new cheap products by the petrochemicals industry led to the rapid demise of the chemurgy movement.

With renewables, assuming that the biomass resource is obtained in a sustainable manner, there should be a constant supply of materials, although there is a finite limit to the amount of material that can be obtained. Figure 1.3 shows a classic growth curve

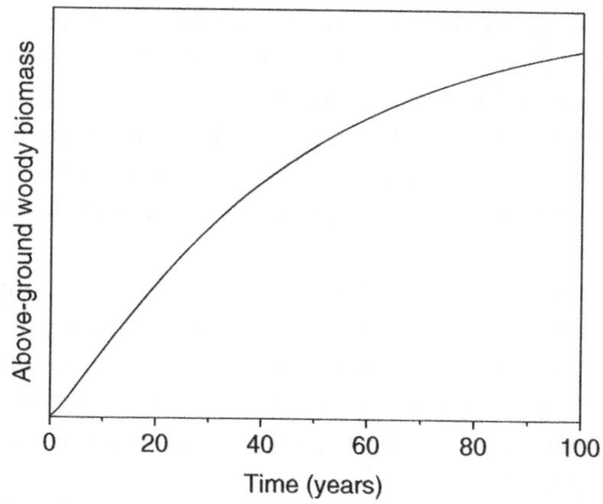

Figure 1.3 *A growth curve showing the increase of biomass in forestry plantation over a 100-year period.*

representing the amount of biomass stored in a crop, which is then available after harvesting to be utilized in products. It should also be noted that, theoretically, there is no downgrading of material quality for a renewable resource.

The biomass in (for example) a forest can be harvested and the sequestered carbon can either be used for energy production (with rapid return of the carbon to the atmosphere) or it can be stored for longer periods in timber products. The example given in Figure 1.4 shows a forest that is clear-felled after 100 years of growth.

The rate of growth of the trees slows as the forest reaches maturity and canopy closure occurs. In addition, the forest eventually establishes equilibrium with the environment, where the rate of carbon sequestration is exactly balanced by the loss of carbon dioxide to the atmosphere due to decay of dead trees and other biomass.

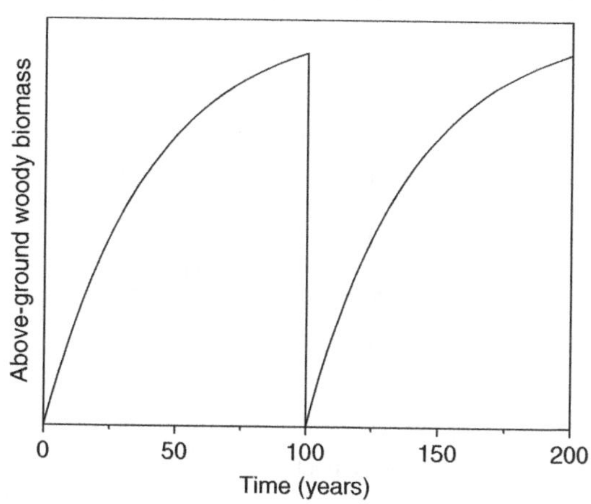

Figure 1.4 *A forest plantation that is clear-felled after 100 years of growth, and subsequent regeneration over the following century.*

The use of wood in long-life products, such as buildings, ensures that this sequestered carbon is held in a materials pool for a longer time. If the use of renewables is encouraged, then more carbon is stored in this manner. Eventually, of course, such systems will establish equilibrium with the environment, as the materials flow into the pool equals the materials flow out into the environment. The use of wood in this way intervenes in a natural cycle, so that wood use and ultimate disposal replaces the natural cycle of wood decay in the forest (Figure 1.5).

The use of forest resources as a feedstock for industrial uses is long established and is, in a sense, superior to the use of agricultural crops, since the supply can be guaranteed well into the future and can be obtained throughout the year, unlike seasonal crops. Although this book is concerned with one small aspect of timber utilization, it should be noted that forest resources can also be used to provide feedstocks for many industrial products, including chemicals.

Timber can be viewed as a classic renewable material. Trees absorb carbon dioxide and utilize water and sunlight to produce a material that can be used in construction, to produce paper or to provide chemical feedstocks, with the production of oxygen as a by-product. Furthermore, at the end of a product life cycle, the material constituents can be combusted, or composted to return the chemical constituents to the 'grand cycles'. In essence, timber use represents a classic example of a cyclic materials flow, mimicking the flows of materials through natural cycles. Provided that we manage our forests well and do not harvest beyond the capacity of the planet to provide timber, we have at our disposal an inexhaustible resource available in perpetuity.

The purpose of this chapter is to briefly consider the environmental credentials of timber utilization and the changes that are now affecting the way in which timber is used. It is not intended to comprehensively cover the topic, which would require an entire book, but it serves to outline the case for timber as a renewable material and to illustrate how environmental considerations are changing the way in which the material is being used. In particular, the use of wood preservatives, which have ensured that this renewable material has continued to remain competitive against nonrenewables, will be discussed. Finally, the importance of wood modification as an emerging technology will be briefly considered.

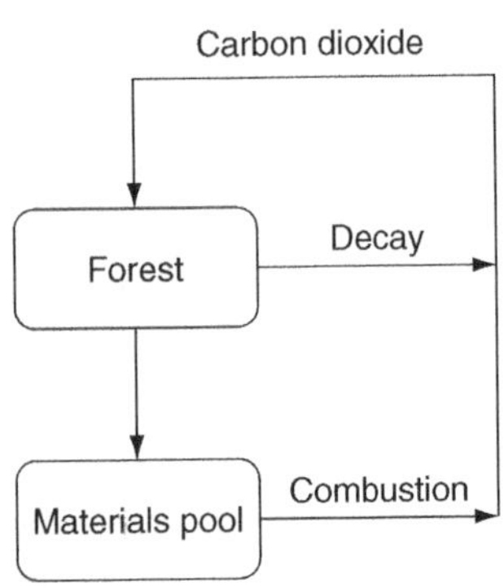

Figure 1.5 The use of wood products stores carbon in a materials pool, eventual disposal returning the sequestered carbon to the grand carbon cycle to provide for growth of new forest.

1.4 The Global Timber Resource

The industrial processing of timber has a long history and provides an excellent case study exemplifying how renewable materials can be processed, utilized and disposed of with minimal environmental impact. It is undeniable that unsustainable harvesting practices were used in the past, resulting in a generally poor perception of the industry by the public. It is also true that, in some parts of the world, harvesting of the timber resource continues to this day with scant regard paid to long-term issues of sustainability. However, there have been significant achievements in addressing the concerns that are often raised. The timber industry is long established, but continues to evolve to meet present and future demands. In response to criticisms, the forestry and timber-processing sector has introduced various certification schemes. The greatest interest in these schemes has been shown by importing countries, especially in Western Europe. According to the Food and Agriculture Organisation of the United Nations (FAO, 2001), the area of certified forests was 90 million ha in 2000, and continues to rise.

Globally, it is estimated that there is 3870 million ha of forests (30 % of the Earth's land area), of which 95 % is natural or semi-natural (Table 1.1). At the present time, in most developed countries, the forest area is increasing and this is likely to continue given the present rates of harvesting. However, in tropical parts of the world, the forest area continues to decline, giving rise to serious concerns. It is estimated that between 1990 and 2000, there was a loss of 9.4 million ha per annum of forest worldwide (an annual deforestation rate of 14.6 million ha and an increase in forest area of 5.2 million ha per year; Table 1.2). The only way to reverse this trend is to place sufficient economic value on forest resources, with incentives to encourage sustainable forest management.

Harvesting operations are moving away from virgin forests, and there is an increasing reliance upon plantations as a source of industrial timber, although the development of such plantations is relatively recent. According to the FAO (2001), half of all established plantations are less than 20 years old. It is predicted that future increases in wood demand will be met largely from plantations.

Table 1.1 *Forest area by region (FAO, 2001)*

Region	Land area (million ha)	Total forests (natural and plantation)			Natural forest (million ha)	Plantation forest (million ha)
		Area (million ha)	% of land area	% of global forests		
Africa	2 978	650	22	17	642	8
Asia	3 085	548	18	14	432	116
Europe	2 260	1 039	46	27	1 007	32
North and Central America	2 137	549	26	14	532	18
Oceania	849	198	23	5	194	3
South America	1 755	886	51	23	875	10
World total	**13 064**	**3 869**	**30**	**100**	**3 682**	**187**

Table 1.2 The annual gross and net change (in million ha) in forest area by region, 1990–2000 (FAO, 2001)

Domain	Deforestation	Increase in forest area	Net change in forest area
Tropics	−14.2	+1.9	−12.3
Nontropics	−0.4	+3.3	+2.9
World	−14.6	+5.2	−9.4

Global production of roundwood was 3335 million m^3 in 1999 (3352 million m^3 in 2000), about 50 % of which was as fuel wood, of which 90 % was consumed in developing countries. Industrial roundwood production (1550 million m^3 in 1999) was dominated by developed countries (79 % of total annual production). This trend will change, in particular with the emergence of China as a major economic force.

1.4.1 Timber and the Carbon Cycle

In recent years, there has been increasing concern over the build-up of greenhouse gases in the atmosphere, particularly CO_2, as a cause of global warming. A number of reports have been issued by the Intergovernmental Panel on Climate Change (IPCC), outlining the scientific basis behind this concern. It is well established that the carbon dioxide derived from anthropogenic emissions arising from the burning of fossil fuels for energy is a major greenhouse gas. Deforestation and land-use change also play a significant role in contributing to the increase in levels of atmospheric carbon dioxide. It is estimated that emissions of CO_2 due to the burning of fossil fuels amount to some 6 gigatonnes (Gt) of carbon per annum, whilst deforestation contributes an additional 2 Gt. It has been calculated that the total amount of carbon emitted up to the year 2000 due to fossil fuel use, plus that due to land-use change, is of the order of 420 Gt of carbon. Projections based upon the work of the IPCC estimate that the amount of carbon that will be emitted in the 21st century range from 690 to 2090 Gt of carbon (Cannell, 2003). However, biological systems also have the capacity to reverse or ameliorate this trend, because the atmospheric carbon can be sequestered in sinks, such as expanding forests. It is estimated that the world's forests presently contain over 4×10^{12} tonnes of biomass (Table 1.3).

Table 1.3 The distribution of above-ground woody biomass by region (FAO, 2001)

Region	Biomass (million tonnes)	%
Africa	70 916	16.8
Asia	45 036	10.7
Europe	61 070	14.5
North and Central America	51 895	12.3
Oceania	12 350	2.9
South America	179 497	42.7
World total	**421 214**	

The process of tree growth utilizes atmospheric carbon in the production of wood biomass. Furthermore, this sequestered carbon can continue to be held in products that are manufactured from wood. Although much research has been done in investigating forests as actual or potential carbon sinks, there has been rather less work looking at the implications of the use of wood products as a medium-term carbon store.

Although forest biomass can be used as a sink for anthropogenic CO_2, as the forest matures, the carbon inventory reaches a plateau as the forest establishes equilibrium. Thus, in order to ensure continued sequestration of atmospheric carbon, it is necessary to harvest the biomass (and replant, or allow for natural regeneration) and utilize this material in long-life products, thereby removing the carbon for longer periods in materials pools. Providing markets for timber ensures the survival of the forestry industry and provides an incentive for further planting. In the EU at present, only 7 % of carbon emissions are offset by growth of wood biomass (Liski *et al.*, 2000). Large-scale afforestation programmes could be used as a means of sequestering much larger amounts of CO_2 than at present, if the political will was present (Booth and Elliott, 1993). If significant afforestation was to be implemented immediately, then an extra 52–104 Gt of carbon could be sequestered in trees over the next 50 years (Cannell, 2003). However, the implementation of forestry programmes to offset anthropogenic carbon emissions can only be viewed as part of an overall strategy to reduce global warming. For example, if *all* of the agricultural land of the EU15 was to be used for forestry, then the amount of carbon sequestered would represent only 20–55 % of the current EU15 emissions (Cannell, 2003). It will clearly not be possible to achieve this, and other measures must also be implemented to reduce CO_2 emissions.

The growth of woody biomass in one year's 'annual increment' represents the quantity of material that can be harvested without affecting the productive capacity of the forest in subsequent years. The gross annual increment (GAI) is the yearly increase in woody biomass, whereas the net annual increment (NAI) is the GAI adjusted for natural losses such as fire, insect damage and so on. The NAI is often referred to as the 'allowable cut'. In boreal and temperate zones, the removal of woody biomass is lower than the NAI, and thus these forests are presently acting as net sinks for carbon dioxide (Figure 1.6). If all of the NAI was harvested, then the forests would no longer act as sinks for CO_2, but would be in balance with the atmosphere.

The material that is harvested can be utilized in products, and when they are disposed of, the sequestered carbon is returned to the atmosphere. This gives rise to the often-quoted property of timber as being 'carbon neutral'. However, this is erroneous, in that harvesting, transport and conversion of timber all result in net carbon emissions, which must be taken into account in carbon balances. But if the timber is used to substitute for products that have higher carbon emissions, then real gains are achieved. The arguments for the utilization of timber are complex and outside the scope of this book, but the following conclusions can be drawn:

- If the forest resource is properly managed, then timber can be harvested indefinitely.
- The use of timber in products represents a means by which atmospheric carbon can be stored in materials pools.
- Extending the life of timber products will result in carbon being stored in a materials pool for longer periods.
- Ultimate disposal of timber products will return the sequestered carbon to the atmosphere, where it is available for continued production of timber.

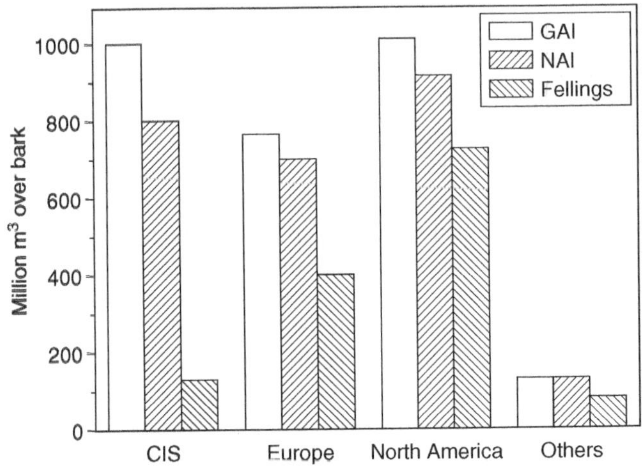

Figure 1.6 The net annual increment and fellings for temperate and boreal zone forested countries.

It is important to emphasize that forestry is a long-term industry. Decisions taken now concerning the way in which we manage our forests will have an effect way into the future, perhaps as much as a century later.

1.5 Timber Production

Wood has been used by mankind for millennia because of its excellent material properties. Although the use of timber in some markets has decreased, the consumption of timber overall continues to rise. Projections have been made until the middle of the 21st century that in most cases show a rise in demand for timber (in all but low economic growth models) and an increase in production (Figure 1.7) (Brooks *et al.*, 1996). There is, however, concern that the supply of timber for industrial purposes may not be able to match demand. For example, Bowyer *et al.* (2003), note that there will be a shortfall in the amount of forest area providing industrial timber by the year 2100, due to the rise in human population during this time (Table 1.4).

This requires the urgent development of new technologies to ensure the more efficient use of the resource by, for example, extending the life of timber-based products.

Increasing quantities of timber are being sourced from managed plantation forests, which are now being established at a rate of over 3 million ha per year. These plantation forests have often been developed to utilize fast grown species (such as *Pinus* and *Eucalyptus*), where yield rather than timber quality is the primary consideration for species choice and management methods. Active management of plantation forests utilizes thinning regimes to maximize yield, sometimes combined with fertilization of the soil to encourage growth. Plantation forests in the temperate zones invariably utilize softwood species chosen for ease of conversion and management.

Fast grown softwood from plantation sources is generally characterized by a high proportion of juvenile wood and often poorly developed heartwood. A fast rate of growth results in wide growth rings, producing low-density timber that exhibits inferior mechanical

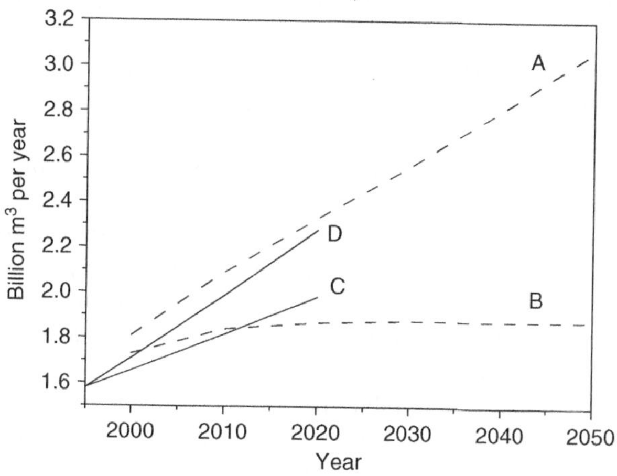

Figure 1.7 *Predictions of growth of timber production and consumption globally until 2050 (Brooks et al., 1996). (A: high demand, B: low demand, C: low production, D: high production).*

Table 1.4 *A comparison of global annual per capita wood consumption for the years 2000 and 2100 (Bowyer et al., 2003)*

Net annual forest growth	1.5 $m^3 ha^{-1}$
Annual per capita wood consumption	0.5 m^3
Forest area required per capita to support consumption	0.4 ha
Forest area available per capita, 2000	0.6 ha
Forest area available per capita, 2100	0.3 ha

properties, when compared with timber sourced from virgin forests. There is also evidence to show that the durability of plantation-grown timber is often inferior to that sourced from natural-growth forests. All of the above factors result in an inferior product compared with the material obtained from virgin forests, requiring some means of upgrading the timber to achieve comparable properties. The harvested timber from semi-natural forests may often be of higher quality, but reduced rates of production place more severe constraints upon the quantity of material that can be extracted in a given time period.

Changes in the quality of feedstock have resulted in the introduction of new timber processing technologies, in particular the production of reconstituted timber and engineered wood products, which do not rely upon the availability of large-diameter logs, which are now in increasingly short supply. The declining quality of timber is a powerful impetus to the development of technologies to upgrade wood quality.

1.6 Wood Preservation

Although wood is a perishable material, this can be seen as an advantage, in that wood can be disposed of into the environment at the end of its useful life, where its molecular constituents are broken down by natural processes and assimilated into nutrient cycles.

However, it is obviously not desirable that this process takes place when wood is used in service situations.

In the past, undesirable properties such as susceptibility to biodegradation were overcome by the use of durable hardwood (in the main) species, particularly tropical hardwoods. This has contributed to tropical deforestation, although factors other than harvesting for industrial timber have had a greater effect. Nonetheless, extraction of target trees from virgin forest leads to a substantial environmental impact and there is a strong tendency for settlement to follow logging roads, with practices such as slash and burn of the forest occurring. This has led to substantial public disquiet regarding tropical forest operations, and this poor public perception has gradually encompassed virtually all forestry operations. Furthermore, the quality and quantity of tropical wood has declined as the resource becomes scarcer and more expensive to extract.

As the availability of naturally durable species has declined, the industry has turned to softwoods, and increasingly to softwoods from managed forests or plantations. In order to achieve acceptable longevity under service conditions, it has been necessary to use preservatives to prevent biological attack. Such preservatives have tended to rely upon broad-spectrum biocidal activity and have become very common, particularly for exterior applications.

The beginning of the modern timber preservation industry can be traced back to the late 1830s, when Bethell developed a method for the pressure impregnation of timber. This was used for the treatment of sleepers and poles for the rapidly expanding railway industry, using creosote and tar oils. The early part of the 20th century saw the development of water-based systems employing arsenates, chromates, fluorides and nitrophenols. In 1933, an Indian government research officer (Dr Sonti Kamesan) developed the first copper–chrome–arsenic (CCA) wood preservative. This proved to be an excellent wood preservative that was used in increasing amounts throughout most of the 20th century.

Although CCA is an exceedingly effective preservative in service, attention has increasingly been focused on the fate of CCA when the treated timber products are disposed of. This has led to concerns especially regarding the ultimate release of arsenic and chromium into the biosphere.

Concerns have also been raised regarding the use of chromium and arsenic in preservatives where there is a high probability of human exposure to the treated products. In this context, it should be noted that, in a study of the consequences of the use of CCA-treated wood in playgrounds, it was concluded that children would have to ingest 10–30 kg of soil in the immediate vicinity of the treated timber on one occasion for it to pose a hazard (Henningsson and Carlsson, 1984). Nonetheless, legislative bodies have adopted a precautionary approach regarding the use of CCA-treated wood where a high probability of human contact is expected.

As a consequence of these concerns, many countries have now either banned the use of CCA outright or have severely limited its use to specific products or market sectors. Even in the latter case, there can be no doubt that an outright ban will follow in time.

Analyses of the quantities of various treated timbers used in various countries have been performed and some examples are given herein to illustrate the scale involved. In 1990, 1 967 600 m^3 of preservative-treated wood and 327 000 m^3 of anti-sapstain treated wood was used in the UK (Jermer, 1990). Of this total, 68 900 m^3 was composed of poles and sleepers, sawn/other wood products comprised 1 484 600 m^3 (75 %) and fencing posts accounted for 414 100 m^3 (21.5 %). Within the 15 member states of the European Union prior to enlargement, approximately 18×10^6 m^3 of timber was preservative treated, with 10.3×10^6 m^3 of this being for construction timbers. In 1991, global annual consumption of CCA was estimated to be 118 thousand tonnes, with Europe consuming something of the order of 39 000 tonnes. Japan used 96 500 tonnes of CCA in 1980, but this had declined to 2 thousand tonnes by 1989. Although the constituents of CCA are fixed in the wood during the lifetime of the product, they cannot be regarded as permanently immobilized, since the treated wood must ultimately be disposed of.

This treated wood will inevitably appear as waste in the future and strategies for disposal will have to be developed. At the present time, procedures are being developed to collect preservative-treated timber, incinerate the material, and recover the flue-stack emissions and ash to prevent dispersal into the environment. A significant concern with the use of incineration is the high volatility of arsenic (Dobbs and Grant, 1978). There have been suggestions that the recovered metals could be reused in preservative treatments, resulting in a cyclic flow of the metals used for wood preservation. Whilst such recovery schemes are highly desirable and should be developed, there is no denying that 100 % recovery of the metals derived from preservatives is not achievable, even theoretically, and there will inevitably be dissipation into the biosphere. Even assuming 90 % recovery, there is an inevitable dissipation of 10 % of the metals into the biosphere with each recovery cycle. It can be readily shown that after only seven recovery cycles, just over 50 % of the material has been dissipated into the environment (Figure 1.8).

Whether the reuse of the metals obtained from incineration as a preservative, or some form of permanent immobilization is preferable requires careful thought. Low-temperature pyrolysis has been suggested as an alternative to incineration, since this would be expected to lead to lower losses of metals (Helsen *et al.*, 1998).

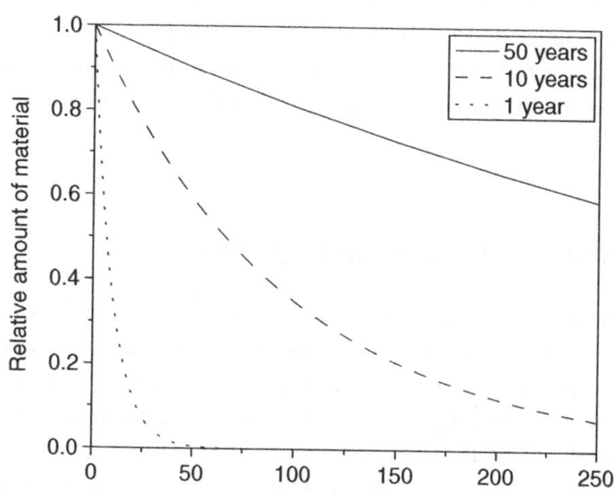

Figure 1.8 *The loss of material into the environment as a result of recycling processes, assuming a recycling efficiency of 90 % and product lifetimes of 1 year, 10 years and 50 years.*

Alternative strategies have also been proposed, in which the metal components are removed from the wood prior to incineration. These include extraction using phosphoric acid (Kazi and Cooper, 2002), ethylenediaminetetracetic acid (Nami Kartal, 2003) or oxalic acid combined with bacterial culture (Clausen, 2004a,b), and electrodialysis (Mateus *et al.*, 2002).

Although both copper and, to a greater extent, chromium (Cox and Richardson, 1978; Richardson and Cox, 1985) have associated environmental impacts, particular concerns have been expressed regarding the use of arsenic.

1.6.1 Arsenic

Arsenic occurs primarily in sulphide minerals associated with copper ores, and to a lesser extent with zinc, lead and gold ores. Arsenic is produced as a by-product of the smelting of these metals. Primary arsenic production has now ceased in the USA and Europe, and most arsenic is now imported from China and Mexico. The volatility of arsenic represents a significant concern, and there is at present no known natural mechanism by which arsenic is immobilized in the environment. Anthropogenic activities account for an input of some 19 000 tonnes into the atmosphere, compared with 12 000 tonnes from natural processes, such as volcanism and forest fires (Ayres and Ayres, 1996).

Until the mid-1970s, the most significant industrial use for arsenic was as general agricultural pesticides, and as desiccants in the cotton production process. According to the US Bureau of Mines, the use of arsenic for agricultural purposes was in the range of 15 000–20 000 tonnes per annum between 1963 and 1973. As agricultural use declined, there was an increase in the use of arsenic for wood preservation, with over 25 000 tonnes of arsenic being used for this purpose in 1998. According to the USGS, it has been estimated that by 2010 the amount of arsenic in discarded treated wood will be of the order of 14 500 tonnes in the USA.

Arsenic is also used in small quantities in the manufacture of lead–acid batteries (which are recycled), in the production of a few nonferrous alloys and in the electronics industry. It has been suggested that rather than importing primary arsenic for industrial uses, this could be recovered from wood waste, although the amounts required are only of the order of one to two thousand tonnes per year in Europe, and similar amounts in the USA (Lindroos, 2002).

1.7 Preservative-treated Wood and Legislation

Concerns about the disposal of preservative-treated wood, as well as perhaps less well founded concerns regarding the safety of preservative-treated wood in service, have resulted in countries introducing legislation to phase out certain classes of preservatives. At present, toxic metal containing preservatives are being banned, or restrictions placed upon their use, with the trend in wood preservation being towards the use of low-toxicity metals (e.g. copper and zinc) and organic nonchlorine-containing biocides. Legislation is also moving towards the registration of wood preservative chemicals, with certain requirements being placed upon determining the environmental impact of using these chemicals. Such restrictions are inevitably going to lead towards increased costs in the development of new biocides, and it is most unlikely that any new biocides will be developed specifically for wood preservation.

In February 2002, the US Environmental Protection Agency announced a voluntary decision by the lumber industry of the United States to replace the sale to consumers of CCA-treated wood with alternative preservative systems by the end of 2003. This voluntary ban affected all residential uses of CCA-treated wood, including decking, picnic equipment, playground equipment, residential fencing and so on. In the EC, Commission Directive 2003/02 (6 January 2003) was published, concerned with restrictions on the use and marketing of arsenic. According to this directive, CCA-treated wood will not be allowed for certain end-uses (Table 1.5). As of 30 June 2004, the net effect of this directive is to severely restrict the use of CCA-treated wood in situations in which there is the potential of human contact. These restrictions apply to imported treated wood and to waste wood containing preservatives. From 2007/8, CCA will require authorization according to the Biocidal Products Directive (BPD).

Similarly, legislation has been, or will be, introduced to deal with the disposal of treated wood waste at the end of a product lifetime. No longer will it be acceptable to dispose of treated wood waste by dumping in landfill. Proper disposal will require the incineration of treated wood waste in appropriate facilities that have the necessary equipment to prevent stack emissions of toxic compounds. This requires expensive investment to build plant that can meet the relevant environmental requirements. Such methods probably represent the best option for the permanent removal of these potential pollutants. The ash generated in these plants may contain high concentrations of arsenic, which will then have to be disposed of as hazardous waste.

The recycling of wood waste is also increasingly being used, where timber products at the end of life are usually reduced by chipping, or fibre production, and the resultant material is then reused in reconstituted wood products. Strict monitoring and control are required in order to ensure that inappropriately treated waste is not included in the feed-stock stream for recycling wood. The use of recycled wood in reconstituted products is rising rapidly; for example, in the UK, the panel products industry used 400 000 tonnes of recycled wood in 1999, and this had risen to 932 000 tonnes by 2002 (UK Forestry Commission data 2003).

Table 1.5 *Allowed and unacceptable uses of CCA-treated wood according to European Directive 2003/02/EC*

Allowed uses	Unacceptable uses
Timber used structurally in public, agricultural, office or industrial buildings (no human contact)	Residential or domestic constructions (all uses)
Bridges and bridgework	Any application where there is a likelihood of repeated dermal contact
Constructional timber in freshwater areas	Marine water installations
Noise barriers, highway safety fencing, earth-retaining structures, avalanche control, livestock fence posts	Agricultural uses, other than fencing or structural uses
Utility poles (electric power transmission, telecommunications)	Applications where the treated wood may come into contact with products intended for human or animal consumption

All of the above environmental concerns are leading to restrictions being placed upon the use and composition of preservatives used to treat timber. Preservative systems of the future will therefore be selected not simply on the basis of criteria such as efficacy and cost, but also environmental impact, both during service and at the end of product lifetime. This will inevitably have economic implications.

1.8 Competition from Nonrenewable Materials

Timber continues to be used in huge quantities in the built environment. However, for some applications there has been significant competition from nonrenewable materials. Concrete, cement, brick, stone, ferrous and nonferrous metals, and various polymeric materials have all been used in place of timber and in some cases have completely replaced timber for some applications. The almost complete loss of the window and door markets to competitors such as unplasticized polyvinyl chloride (uPVC) has been a serious blow to wood processors in some countries. This change has been brought about mainly due to public demand, with wooden products being perceived as having high maintenance requirements and poor performance in service. Aggressive marketing has also attempted to portray timber as an environmentally unfriendly material, with the use of nonrenewables being a better option, since this leaves the forests intact. This publicity has also been backed up by life cycle analysis (LCA) studies purporting to indicate the superior environmental credentials of certain nonrenewables. The timber sector was slow to respond to this threat and has only recently begun to reclaim lost markets. In some cases, the appropriate response has been to design better products, or to use more sophisticated manufacturing methods. However, it is much more complex to counter the negative environmental arguments.

The whole area of LCA is very complicated and it is not possible to do sufficient justice to the topic within the space limitations of this volume. However, contributions to the apparent negative environmental impacts of timber could include unrealistically high maintenance requirements (e.g. regular applications of paint or varnish, and regular inspection), the use of preservatives and the assumption of a relatively short product lifetime due to deterioration in service. Favourable LCAs for a material such as uPVC can be obtained by assuming that all material at the end of a product life will be recycled (thus ignoring ultimate disposal), assuming that the product life is equal to the life of the material, or assuming low maintenance requirements when compared with timber.

The replacement of timber products by nonrenewable materials is an unfortunate development, since it has been repeatedly shown that the use of timber does have associated environmental benefits compared with the use of nonrenewables (e.g. Marcea and Lau, 1992; Hillier and Murphy, 2000; Bowyer *et al.*, 2003; Lippke *et al.*, 2004). Timber has a lower embodied energy content (and hence a more favourable carbon emission profile) compared to most other building materials and can provide other benefits, such as improved thermal properties. It and the products made from it (in common with other renewable materials) can be used as a repository for atmospheric carbon dioxide. Wood is derived from a renewable resource, albeit potentially an exhaustible one unless it is managed correctly. Disposal of wood can be readily achieved with little environmental impact (subject to how the wood has been treated prior to disposal).

Unfortunately, one consequence of the ever more stringent controls upon the use of preservatives and upon preservative-treated wood will be the rejection of wood as a construction material.

1.8.1 Polyvinyl Chloride

The estimated annual production of PVC for all applications is around 22 million tonnes (Jaksland *et al.*, 2000). In many ways, uPVC is an excellent material. It is very stable and exhibits good durability, making it ideal for use in long-lived low-maintenance applications (such as underwater drainage). However, it should be noted that uPVC usage carries with it significant environmental impacts. Although PVC production waste is recycled mechanically, most of the post-consumer waste is not, with much of it being disposed to landfill at present. PVC exhibits a degree of thermal and UV instability, requiring the use of metal-containing compounds, such as cadmium soaps, as stabilizers. Degradation of uPVC products in exterior service has resulted in paint manufacturers producing products specifically for application to uPVC.

Production of elemental chlorine for the synthesis of vinyl chloride monomer (the biggest single use of chlorine worldwide) can have significant environmental impacts, particularly if a mercury cell is used for chlorine production. Ultimate disposal of uPVC does not appear to have been seriously addressed until now, with some analyses even assuming 100 % recycling of the material into new products. Even if this was technically (or even thermodynamically) possible, it still does not address the issue of *ultimate* fate. With present technologies, this can only be dissipation into the environment, landfill or incineration. There is research under way investigating alternative means of recycling PVC waste, other than mechanical grinding, although the costs of such processes are not yet known (Jaksland *et al.*, 2000).

1.9 The Need for Wood Modification

Due to environmental concerns regarding the use of certain classes of preservatives, there has recently been a renewed interest in wood modification. Wood modification represents a process that is used to improve the material properties of wood, but produces a material that be disposed of at the end of a product life cycle without presenting an environmental hazard any greater than that associated with the disposal of unmodified wood. Although wood modification has been the subject of a great deal of study at an academic level for over 50 years, it is only comparatively recently that there has been significant commercial development.

1.10 Conclusions

Although timber production and utilization can result in substantial environmental benefits compared with materials extracted from nonrenewable sources, timber utilization and forestry have become associated with negative environmental impacts. Tropical deforestation continues to be a cause of concern, although there are many factors associated with this other than timber production. The demand for timber will continue to rise, leading to concerns of a timber shortage by the end of the 21st century. In the future, increasing quantities of timber will be sourced from managed forests and there will be an associated reduction in timber quality in some cases. Environmental concerns have led to legislation that has resulted in significant reductions in the quantities of timber treated with conventional preservatives (especially CCA). These trends will continue in the future, with the cost of preservative-treated timber rising. There has been a significant increase in the use of materials obtained from nonrenewable sources as substitutes for timber in some markets. These major changes in the timber sector have resulted in a substantial increase in interest in wood modification as a means of improving the properties of the material.

2

Modifying the Properties of Wood

2.1 Introduction

Although wood continues to be used for many applications because of its many excellent material properties (such as a good strength to weight ratio, aesthetic appearance etc.), it also suffers from a number of disadvantages. Dimensional changes in response to altering atmospheric conditions, susceptibility to biological attack and changes in appearance when exposed to weathering place restrictions on the potential end-uses of wood. The origin of the wood properties can be understood by referring to its structure at the macroscopic, microscopic and molecular levels. This chapter gives a brief overview of the properties of the material and how modification can be used to alter them. Although, necessarily in a book of this length, it is not possible to go into many of the aspects of wood structure, chemistry, biology or material properties in any great depth, there are a great many excellent texts that deal with these subjects, to which the interested reader is directed (e.g. Stamm, 1964; Sjöström, 1981; Wilson and White, 1986; Fengel and Wegener, 1989; Zabel and Morrell, 1992; Eaton and Hale, 1993; Walker, 1994; Desch and Dinwoodie, 1996; Dinwoodie, 2000; Hon and Shiraishi, 2001; Barnett and Jeronimidis, 2003; Bowyer *et al.*, 2003; Goodell *et al.*, 2003). In some cases, this chapter does go into more detailed discussions of some of the properties of wood that are relevant to the material covered in the other chapters in this book.

2.2 Wood Properties and Wood Modification

Wood is a natural material, and as such it exhibits great variability in its properties. Variations exist between species, between trees and within the tree itself. This can cause difficulties for any study of the material, making it difficult to compare results. In any review of the wood science literature (or indeed any branch of science), it is readily apparent that there are sometimes disagreements between different studies, and it is not

Wood Modification: Chemical, Thermal and Other Processes C. Hill

always possible to reach a consensus. The wood modification literature is no different in this respect!

Since many of the properties of wood are ultimately determined by its chemical constituents, wood modification often seeks to make changes at this level, in order to produce a material that has the desired properties. However, there are also wood modifications that do not involve alteration of the chemical composition of the material.

Wood modification is a means of altering the material to overcome or ameliorate one or more of its disadvantages. The aim of a modification may be to bring about an improvement in decay resistance or dimensional stability, to reduce water sorption, to improve weathering performance and so on. The term 'wood modification' applies to the application of a process that alters the properties of the material such that during the lifetime of a product no loss of the enhanced performance of the wood should occur.

For example, using conventional technologies, the decay resistance of wood can be increased by the application of wood preservatives (e.g. CCA, creosote etc.). To be sure, the wood property has been improved and the wood can, in a sense, be considered to be modified. In addition, the property enhancement can be considered to occur for the lifetime of the product. However, although these changes have been made, the mode of action of the preservatives is due to the biocidal action of these substances on decay organisms. Such activity is often nonspecific, in that although the attack by the target organisms is prevented or retarded, the property may also be extended to nontarget organisms, such as mammals. Thus, although during the lifetime of the treated product there may be a very low hazard associated with in-service use, this may well change at the end of life of the product. Therefore, potential loss of the preservative into the environment may become a problem in disposal or recycling situations. Where this occurs, the biocidal action of the preservative remains and this can have a negative environmental impact. Of course, this problem can be addressed by choosing a substance that breaks down rapidly in the environment when released from the wood. However, if the mode of action of such a substance whilst in the wood is biocidal in nature, then a treatment of this kind is not considered to be a wood modification.

To give another example, it may be possible to improve the dimensional stability of solid wood by applying a coating, or in a reconstituted wood product by adding wax during the forming process. However, such treatments are not permanent, in that a coating may fail by cracking, or wax may leach or exude during the lifetime of the product. This loss of performance occurs because the treating agent is not chemically bonded to or otherwise immobilized within the wood structure. However if, for example, the coating were to be applied so that there exists a covalent bond between it and the wood, an improvement of the performance of the coating would be expected. Such a treatment would be a surface modification of the substrate. Such a surface modification is not restricted to direct chemical bonding to the surface, but may involve the action of chemicals, biological agents (e.g. enzymes) or some other agent (e.g. electromagnetic radiation or electron bombardment) to alter the surface chemistry of the substrate to improve the compatibility between the coating and the wood surface.

In consequence, the following definition can be applied:

Wood modification involves the action of a chemical, biological or physical agent upon the material, resulting in a desired property enhancement during the service life of the modified wood. The modified wood should itself be nontoxic under service conditions and, furthermore, there should be no release of any toxic substances during service, or at end of life following disposal or recycling of the modified wood. If the modification is intended for improved resistance to biological attack, then the mode of action should be nonbiocidal.

Note that the above does not necessarily preclude the use of a hazardous chemical in the preparation of the modified wood, provided that no hazardous residues remain once the wood modification process is complete.

2.3 Wood Modification Methods

Modification of wood can involve active modifications, which result in a change to the chemical nature of the material, or passive modifications, where a change in properties is effected, but without an alteration of the chemistry of the material. Some modifications have been classified by Norimoto and Gril (1993), by referring to changes taking place at the cell wall level, from which the scheme given in Figure 2.1 is derived. The subdivisions of the different wood modification methods, as defined in this text, are summarized in Table 2.1, with reference (where relevant) to Figure 2.1.

2.3.1 Chemical Modification

Most of the chemical modification methods investigated to date have involved the chemical reaction of a reagent with the cell wall polymer hydroxyl groups. This can result in the formation of a single chemical bond with one OH group (Figure 2.1d), or cross-linking between two

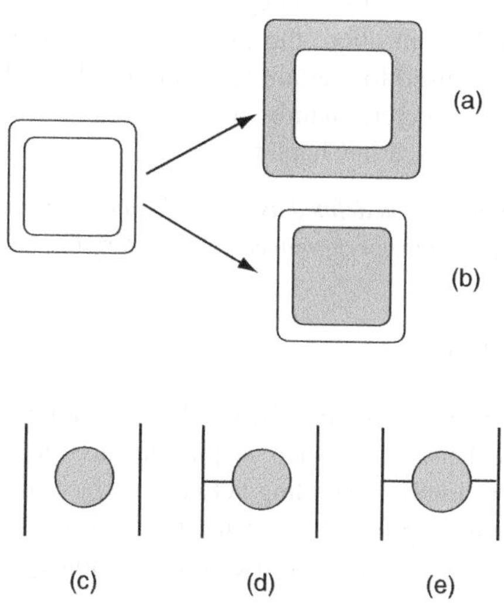

Figure 2.1 A diagram illustrating different types of wood modification at the cellular level (derived from diagrams of Norimoto and Gril, 1993).

Table 2.1 *A classification of wood modification methods (see Figure 2.1)*

Division	Type	Class	Illustration
Active	Chemical modification	Cell wall	d, e
		Surface	—
	Thermal modification	Cell wall	—
	Enzymatic modification	Surface	—
Passive	Impregnation modification	Cell wall fill	a, c
		Lumen fill	b

or more OH groups (Figure 2.1e). The chemical nature of the cell wall polymers is thus changed, which *may* be responsible for the new properties. Chemical modification of wood has been the subject of a number of reviews (Rowell, 1977a, 1983a,b; Banks, 1990; Takahashi, 1993; Banks and Lawther, 1994; Kumar, 1994; Sasaki and Kawai, 1994; Hon, 1996; Militz *et al.*, 1997; Suttie, 1997; Birkinshaw 1998). At the time of writing this book, there is no process currently operating that uses chemical modification, but plans are well advanced in The Netherlands to launch a commercial acetylation process. Chemical modification of the cell wall is dealt with in Chapters 3 and 4, and surface chemical modification in Chapter 6.

Chemical modification of wood is defined as the reaction of a chemical reagent with the wood polymeric constituents, resulting in the formation of a covalent bond between the reagent and the wood substrate.

2.3.2 Thermal Modification

The application of heat to wood results in degradation associated with chemical changes in the material. If carefully controlled, the property changes that are obtained due to thermal modification can be used for certain applications. There has been a great deal of commercial activity in this sector, mainly in Finland, France and The Netherlands. Thermal modification is discussed in Chapter 5.

The thermal modification of wood is defined as the application of heat to wood in order to bring about a desired improvement in the performance of the material.

2.3.3 Surface Modification

Changes to the wood surface can involve chemical modification, biological modification using enzymes or physical processes such as plasma modification. The main areas of interest are modification to improve bonding between surfaces directly, or between wood surfaces and coatings, or to improve the weathering performance of wood. There is, as yet, no commercial process operating that uses this technology. Surface modification is the subject of Chapter 6.

The surface modification of wood is defined as the application of a chemical, physical or biological agent to the wood surface in order to effect a desired performance improvement.

2.3.4 Impregnation Modification

Impregnation of the wood cell wall with chemicals of various types is a very broad area, which continues to attract interest. In most of the studies reported here, the impregnation involves the treating of wood with a monomer solution that diffuses into the cell wall, followed by subsequent polymerization. Property improvements occur primarily due to bulking of the cell wall by the impregnant. Although lumen-fill treatments have also been reported over the years, this has not been covered in this book. There are examples of commercial impregnation processes that have been running for many years. Impregnation modification is covered in Chapter 7.

Impregnation modification of wood is defined as any method that results in the filling of the wood substance with an inert material (impregnant) in order to bring about a desired performance change.

2.4 The Cell Wall of Wood

For a detailed description of the ultrastructure of wood and the cell wall, the reader is referred to the comprehensive texts listed above. Briefly, the cell wall of wood is composed of a number of discernable layers (Figure 2.2). These are divided into the primary (P) and secondary (S) layers; the secondary layer is further subdivided into the S_1, S_2 and S_3 layers. The primary layer is the first to be laid down when the cell is formed and is composed of microfibrils, which have an essentially random orientation that allows for expansion of the cell to occur as cell growth takes place. The secondary layer is subsequently formed, with each of the sub-layers exhibiting different patterns in the way the microfibrils are oriented, as illustrated in Figure 2.2. Of these, the S_2 layer occupies the

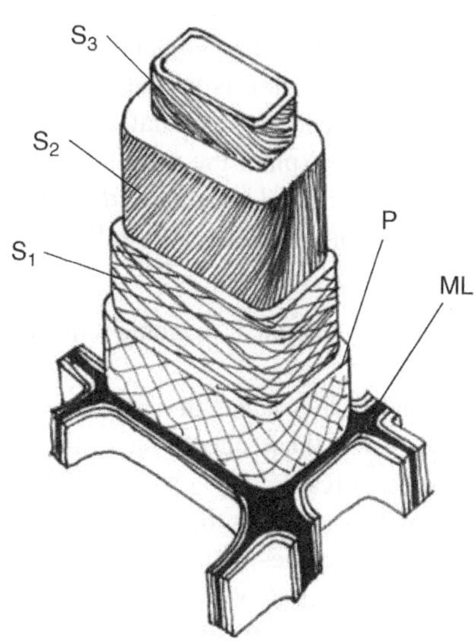

Figure 2.2 A schematic of the ultrastructure of the wood cell wall, showing the middle lamella, the main cell wall layers and the associated microfibrillar orientation.

greatest volume of the cell wall and, as a consequence, is the layer that has the greatest influence on many of the properties of the cell and hence the macroscopic properties of the wood. The S_2 layer exhibits a definite microfibrillar orientation, and is itself composed of many lamellae consisting of a great many closely associated microfibrils that exhibit a helical winding pattern.

2.4.1 Cell Wall Microporosity

The space between the microfibrils is occupied by the hemicelluloses and by lignin. However, the incomplete filling of the intermicrofibrillar region results in the existence of what are usually referred to as micropores (or microvoids) in the cell wall. These have diameters of the order of nanometres and thus technically should be referred to as nanopores, but since the term 'micropores' is the most commonly used in the literature, it will be used throughout this book.

When the cell wall is fully swollen, the micropores are open and the interior of the cell wall can be accessed by entities that are smaller than the diameter of the micropores. When wood is dried from a water-saturated condition, as water is removed the lumen and other macrovoids, and then subsequently the cell wall, lose moisture. As the water is removed, the micropores begin to collapse, and this process continues until the wood is dry.

Evidence for the existence of the cell wall micropores has been obtained from various experimental studies (Hill and Papadopoulos, 2001). The presence of the micropores can be inferred by pycnometric determinations of the density of wood. If wood is immersed in a nonswelling solvent, such as a nonpolar hydrocarbon, then cell wall densities of the order of 1.44–1.47 $g\,cm^{-3}$ are obtained, whereas with a swelling solvent such as water, higher densities of the order of 1.50–1.55 $g\,cm^{-3}$ are found (Siau, 1984). This difference can be explained by postulating that there are micropores inside the cell wall that can only be accessed when the cell wall is swollen. Since the cell wall density is found to be lower with nonswelling solvents, or even when helium is used as the displacement medium, it can be further inferred that collapse of the micropores in the cell wall interior is incomplete when the wood is dried, but that this region is nonetheless sealed off from the exterior during the drying process. The existence of residual microporosity in the cell wall of oven-dry wood has been proposed to explain the results obtained in the measurement of the dynamic mechanical properties of wood (Akitsu *et al.*, 1993a).

The determination of the accessibility of the cell wall interior is of importance for a number of reasons:

- In many cases, modification with chemical reagents requires reaction with OH groups inside the cell wall.
- In some classes of modification, it is necessary for the modifying agent to penetrate the cell wall, which requires that these agents have dimensions no greater than the diameter of the cell wall micropores.
- Improvement of some properties (such as dimensional stabilization) requires that, in some cases, the modifying agent resides within the cell wall micropore structure.
- Blocking of the cell wall micropores may be a mechanism in explaining some of the properties of modified wood (e.g. decay resistance).
- Modification resulting in the formation of cross-links within the cell wall prevents the micropores from opening when the wood is exposed to moisture.

Accessibility is defined in terms of what is accessing the cell wall interior. This requires the movement of a molecule or some other entity from the lumen into the cell wall via the cell wall micropore network. Methods for probing the accessibility of the cell wall require that the cell wall be swollen, essentially restricting such studies to aqueous environments. The most commonly employed technique is that of solute exclusion, where water-saturated wood is immersed in a solution of probe molecules and allowed to reach equilibrium with the solution. Probe molecules enter the macropores of the wood (lumens, pits etc.) and if they are small enough they can also penetrate the cell wall. When probes enter the pores of the wood, they replace the water in them and this water dilutes the probe solution. The decrease in the concentration of the probe solution is a measure of the pore volume of the wood. By using probe molecules of different sizes, it is possible to obtain data for the pore size distribution of the wood. As the diameter of probe molecules increases, a point is reached at which there is little further change in the solute concentration, and this indicates the limiting size for penetration of the cell wall (Hill *et al.*, 2005). Most of these studies report a maximum size for cell wall micropores in the region of 2–4 nm. This technique is of value in determining size criteria for determining cell wall accessibility, although it is not an absolute determinant of the geometric distribution of micropores.

2.5 The Chemical Constituents of Wood

Many of the physical, chemical and biological properties of wood can be understood by referring to the polymeric chemical constituents. In many cases of wood modification, these polymeric components are modified to some extent. The three structural polymeric components of the wood cell wall are cellulose, hemicellulose and lignin. There are many excellent texts describing the structure and function of these components, and only a brief account is given here.

2.5.1 Cellulose

The cellulose content of wood varies from about 40 % to 50 % (Fengel and Wegener, 1989). Cellulose is composed of β-D-glucopyranose monomeric units (Figure 2.3a) that are alternately inverted to form cellobiose dimeric units, which are 1.03 nm in length (Figure 2.3b). This results in the cellulose backbone being linear (Figures 2.3c, d). The cellobiose units link together via glycosidic linkages to form the polymer cellulose. The degree of polymerization (DP) of cellulose has been determined as about 10^4 in wood (indicating that a cellulose macromolecule is about one micron in length), although this is likely to be an underestimate, since some degradation will inevitably occur when cellulose is isolated from the cell wall. A number of cellulose chains are closely associated via extensive H-bonding networks to form the microfibril, which is the primary reinforcing element in the cell wall. The exact number depends upon the cross-sectional area of the microfibril (about 10–25 nm). There are numerous models describing the structure of microfibrils, and a good account can be found in Fengel and Wegener (1989). The microfibrils have associated with them crystalline and amorphous components with associated OH groups, which can be detected using a variety of methods, such as X-ray diffraction,

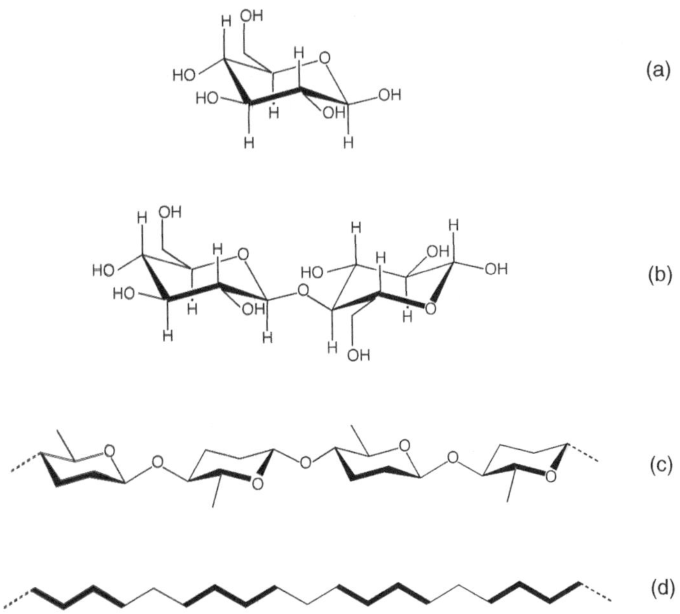

Figure 2.3 The molecular structure of β-D-glucopyranose (a), cellobiose (b) and the glucopyranose backbone of cellulose (c), and (d) a schematic of the linear arrangement of the glucopyranose units.

nuclear magnetic resonance (NMR) and infrared (IR) spectroscopy. Because of the highly crystalline nature of the microfibrils, the cellulose component of wood is relatively unreactive and thermally stable. It appears that the crystalline region is associated with the core of the microfibril, with the exterior corresponding to the amorphous cellulose content (Chanzy, 1990).

2.5.2 Hemicelluloses

Like cellulose, hemicelluloses are polysaccharides, but they are composed of a number of different sugar units. The DP of hemicelluloses is of the order of 200–300, and they are generally less ordered than cellulose, although some can form crystalline units (indeed, microfibrils composed of hemicelluloses have been identified (Savage, 2003). They also differ from cellulose, in that some of the OH content is naturally acetylated, and there are also carboxylate groups associated with the structures. Unlike cellulose, the main backbone of hemicellulose structures also has short branches of sugar units attached (Figure 2.4). Because of the generally amorphous nature of the hemicelluloses, they contain the greatest proportion of the accessible OH content of the cell wall, react more readily, and are less thermally stable than cellulose or lignin. Hemicelluloses appear to act as interfacial coupling agents between the highly polar surface of the microfibrils and the much less polar lignin matrix. Hemicellulose degradation results in wood becoming brittle and

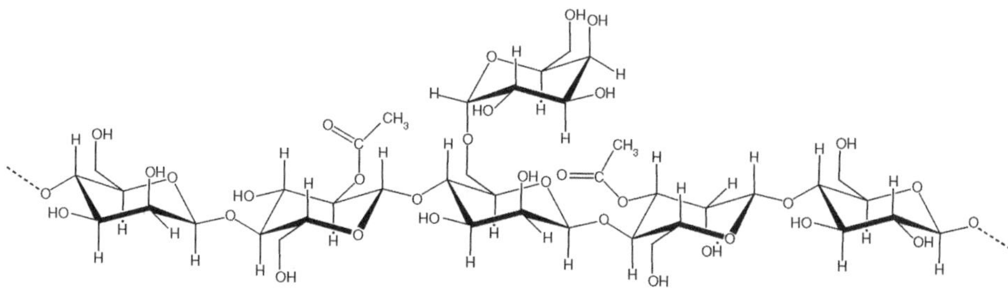

Figure 2.4 The molecular structure of a hemicellulose (O-acetylgalactoglucomannan).

rigid, indicating the important role that they have in imparting viscoelastic properties to wood. The hemicelluloses form H-bonds with the surface of the microfibrils and covalent linkages with the lignin matrix (Koshijima and Watanabe, 2003). Hardwood hemicelluloses are generally found in higher proportions compared with softwoods, with a higher proportion of pentosans and a higher degree of acetylation.

2.5.3 Lignin

Lignin is a highly amorphous phenolic polymer of indeterminate molecular weight. Lignification of the wood cell wall involves the diffusion of the phenylpropane monomer units and polymerization to produce a random three-dimensional network, via a free-radical mechanism. Due to the random nature of the polymerization reaction, there is no definitive structure to lignin, although the frequency of individual bond types is well established. A representative structure of lignin is shown in Figure 2.5. As with all such structures, the two-dimensional representation does not give an accurate impression of the true complexity of the structure. The random nature of the lignin polymer network is the main factor in determining the complex geometry of the cell wall micropores. Lignin is responsible for providing stiffness to the cell wall and also serves to bond individual cells together in the middle lamella region. Although lignin is relatively rigid at room temperature, it undergoes a glass transition at around 140 °C, and the presence of moisture in the cell wall additionally serves as a plasticizer for the lignin network. The presence of moisture in the cell wall opens up the structure of lignin. Lignin has a low concentration of OH groups compared to the polysaccharide components. Hardwood lignins have a syringyl content varying from 20 % to 60 %, whereas softwood lignins have very low syringyl contents (Fengel and Wegener, 1989).

2.5.4 Extractives

Depending on the species, wood can contain levels of extractible material ranging from approximately 0.5 % to around 20 % by weight. This can present problems when wood modification processes are attempted. In chemical modification reactions, leaching of extractives into the modification medium can lead to inaccurate determinations of weight gain due to modification. With impregnation modifications, leaching can also present a

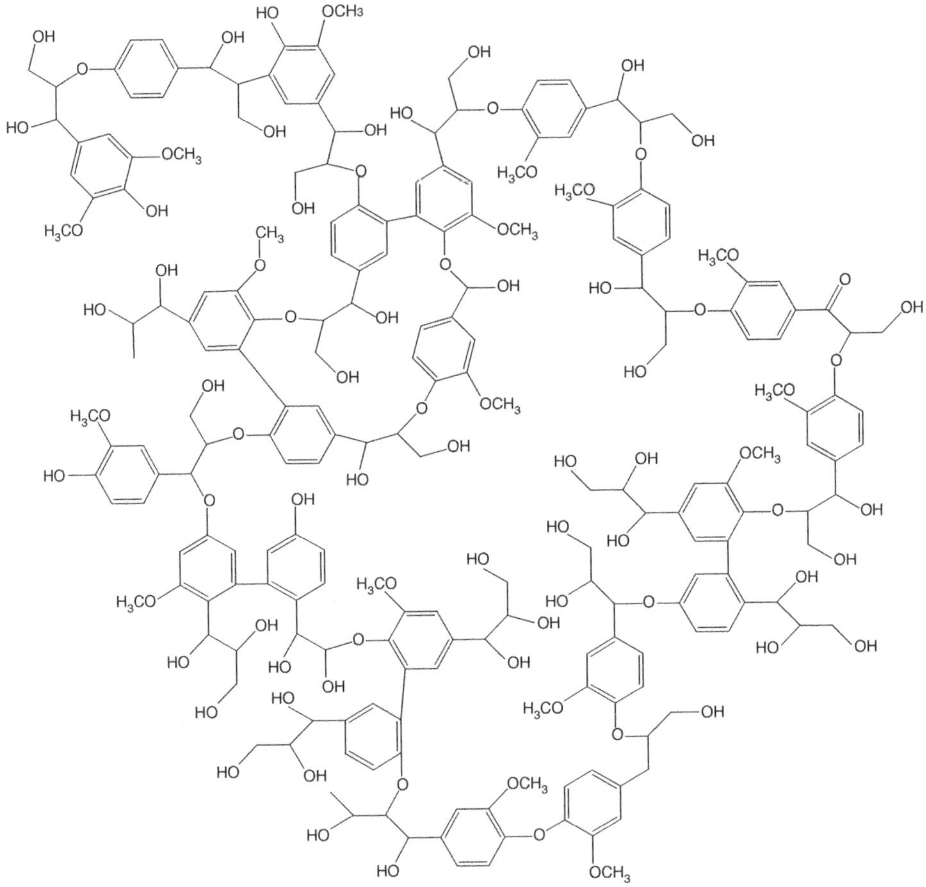

Figure 2.5 A representative molecular structure of a softwood lignin.

problem to a lesser extent. In the absence of a solvent, migration of extractives can occur with treatments that involve the application of heat or vacuum to the wood. In many laboratory-based studies, extractives are removed prior to modification, but this is not possible when scale-up is attempted.

2.5.5 Hydroxyl Accessibility

The OH groups associated with the cell wall polymeric constituents are very important in determining many of the properties of wood. Many of the material properties of wood depend upon the presence of these groups and, in particular, hydrogen bonding interactions between the molecular components of the cell wall. Moreover, the OH groups are also responsible for absorption of moisture (leading to dimensional instability) and permit biological attack of the material. On a molecular level, the cell wall OH groups represent the most important component influencing the performance of wood.

Hydroxyl groups are associated with cellulose, hemicellulose and lignin in the cell wall. Cellulose is the molecular component of the reinforcing elements of the cell wall (the microfibrils) and a proportion of the cellulosic OH groups are involved in hydrogen bonding between cellulose polymers within the crystalline core of the microfibrils. These OH groups are considered to be inaccessible, in that they are not involved in hydrogen-bonding interactions with sorbed water molecules. Hydroxyl groups on the surface of the microfibrils are able to interact with sorbed water and are thus considered to be accessible. The majority of the accessible OH groups of the cell wall are associated with the hemicellulosic constituents; the lignin less so, due to its relatively low abundance of OH groups.

The number of available (accessible) OH groups in wood is not known precisely, but there have been studies of wood exposed to deuterated or tritiated water, where estimates of accessible OH content have been made. Tritium exchange with microtomed sections of black spruce (*Picea mariana*) and white birch (*Betula papyrifera*) have yielded OH accessibility values of 7.97×10^{-3} and 6.90×10^{-3} moles g^{-1}, respectively (Sumi *et al.*, 1964). Deuterium exchange using never-dried microtomed sections of Douglas fir indicate that in the sapwood, approximately 45 % of the total OH content of the wood was accessible in the latewood, but only about 24 % in the earlywood at an EMC of 40 %, with accessibility decreasing in direct proportion with the cell wall MC (Chow, 1972). Such studies are complicated by the fact that the extent of OH accessibility depends upon the history of the samples studied and, in particular, whether such samples have been previously dried (Chow, 1972). For example, an estimate of the number of accessible OH sites in wood, as determined by deuterium exchange, gave a value of 3.3×10^{-3} moles g^{-1} when Japanese cypress was exposed to D_2O from the vacuum dry state (Taniguchi *et al.*, 1978). This appears to be a low value compared to other studies and theoretical calculations. Similarly, D_2O exchange with OH protons of a birch xylan dried at 75 °C for 48 hours gave a value of hydroxyl accessibility of only 52 % (Sepall and Mason, 1961). This low level of accessibility would be due to hornification of the hemicellulose upon drying directly from a water-saturated state.

The number of OH groups per unit mass of dry wood can be calculated on the basis of an estimate of the number of the OH groups associated with each of the cell wall polymeric constituents (Table 2.2) (Rowell, 1980). An example is given for Scots pine (*Pinus sylvestris*). This gives the number of theoretical OH groups according to

$$(0.26 \times 1)/180 + (0.52 \times 3)/162 + (0.14 \times 3)/162 + (0.08 \times 2)/132 \qquad (2.1)$$

This yields a value for the total OH content of approximately 14.8×10^{-3} moles per gram of dry wood mass. This value is relatively insensitive to the polymeric composition, with

Table 2.2 *The distribution of OH groups and composition of Scots pine*

Component	% Composition	OH groups per unit	Mol. wt of units
Lignin	26	1/C9	194
Cellulose	52	3/C6	162
Hexosan	14	3/C6	162
Pentosan	8	2/C5	132

a calculation for European beech (*Fagus sylvatica*), based upon the composition given by Fengel and Wegener (1989), giving a total OH content of approximately 19.8×10^{-3} moles g^{-1}. Although such figures represent a crude estimate, they are nonetheless acceptably close to the values obtained from isotopic exchange experiments. The accessible OH content will be lower than the total OH content, since a proportion of the OH groups will be located in inaccessible regions, such as the crystalline core of the microfibrils. The accessible OH content can be crudely estimated by using a value of 65 % for the proportion of cellulosic OH groups located in crystalline regions (Fengel and Wegener, 1989), giving the accessible OH content for Scots pine as 8.6×10^{-3} moles g^{-1}.

2.6 The Wood–Water Relationship

Wood is a hygroscopic material, due to the fact that the cell wall polymers contain hydroxyl groups. In an environment containing moisture, dry wood will absorb moisture until it is in equilibrium with the surrounding atmosphere. Similarly, saturated wood, when placed in an atmosphere of lower relative humidity (RH), will lose moisture until equilibrium is attained. If the wood is placed in an environment where the RH is stable, it will attain a constant moisture content (MC), known as the equilibrium moisture content (EMC). At this point, the flux of water molecules into the cell wall is exactly balanced by the outward flux into the atmosphere.

Wood responds to its environment; in particular, it changes dimension in response to changes in atmospheric RH. As water vapour enters the wood cell wall, it occupies space, causing the dimensions of the cell wall to increase, and as a consequence the wood increases in dimension. These changes in dimension are anisotropic and are greatest in the tangential direction, then the radial and least in the longitudinal direction.

2.6.1 Moisture Sorption and Desorption

The sorption and desorption behaviour of wood is studied by determining the relationship between EMC and RH at constant temperature (an isotherm). Such a plot produces the classic sigmoidal relationship shown in Figure 2.6. In practice, it is not possible to study the relationship to 100 % RH and, generally, experiments of this type are terminated at around 95–98 % RH. In this region of high RH, the EMC of the wood is increasing rapidly. The relationship between EMC and RH in the sorption and desorption modes differs, giving rise to the phenomenon of hysteresis.

Many models have been developed that deal with the sorption properties of wood in the presence of moisture: these have been discussed in a number of works (e.g. Skaar, 1972; Siau, 1984). They can be approximately divided into sorption models, such as the Brunauer–Emmett–Teller (BET) model, or solution models (such as the Hailwood–Horrobin, H–H, model). The sigmoidal shapes of sorption or desorption isotherms can be deconvoluted into two components. These are often taken to represent a monomolecular water layer (associated with the primary sorption sites, OH groups), and a multilayer component where the cell wall bound water molecules are less intimately associated with the fixed cell wall OH groups.

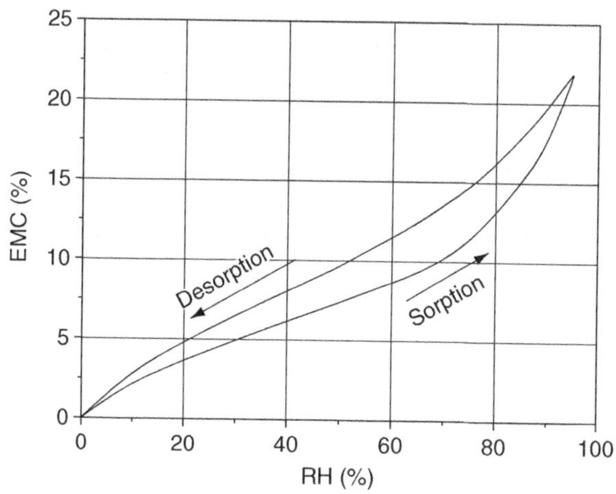

Figure 2.6 *A sorption and desorption isotherm, showing hysteresis.*

2.6.2 The Fibre Saturation Point

Many workers extrapolate the line of best fit through the sorption isotherm data points to determine the theoretical EMC at 100 % RH. This EMC is taken to represent the fibre saturation point (FSP), which represents the wood moisture content at which the cell wall cannot increase its moisture content any further. Any further water that is sorbed into the wood structure will now be located in the lumen, or in other macrovoids. As a consequence of this, the physical properties of the wood will now no longer change with any further increase in wood moisture content. Thus, if the modulus of elasticity (MOE), modulus of rupture (MOR), tensile strength, swelling and so on are determined as a function of wood MC, then they will exhibit a constant value above the FSP. This then gives several methods to determine the FSP, and these various techniques are discussed in Stamm (1964), Skaar (1972) and Siau (1984), amongst others. This was first noted in the work of Carrington (1922), where changes in MOE, MOR and so on of sitka spruce were determined as a function of wood MC. Estimates for the value of the FSP generally lie between 21 % and 32 %, and Siau (1984) assumes an average value of 30 %.

It is worth considering the measurement of the FSP in some detail, since many wood modification techniques result in a reduction of EMC at a given RH and there have been studies in which attempts have been made to correlate the FSP with other properties, such as decay resistance.

There is some controversy regarding the absolute determination of the FSP and there are differences in the value depending upon the methods used to determine it. It has been argued that the use of extrapolated sorption isotherms is untrustworthy, since in the region of 98–100 % RH the sorption isotherm is rising very rapidly and slight errors in measurement can lead to large errors in extrapolation. Experiments using isotopic exchange have suggested that the FSP may be higher than 30 % (Chow, 1972). Values for the FSP as determined by solute exclusion are found to be around 40 %, which Siau

(1984) attributed to the use of microtomed sections, resulting in there being less mechanical restraint of the cells. However, values of this order are also found for solid wood samples when the solute exclusion technique is used (Hill *et al.*, 2005).

2.6.3 Determining the Moisture Relationship of Modified Wood

In any study of the moisture sorption–desorption properties of modified wood, the moisture content of the samples is determined using gravimetric methods. Invariably, the EMC of the sample is based upon the oven-dry weight of the modified wood:

$$\text{EMC}\ (\%) = [(M_2 - M_1)/M_1] \times 100 \qquad (2.2)$$

In many cases, this is not explicitly stated and can be a source of error. If the weight of a sample of wood is increased by 20 % and it is then exposed in an atmosphere at a given relative humidity (RH) alongside a sample of unmodified wood, there will be an apparent reduction in EMC simply due to the increased weight of the sample due to modification (Figure 2.7). In order to understand the effect of wood modification upon the sorption–desorption properties of modified wood, the EMC values should be based upon the oven-dry weight of the wood substance rather than the actual weight of the sample (weight of wood substance plus weight of modificant). This point has been noted previously, and in some studies a reduced equilibrium moisture content (EMC_R) is used rather than measured moisture content (e.g. Akitsu *et al.*, 1993a):

$$\text{EMC}_\text{R}\ (\%) = [(M_2 - M_1)/M_0] \times 100 \qquad (2.3)$$

where M_0 is the oven-dry weight of the specimen before modification, M_1 is the oven-dry weight of the specimen after modification and M_2 is the weight of the modified specimen at equilibrium with the atmosphere at a given RH.

It is therefore important to ensure that the correct method is used to calculate the moisture content of modified wood. For example, if it is necessary to determine the relationship

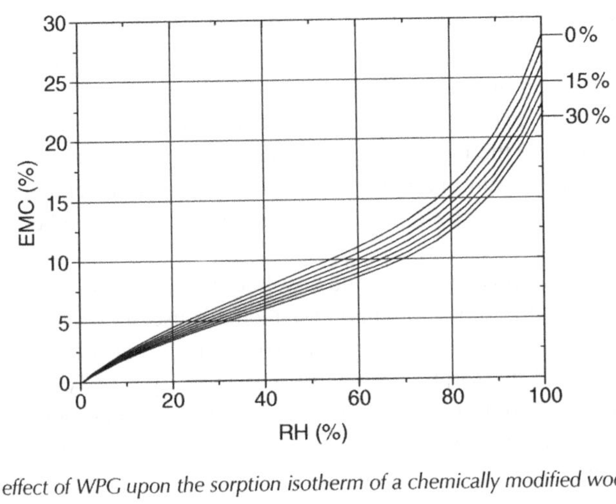

Figure 2.7 *The effect of WPG upon the sorption isotherm of a chemically modified wood.*

between the reduction in wood MC and bulking of the cell wall, then it is essential that EMC_R is the parameter used.

Notwithstanding, even when this potential source of error is taken into account, there is still a genuine reduction in EMC as a result of, e.g. acetylation. Where cross-linking is not the mechanism, this reduction is normally attributed to a combination of cell wall bulking and OH substitution, and it can be difficult to deconvolute the two effects. Finally, reductions in EMC are also accompanied by an increase in the time taken to achieve equilibrium and this can increase considerably at higher WPGs in particular, requiring great care to ensure that true equilibrium has indeed been achieved.

The reduction in EMC as a result of wood modification is sometimes reported in terms of a parameter referred to as the Moisture Exclusion Efficiency (MEE), which is defined as follows:

$$\text{MEE}\ (\%) = [(E_u - E_m)/E_u] \times 100 \tag{2.4}$$

where E_u is the EMC of the unmodified wood and E_m is the EMC of the modified wood.

2.6.4 The Dimensional Instability of Wood

As wood absorbs moisture, the cell wall swells significantly more in the radial/tangential plane than in the longitudinal direction, due to the winding angle of the microfibrils in the S_2 layer being rather low. This increase in the outer dimensions of the cell wall results in a gross swelling of the wood material. Such swelling tends to be greater in the tangential direction, since the rays tend to have a restraining effect upon swelling in the radial direction. When the FSP is reached, there is no further change in the dimensions of the cell wall and water now accumulates in the lumen rather than the cell wall. Dimensional instability gives rise to problems in the satisfactory operation of components such as doors and windows, and can also result in the distortion of items such as cladding.

In a laboratory environment, the dimensional (in)stability of the wood is determined by measuring the external dimensions (longitudinal (l), radial (r) and tangential (t)) of small wood samples. Measurements are performed on relatively small samples, and it is very important to ensure that the grain orientation of the samples is straight and parallel to one of the sides, to minimize inaccuracies caused by distortion of the samples. Changes in dimension are then recorded under different conditions. In one method (Rowell and Ellis, 1978), the dimensions of the wood are first recorded in the oven-dry condition: the wood samples are then vacuum impregnated with distilled or deionized water and then left submerged in the water for several days. The dimensions of the water-impregnated samples are then recorded, and the samples then oven-dried again. This procedure is carried out over a number of oven-dry water soak cycles. Dimensional stability is then reported as the swelling coefficient (S), calculated as follows:

$$S\ (\%) = [(V_{ws} - V_{od})/V_{od}] \times 100 \tag{2.5}$$

where V_{ws} is the water-swollen volume of the wood and V_{od} is the oven-dry volume of the wood.

The effect of a wood modification where stabilization is due to cell wall bulking is illustrated in Figure 2.8.

When chemically modified wood samples are solvent extracted and then oven-dried before experiments, this usually results in them being slightly swollen compared to samples that are oven-dried from a water-saturated state, as occurs with solvent-exchange drying (Hill and Papadopoulos, 2001). This can then affect subsequent experiments, such as dimensional stability tests, when the samples are subsequently subjected to water-saturation, or exposed to high RH, followed by oven-drying. The first cycle can also be affected by the presence of nonbonded but bulking chemical (Rowell and Ellis, 1978). For this reason, in such tests it is good practice to ignore the results from the first cycle, since these are generally unrepresentative (Rowell and Ellis, 1978; Hill and Jones, 1996a).

If the modification imparts dimensional stability to the wood, the difference between the water-saturated and oven-dry volume of the wood sample is reduced, resulting in a lower value for S. The increase in dimensional stability can then be evaluated in terms of an anti-swelling efficiency (ASE):

$$\text{ASE}\ (\%) = [(S_u - S_m)/S_u] \times 100 \qquad (2.6)$$

where S_u is the swelling coefficient of unmodified wood and S_m is the swelling coefficient of modified wood.

The use of a water-soaking, oven-drying series of cycles for the determination of dimensional stability of wood is a severe test (although it may produce useful data) and it does not necessarily reflect the conditions that wood will encounter in service conditions. As a result, some workers determine dimensional stability by subjecting samples to different relative humidities. It can be readily understood that dimensional stability values determined using different methods will not be comparable and it therefore needs to be explicitly stated how these values were obtained.

Since these volumes are determined from external dimensions of the samples, they do not reflect actual changes in the cell wall volumes, and where this is claimed or assumed, the data needs to be treated with caution. Sample geometry is a crucial factor affecting the results obtained. In particular, the orientation of growth rings with respect to the sample

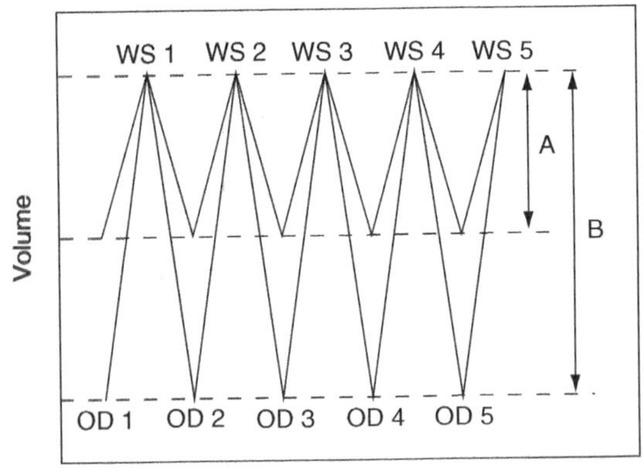

Figure 2.8 *Dimensional stabilization due to cell wall bulking. (A represents a modified wood and B unmodified).*

geometry and the number of growth rings within the sample are extremely important. It is important that the samples are prepared so that the growth rings are exactly parallel to the tangential longitudinal face. In practice, this can be difficult to achieve and some (slight) variability has to be accepted. Samples containing defects (such as small knots) must be rejected due to the variability that these introduce into the data obtained.

Wood Modification and Dimensional Stability

Wood modification can improve the dimensional stabilization of wood by two mechanisms. Where the cell wall is filled in some way by the reagent (whether covalently bound or not), the cell wall is swollen. When the dimensional stability of the modified wood is subsequently determined, the wood can then only swell by an additional amount, which is dependent upon the bulking of the cell wall due to the volume occupied by the modifying agent. This is illustrated in Figures 2.8 and 2.9b.

Where the reduction in ASE is attributable to bulking of the cell wall by the modifying agent, some workers calculate the influence of this in terms of a bulking coefficient (BC), defined as follows:

$$\mathrm{BC}\ (\%) = [(V_m - V_u)/V_u] \times 100 \tag{2.7}$$

where V_u is the volume of the unmodified wood sample and V_m is the volume of the modified wood sample. Since these volumes are invariably determined by measuring the external dimensions of the wood sample, it cannot be assumed that BC is related to bulking of the cell wall, but only to the gross wood structure.

When dimensional stability is achieved due to cell wall bulking, the dimensional stabilization achieved is equal to the volume of the water-saturated sample minus the volume of the modified wood. Another class of modification reaction is due to cross-linking between the cell wall polymeric components. In this case, dimensional stability is imparted to the modified wood because movement of the cell wall is restrained, although the volume of the cell wall occupied by the modifying agent may still have an influence (Figure 2.9c). Ohmae *et al.* (2002) have suggested a method by which the various mechanisms can be

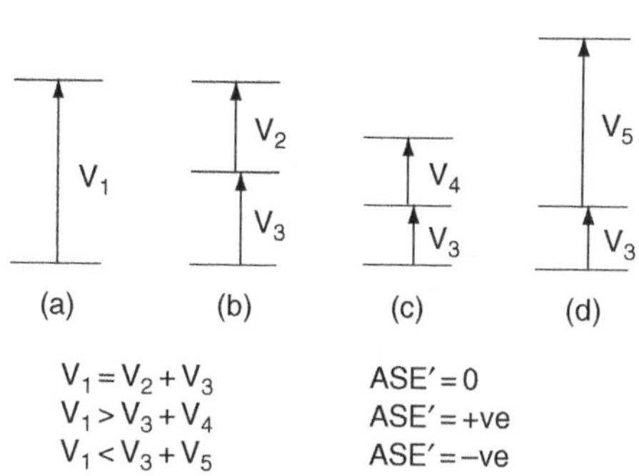

Figure 2.9 *Changes in ASE′ due to different forms of wood modification, according to Ohmae et al. (2002). Modification due to bulking (b) results in an ASE′ of zero, and modification with cross-linking (c) yields a positive ASE′, whereas cell wall damage produces a negative ASE′ (d).*

determined. In addition to the conventional means of determining ASE, an additional measure (ASE′) has been proposed, which is calculated as follows:

$$\text{ASE}'\ (\%) = [(S_u - S_m')/S_u] \times 100 \tag{2.8}$$

where S_m' is calculated by subtracting the oven-dry volume of the wood sample prior to modification from the water-saturated volume of the same sample after modification.

This means that if the dimensional stability of the modified wood is due solely to a bulking phenomenon, then ASE′ will be zero (Figure 2.9b). If, however, cross-linking is also a factor, then the water-saturated volume of the modified wood will be lower than would be calculated by considering bulking only (Figure 2.9c), resulting in a lower value for S_m' and hence a positive value for ASE′. If the modification results in cell wall degradation occurring, which leads to an anomalous increase in volume, then a negative value of ASE′ will be obtained (Figure 2.9d). This is summarized in Figure 2.10.

In determinations of the dimensional stability of wood, it is essential to determine the volume changes over a number of cycles. This is particularly important when the water soak/oven-drying method is used, since if the modifying chemical is susceptible to leaching by water, this will become evident as the test proceeds. Unfortunately, in reports in the literature where only one cycle was used for such tests, it cannot be assumed that the figures reported would look so good had the test been continued. Results from the first cycle should be discarded, since they invariably prove to be unrepresentative of values obtained in subsequent cycles. Ideally, over a number of cycles, a plot such as that illustrated in Figure 2.8 is obtained. In practice, this is seldom observed, with results usually obtained that are similar to those of Figure 2.11. The plots in Figure 2.11 have been exaggerated for clarity, but illustrate typically observed trends.

Figure 2.11a illustrates the situation in which both the water-saturated and oven-dry volumes decrease at an equal rate. In this case, there is no decrease in the dimensional stability (S). In most cases, this can be attributed to the loss of a thermally labile component

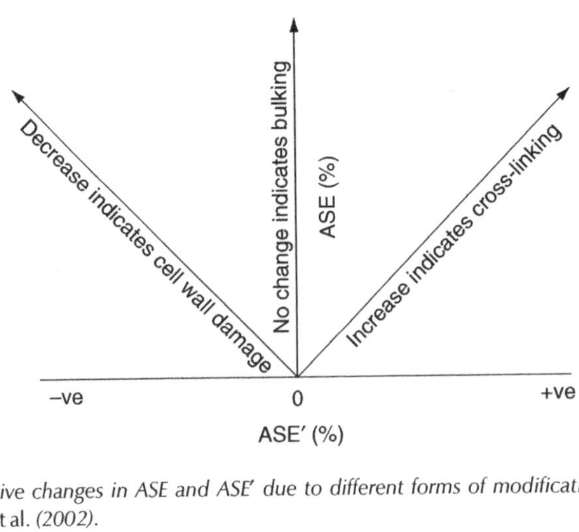

Figure 2.10 Relative changes in ASE and ASE′ due to different forms of modification, according to the criteria of Ohmae et al. (2002).

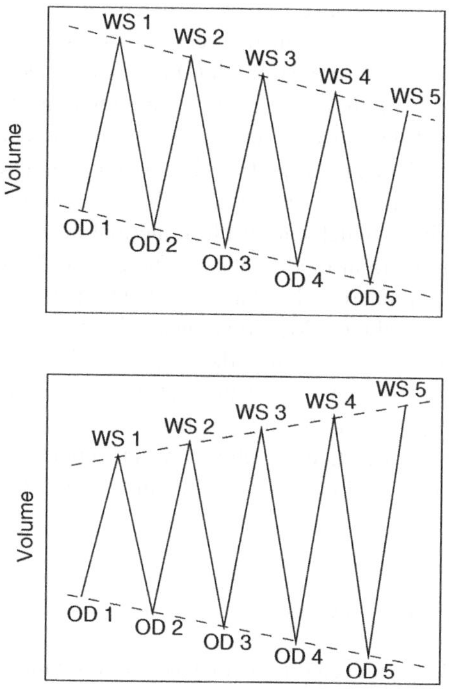

(a)

(b)

Figure 2.11 *The change in volumetric swelling behaviour due to (a) loss of cell wall material but not cell wall bulking agent, and (b) loss of cell wall cross-linking agent.*

of the wood material (probably the hemicellulose fraction), but no loss of the modifying agent. In Figure 2.11b, there is a decrease in the oven-dry volume and an increase in the water-saturated volume, accompanied by a decrease in dimensional stability. This can probably be attributed to loss of the modifying agent responsible for the dimensional stabilisation. In all studies, it is good practice to determine the oven-dry weight of the modified wood for each cycle.

2.7 The Mechanical Properties of Modified Wood

Wood is a material that has evolved to fulfil a number of structural roles in supporting the tree canopy, such as resisting wind loads, and so on. These mechanical properties can be exploited and used in various structural roles by mankind. It is, however, important to note that wood, as it performs in the natural environment, is fully saturated with water. When used in man-made structures, wood is invariably dried to a relatively low moisture content (below the FSP) and as a consequence exhibits physical properties that are subject to change as the moisture content of the cell wall changes.

In most cases, wood modification results in a change in the way in which the wood responds to moisture and, with a great many wood modifications, a reduction in EMC at a given RH is observed. This will in itself alter the mechanical properties of the wood, since it is well known that a reduction in the cell wall moisture content results in an increase in the MOE, and an increase in strength (Dinwoodie, 2000). Creep deformation is also influenced by cell wall moisture content, although the relationship is more complex, and especially so if the wood sample is subjected to cyclic humidity changes. Apart from changes in mechanical properties due to the influence of the cell wall moisture content, wood modification also affects these properties when the process involves the application of heat, which can result in thermal degradation of the substrate (the extreme example being thermal modification). The use of chemicals can also result in degradation of the cell wall macromolecular components (particularly if acidic chemicals are used or occur as by-products). Modification can involve the removal or masking of the cell wall OH groups, which can affect the properties. Modification will, in many cases, result in changes to the dimensions of the material being studied, which must be accounted for when the properties are being calculated. The way in which dimensional change is accounted for must be stated explicitly when reporting measurements, and will lead to erroneous figures if this is not done. This is discussed in more detail in Chapter 3.

2.7.1 Dynamic Mechanical Properties

In these studies, the sample is subjected to a sinusoidal deformation and the response of the sample to this deformation is measured. This provides information regarding the dynamic Young's modulus (E'), which is related to the sound velocity in the sample, and $\tan \delta$ measurements that represent energy loss due to internal friction (damping) in the sample. This property can also be related to other mechanosorptive phenomena, such as creep deformation. Ono and Norimoto (1984) have shown that there is an exponential relationship between the specific longitudinal dynamic Young's modulus (E'/γ) (where γ is the specific gravity of the sample) and damping:

$$\tan \delta = C(E'/\gamma)^{-k} \tag{2.9}$$

where C and k are constants. In a plot of $\log(\tan \delta)$ against $\log(E'/\gamma)$, a linear regression (which will be called the 'reference line') can be drawn through the data at a given cell wall MC, as shown in Figure 2.12. If the EMC of the wood increases, the reference line shifts upwards; and, conversely, downwards if the EMC decreases (Sasaki *et al.*, 1988). This relationship of experimental data to the reference line is affected by the type of modification that is performed on the wood. For treatments where there is no penetration of the cell wall, there is no change in the relationship and the results fall on the reference line (Akitsu *et al.*, 1993a,b). Where there is a cell wall modification either involving cross-linking of the cell wall polymers, or the filling of the cell wall with hydrophobic moieties, then the results fall below the reference line. Where hydrophilic agents are present in the cell wall and cause an increase in the chain mobility of the cell wall polymeric components, the results fall above the reference line at a given RH. Results for acetylation fall into the middle of these three categories.

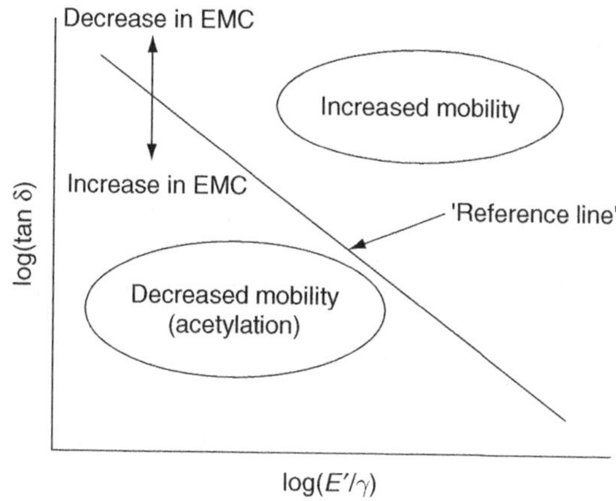

Figure 2.12 *The change in dynamic mechanical properties due to different forms of modification (Ono and Norimoto, 1984).*

2.8 Modified Wood and Biological Degradation

At the end of the life of a tree, all of the chemical components are broken down and the constituent atoms are returned to the environment, where they are available as the building blocks for new organisms. This cyclic materials flow is characteristic of biological processes and can be considered a positive aspect of the use of wood, in that at the end of its life, a wood product can be returned to the natural cycles. However, although this is advantageous from an ecological perspective, the degradation of wood by biological organisms is a distinct disadvantage when wood is used in service.

In order to ensure that timber has a reasonably long life in service, it is necessary to resort to the use of wood preservation techniques to prevent, or at least slow, attack by biological organisms. Simplistically, such attack can be divided into macrobiological (insects and mammals) and microbiological (fungi, staining fungi and moulds, and bacteria).

2.8.1 Wood Degradation by Decay Fungi

Subject to the necessary conditions, wood is susceptible to degradation by microbiological organisms. In order for fungal attack to occur, three components are required: water, oxygen and a source of nutrients. One of the best ways of preventing fungal attack is to prevent the wood from reaching a sufficiently high moisture content, and good design of wood structures should ensure that this is the case. However, there are situations where wood is exposed to conditions in which fungal attack will occur, either through poor design, or because of the application for which the wood is used, and in these circumstances a means of preventing fungal attack is required. Fungal attack is divided into three classes, which are named according to the appearance of the wood following degradation: brown rot, white rot and soft rot.

Brown Rot

Brown rot decay is characterized by the removal of the polysaccharide components of the cell wall, to leave a lignin-rich material. As a result, brown-rotted wood is a reddish-brown colour and in the late stages of decay exhibits deep cross-cracking when the wood is dried; longitudinal cracking may also be observed. Brown rot fungi are unable to degrade lignin, although extensive demethoxylation of the aromatic lignin fraction occurs. High strength losses are found with wood during early stages of decay with brown rot, due to attack of the polysaccharide components of the cell wall.

White Rot

White rot fungi selectively attack the lignin component of the cell wall and utilize the lignin fragments as an energy source, resulting in a bleaching of the wood. White rot fungi can be subdivided into simultaneous (where the lignin and polysaccharide are degraded at similar rates) and preferential white rot (where the lignin component is degraded prior to the polysaccharide). White-rotted wood does not exhibit the cubical cracking that is characteristic of brown rot attack, but the surface becomes softened; with shrinkage of the wood occurring in advanced stages of decay. Hardwoods are generally more susceptible to white rot than softwoods. In the early stages of attack by white rot, only slight strength losses are found in the wood.

Soft Rot

Soft rot is a term used to describe a form of attack of wood that differs from the previous divisions. Soft rot attack is favoured in conditions in which the activity of the generally dominant basidiomycete fungi is retarded in some way. Soft-rotted wood is cheesy soft and darkened on the surface, but below this layer of degradation the wood is sound. When the wood is dry, the surface shows fine cracking in the longitudinal and cross directions. The appearance differs from brown-rotted wood because the cracking is denser and shallower. Soft rot is generally the dominant form of attack where wood is exposed in conditions in which there are high levels of moisture. High strength losses can be found in wood during early stages of decay by soft rot fungi.

Mechanisms of Decay

A range of enzymes are generated by fungi that participate in the breakdown of the cell wall polymer components. These are the polysaccharide-degrading enzymes (endocellulases, exocellulases, hemicellulases, β-glucosidases and oxidases) and lignolytic enzymes, produced by those fungi capable of degrading the lignin component. It has been recognized that these enzymes are too large to penetrate the cell wall of sound wood or during the early stages of decay, because they are larger than the cell wall micropores (Srebotnik and Messner, 1990; Eaton and Hale, 1993; Goodell *et al.*, 2003). For this reason, it has been proposed that the fungi generate a range of low molecular weight diffusible agents (LMWDAs), that have dimensions allowing for penetration of the cell wall and that initiate degradation of the cell wall polymeric constituents. This initial attack results in an increase in the dimensions of the cell wall micropores, so that subsequent ingress by enzymes is presumably facilitated (Flournoy, 1991). In order that these LMWDAs and subsequently the enzymes can penetrate the cell wall, it is essential the moisture content thereof is sufficiently high to allow for efficient diffusion. At the present time, this process is not well understood and is the subject of much research activity.

Testing of Decay Resistance

Testing of decay resistance can be performed in a laboratory environment or in outdoor field trials, and there are many standards defined for these tests. The first objective of a laboratory-based test is to provide a methodology for the rapid screening of a candidate wood preservative, treatment or modification in order to assess which ones exhibit decay resistance. Broadly speaking, laboratory-based tests can be divided into sterile (pure culture) tests and unsterile tests (such as fungal cellar, soil burial etc.).

In sterile tests, the wood sample is introduced into a container in which there is an active mycelium of the test fungus growing on a suitable medium. The container is sealed in such a way so as to prevent contamination, but to allow for the transport of moisture and gases into and out of the vessel. It is essential to ensure that the wood, container and growth medium are sterile, so that the wood is exposed only to the chosen fungus. Decay resistance is determined after a period of exposure by determination of a change in a chosen parameter, such as a mechanical property, or weight loss. These tests usually involve testing of wood at different levels of treatment (e.g. WPG for chemically modified wood), and the relationship between this and, for example, weight loss due to decay is plotted to determine a decay protection threshold (Figure 2.13). In some sterile tests,

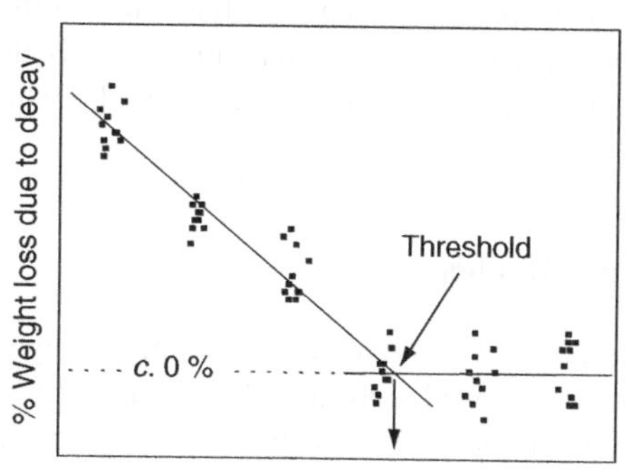

Figure 2.13 *An idealized relationship between weight loss due to decay in a laboratory test and WPG for a modified wood.*

samples are buried in an inert medium, such as vermiculite, that has been innoculated with a range of test fungi.

There can be a great deal of variation in the results reported from such tests, and it is important to try to minimize the variation as much as possible. Since wood itself is inherently variable, it is often recommended that matched wood samples are used for the control and treated specimens. Size of wood sample is an important variable that is defined in the standards, but may not be adhered to by some workers for various reasons. The activity of the fungus can differ greatly, between laboratories, between containers in the same test and even within the containers themselves. This can be minimized by adopting protocols that take some account of this, by using only one fungal strain, or by performing a sufficiently robust test or series of tests, so that some confidence can be placed in the results obtained. However, it has to be acknowledged that in dealing with natural systems, there will always be variability and there will always be tests that produce unexpected, or contradictory, results. The tests that have been developed have been concerned with the screening of more conventional wood protection systems, and may not necessarily be appropriate for the testing of modified wood. For example, modified wood often exhibits lower EMCs compared to unmodified wood, and in conventional decay tests, the wood may not attain sufficiently high moisture contents to facilitate fungal attack. In a situation in which screening is the objective, this may not prove to be a problem, but if some understanding of the mechanism of decay resistance was the objective, then a modified test would be necessary. As a consequence, there is a tendency to find that many studies of modified wood have used nonstandard protocols and there is, as yet, no consensus as to what tests are the most appropriate, although this has been recognized and is the subject of debate (e.g. Van Acker and Stevens, 2000; Van Acker and Jones, 2002).

With laboratory-based unsterile tests, samples (usually stakes) are placed into soil containing a range of organisms. Some tests specify a commercial plant growth product as the soil medium. In some cases, the soil is inoculated with additional materials, such as litter from a forest, which has a specific type of fungal activity. The moisture content of the soil bed is maintained within certain limits, in order to encourage a specific form of attack (soft rot, for example). The progress of the decay can be measured by removing stakes at intervals, and determining the static bending modulus, and the stakes can be replaced. At the end of the test, the stakes are removed and the weight loss determined, and/or some other property measured, such as the dynamic modulus. Although such tests are closer to what might be expected in service conditions, there is clearly more variability to be expected in terms of what type of attack will occur and by what species, compared to pure culture tests.

2.8.2 Staining Fungi and Moulds

Fungi that cause discolouration of wood are called staining fungi, whilst those that grow superficially on the wood are referred to as moulds. Technically, staining is referred to as sapstain in freshly felled wood and as blue stain in wood in service. The main source of energy of the fungi is from freely occurring sugars and starch, not the cell wall structural macromolecules. Such attack does not result in significant losses of strength or mass of the wood unless staining is very substantial, but has great economic importance because of the degradation in the aesthetic qualities of the material. There are a variety of tests used to evaluate sapstain, with no one test having received general acceptance. Standardized procedures have been developed for the testing of blue stain. Overall, very little work has been done on the resistance of modified wood to blue staining.

2.8.3 Wood Degradation by Bacteria

Bacterial attack is an early stage in the degradation of wood exposed in wet or moist conditions. Bacteria can be the dominant form of attack when fungal decay is suppressed by a wood-preserving treatment. Bacteria can attack the cell wall of wood by tunnelling, cavitation or erosion mechanisms (Eaton and Hale, 1993).

2.8.4 Wood Degradation by Insects

Although wood-destroying insects can attack wood at a lower moisture content than that required by other degrading organisms, there are also a variety of insects that will attack wet wood. An examination of the range of insects and forms of attack is not possible within this text, but a very good coverage of this subject is given in Eaton and Hale (1993). Most of the studies of insect attack on modified wood have been concerned with termites, and are clearly of great economic importance where such species are active.

2.8.5 Attack by Marine Organisms

In circumstances in which wood is exposed in a marine environment, attack by marine organisms, such as fungi, bacteria and marine boring animals, can occur. Of these, the marine boring organisms (molluscs and crustaceans) are responsible for most of the damage. There is relatively little literature on this subject in connection with modified wood.

2.9 Wood and Weathering

Wood exposed in exterior conditions is subject to degradation by the UV component of the solar spectrum. This degradation is essentially confined to the lignin component of wood, resulting in the gradual release of polysaccharide-rich wood cells, which are subsequently removed from the wood surface by wind and rain erosion. The lignin fragments produced by UV irradiation are subsequently metabolized by blue stain fungi, resulting in the production of what is often optimistically referred to as a silver-grey (or driftwood) exterior, but what is in reality an aesthetically inferior surface appearance.

2.10 Proof of Bonding

With chemical modification of wood, it is necessary to prove that a chemical bond has been formed with the wood cell wall polymers. One simple test involves determining the weight and dimensions of wood before and after modification. Weight gain due to reaction is commonly reported in the wood modification literature as weight percentage gain (WPG), which is calculated as follows:

$$\text{WPG}\ (\%) = [(M_\text{m} - M_\text{u})/M_\text{u}] \times 100 \tag{2.10}$$

where M_m is the oven-dry mass of the modified wood and M_u is the oven-dry mass of the unmodified wood.

Similarly, the volume change (VC), or bulking coefficient (BC), can be reported as a percentage increase compared to the original volume (Equation (2.7)).

Where the wood has gained weight and is swollen after modification, this probably indicates that the reagent has penetrated the cell wall. However, this does not constitute proof that there is any chemical bond between the wood and the reagent. If the wood is subjected to rigorous solvent extraction after modification and a weight gain and swelling of the material are still observed, then this indicates that the modification is stable, but again does not prove the existence of a chemical bond.

Similarly, leaching of the wood in water after an appropriate solvent extraction does not prove that a chemical bond exists between the modificant and the wood.

In circumstances in which the modifying agent cannot polymerize and a stable weight gain and volume change (after solvent extraction and leaching) are obtained, then this is a strong indication that a chemical bond has been formed. However, absolute proof requires confirmation by the use of another technique, such as IR or NMR spectroscopy, to identify a specific bond type associated with the modification. Where polymerization of the reagent is possible, then it is absolutely essential to obtain unambiguous confirmation using a reliable technique. Although a chemical bond may be the objective, provided that the modification is stable and provides the desired performance benefits, it is of little consequence whether or not a bond has been formed. However, in the absence of definite proof, it is not possible to claim that a chemical bond has been formed. Unfortunately, the wood modification literature contains more than a few examples where a chemical bond is claimed, in the absence of definitive scientific proof.

2.11 Conclusions

It is recognized that as wood preservative systems that exhibit broad-spectrum activity are phased out, there will be an increase in the use of systems that are designed to be 'fit for purpose'. In the area of wood modification, it is well known that some systems perform well in some situations and less well in others. There will need to be a framework of tests that recognize this fact and allow products to be licensed that perform entirely satisfactorily in service, even though they may fail the current tests as they stand at the moment.

Until recently, the lack of uniformly adopted standard test procedures specifically designed for evaluating modified wood has been little more than an inconvenience: indeed, the use of different test methods helps to determine which of the tests adopted may be the most appropriate. Due to the rapid changes in this area, with many commercial processes now being adopted, it is now a matter of great urgency to ensure that the appropriate protocols are agreed at an international level.

3

Chemical Modification of Wood (I): Acetic Anhydride Modification

3.1 Introduction

This chapter reviews research into the chemical modification of wood with acetic anhydride. Modification with higher molecular weight anhydrides and cyclic anhydrides are considered separately in Chapter 4. The surface properties of anhydride-modified wood are also considered separately in Chapter 6. Reaction of acetic anhydride takes place with the cell wall polymeric hydroxyl groups to form an ester bond, with acetic anhydride produced as a by-product (Figure 3.1). Of all the wood chemical modification reactions, acetylation has the longest pedigree, with Rowell (1983a) citing work dating back to 1928 in his review. Acetylation has been the subject of extensive research and is currently in the process of being commercialized.

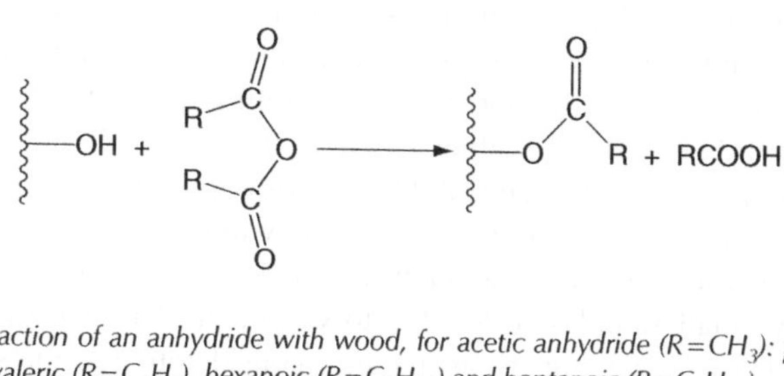

Figure 3.1 *The reaction of an anhydride with wood, for acetic anhydride (R=CH_3): propionic (R=C_2H_5), butyric (R=C_3H_7), valeric (R=C_4H_9), hexanoic (R=C_5H_{10}) and heptanoic (R=C_6H_{11}).*

Wood Modification: Chemical, Thermal and Other Processes C. Hill

3.2 Reaction Protocols

The reaction of an acid anhydride with wood is complex, with many factors affecting the reproducibility of the results. The variables include the following:

- sample and sample pre-treatment (e.g. presence of extractives, size of samples, wood species, wood density and moisture content of samples);
- reaction medium (e.g. neat anhydride or solution, use of catalyst and/or swelling agent, vapour-phase or liquid-phase reaction, and presence of acetic acid);
- reaction variables (e.g. scale of reaction, temperature of reaction, time of reaction, pre-impregnation of samples, method of applying heat to the reaction medium/vessel, ambient pressure or pressure vessel, and quenching of reaction);
- clean-up procedure (e.g. solvent extraction, vacuum with heating, water-soaking, steam stripping, and solvent stripping).

Many of the above are generic and apply to all chemical reactions taking place within the cell wall.

3.2.1 Sample Variables

Extractives

The presence of extractives in the wood can have a significant effect upon the results obtained during acetylation studies, with leaching of extractives into the acetylation medium leading to lower WPG values being recorded than have actually been obtained. For this reason, extractive-free wood is often used in laboratory studies. It would clearly be impractical to use extractive-free wood on an industrial scale, and where laboratory studies do use wood containing extractives, alternative methods for determining the extent of the reaction (i.e. not gravimetric) are required.

Sample Dimensions

The efficiency of the wood acetylation reaction depends on the accessibility of the reagent to the bulk of the wood and ultimately in the cell wall interior. Reactions occurring within the wood cell wall require consideration of the accessibility of the cell wall interior to the reagent molecules. It is known that the rate of reaction is promoted by wood-swelling agents (such as pyridine). This swelling expands the cell wall micropore network, improving access of reagent to the cell wall OH groups (Patscheke and Dilg, 1958).

Macroscopic distribution of reagent is an exceedingly important issue when considering the properties of acetylated timber on an industrial scale, but also needs to be considered even for relatively small samples in a laboratory context. The penetration of reagent molecules into timber of large dimensions requires consideration of the same issues that affect treatability of larger timber specimens with wood preservative. Drying of wood of larger dimensions can affect the permeability of the timber, with penetration problems occurring due to pit aspiration, for example. Rowell (1977b) addressed this problem by using solvent vapour drying to prevent pit aspiration, but no improvement in reactivity was observed. It is an important concern to ensure that distribution of reagent in samples of large dimension is uniform. Uneven distribution can result in 'envelope' treatments, where the core of the material is not protected to the same extent as the interior and as a result may attain a higher MC in service, resulting in greater dimensional instability and reduced decay resistance. Nonuniform reagent penetration is also very likely to lead to distortion of the wood when the modification process takes place. Refractory species of large dimension, such as spruce and Douglas fir, present impregnation problems for the acetylation process, and a fungal pre-treatment has been investigated as a possible means for overcoming this difficulty (Messner *et al.*, 2003). Pre-steaming has also been used to improve the treatability of wood for conventional preservatives (Bergervoet, 1983), and for formaldehyde modification (Burmester, 1971b), but does not appear to have been studied for anhydride modification. Torgovnikov and Vinden (2000) noted that the use of pre-steaming to increase wood permeability can result in strength losses, although they report that this method is commonly used to condition radiata pine prior to preservative treatment. They have developed a microwave technique to improve wood permeability, which may have application as a pre-conditioning process prior to acetylation.

The Effect of Wood Moisture Content upon Reactivity

The presence of moisture within the cell wall of the wood will improve both cell wall and gross accessibility of the reagent to the wood interior, because the cell wall is swollen. However, reaction of cell wall water with the anhydride also results in the loss of reagent as the corresponding carboxylic acid. The presence of water in the lumens confers no advantage whatsoever. The influence of wood moisture content (MC) upon the reactivity of wood to acetic anhydride has been considered in several studies. Rowell *et al.* (1990) found that the rate of acetylation was the same at wood MCs of 0, 5 and 7 %. Beckers and Militz (1994) noted that modification of wood at low MCs was not likely to be viable on an industrial scale, and studied the effect of a wide range of MCs (from zero to 26 %) upon the degree of acetylation. A decrease in WPG after reaction was found as wood MC increased, but even above wood MCs of 20 %, reaction with the wood still occurred, with WPGs of the order of 8 % being found.

The Influence of Wood Type and Species upon Reactivity and Distribution

It is possible that there will be differences in the reactivity of heartwood and sapwood, or between juvenile and mature wood in acetylation reactions. As a consequence, larger-dimension wood specimens containing both heartwood and sapwood, or mature and juvenile wood, could be liable to distortion when modified. Considering the importance of this to any commercial acetylation operation, it is surprising to see that there is, in fact, very little literature dealing with this issue.

In a study of acetylation of loblolly pine oriented strand board (OSB) flakes, it was found that reactivity of the wood increased in proportion to the juvenile wood content, when moving from the base to the top of the tree (Hon and Bangi, 1996). However, no difference in reactivity was found in moving from the core to the outer shell (i.e. radially). Rowell and Plackett (1988) found no difference in the reactivity of the heartwood and sapwood when *Pinus radiata* flakes were acetylated using a dip procedure. Rowell *et al.* (1990) also found that the reactivity of pure heartwood pine flakes was the same as that of commercially produced particles, which consisted of a mixture of heartwood and sapwood. However, for reaction of larger-dimension specimens, it was found that pine sapwood exhibited a higher rate of reaction with acetic anhydride than heartwood and also reacted to higher ultimate WPGs (Beckers and Militz, 1994). This indicates that reduced reactivity of the heartwood is a function of permeability and is controlled by macroscopic features (e.g. pit aspiration), rather than being due to an inherent difference in the reactivity of the cell wall. In view of the lack of literature on this subject, there is clearly a need for further research with respect to the acetylation of solid wood.

Differences in reactivity between wood species have been observed, which can be attributed to a number of factors. Rowell *et al.* (1986c) acetylated southern pine and aspen flakes under identical reaction conditions, and found that the softwood species exhibited a higher rate of reaction. Beckers and Militz (1994) performed a pilot-scale study of the acetylation of beech, eucalyptus, poplar, pine, Douglas fir and spruce, finding considerable differences in reactivity between the woods studied.

3.2.2 The Reaction Medium

The Use of Catalysts

There has been extensive research into the use of catalysts in anhydride reactions with wood. The earliest work studying acetylation of wood used mineral acid catalysts in reactions with sawdust or wood flour. Generally, the use of an acid as a catalyst is to be discouraged, since the acid will also degrade the polysaccharide components of the wood. The first recorded acetylation reaction of whole wood is attributed to Stamm and Tarkow (1947), who used pyridine as a catalyst. Pyridine functions as a solvent, as a swelling agent for the wood and also as a catalyst for the reaction (Figure 3.2). Many other catalysts have been investigated besides pyridine (Rowell, 1983a).

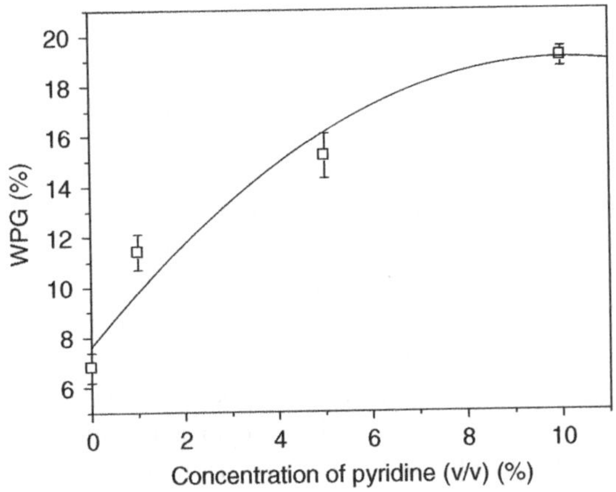

Figure 3.2 *The effect of increasing the concentration of pyridine catalyst upon the reactivity of acetic anhydride with wood, for reaction at 100°C for 30 minutes (from data of Çetin, 1999).*

Most acetylation reaction studies involve reactions at temperatures ranging from 100 °C to 140 °C. However, WPGs of the order of 16–30 % were achieved when reacting acetic anhydride with wood at temperatures as low as 25–45 °C by the use of trifluoroacetic acid and magnesium perchlorate as synergistic catalysts (Arni *et al.*, 1961b). Although the use of catalysts and swelling agents in acetylation reactions has been of interest in academic studies, their use is not considered suitable for large-scale industrial processes.

The Influence of Acetic Acid on Reactivity

Since acetic acid is produced as a by-product in acetylation reactions using acetic anhydride, there has naturally been interest in studying the effect of the presence of acetic acid in the anhydride upon reactivity. There is also the possibility that the presence of acetic acid in the cell wall at high temperatures can lead to degradation occurring. Acetic acid can serve to swell the cell wall and hence promote the rate of reaction, although the presence of acid will also dilute the active anhydride reagent, thereby reducing the rate of reaction. Acetic anhydride itself is not capable of swelling the cell wall to any significant extent (Rowell, 1983a). Acetic acid in the anhydride was found to result in a slight increase in the reaction rate up to an acetic acid content of 15 %, and a decrease thereafter (Rowell *et al.*, 1990; see Figure 3.3). The degree of acetylation at 30 % acetic acid content was the same as with pure acetic anhydride.

It is therefore important to monitor the acetic acid content in the anhydride, in order to control the process efficiency. Recovery of acetic acid from the anhydride following reaction is an important part of the economics of the whole process.

Reactions in the Vapour Phase

Although most of the studies of acetylation of wood have used liquid phase reactions, there have been some reports of vapour phase reactions. Arora *et al.* (1979, 1980) and

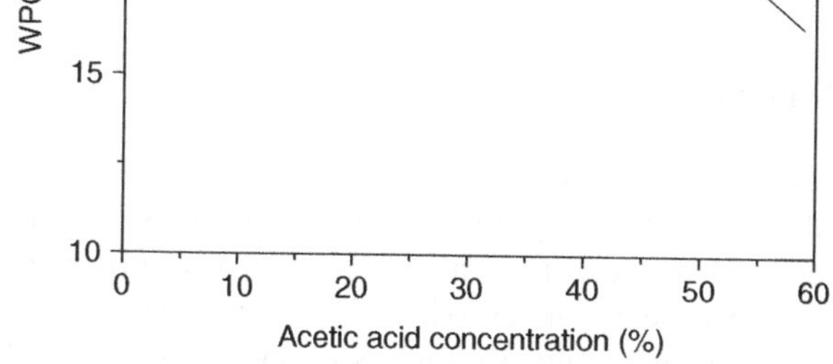

Figure 3.3 *The effect of acetic acid concentration on reactivity of acetic anhydride with wood, for reaction at 120 °C for 5 hours (redrawn from data of Rowell* et al.*, 1990).*

Nishino (1991) noted the presence of an induction period in the kinetic profile of reactions of solid wood samples with acetic anhydride vapour. The duration of this induction period decreased and the rate of reaction increased with both increasing temperature and concentration of acetic anhydride vapour in the reaction vessel (Nishino, 1991). The rate of reaction was diffusion limited, with the rate of diffusion being greatest in the longitudinal direction of the wood samples. This phenomenon is only found in solid wood samples, since no induction period was noted for the reaction of wood fibres with acetic anhydride vapour (Loras, 1968).

3.2.3 Reaction Variables

The rate of reaction is related to the temperature of reaction, with higher temperatures resulting in faster reaction. Reaction profiles are usually reported as WPG against time (Figure 3.4a). The upper limit for the reaction temperature is around 120 °C, since cell wall degradation becomes more significant above this. If the wood is pre-impregnated with solvent but not reagent prior to reaction, then the reaction profile is diffusion limited until such a time when most of the available OH groups are reacted (Figure 3.4b). Many studies, however, involve pre-impregnation of the wood with anhydride reagent prior to heating, resulting in penetration of the cell wall by acetic anhydride. Where reagent is present in the cell wall, there is a rapid initial reaction of anhydride molecules with the cell wall OH groups. Once this stage of the reaction is complete, the kinetics then become diffusion limited, as it is necessary for anhydride molecules to diffuse to reactive sites in the cell wall (Hill *et al.*, 2000). If the wood is placed in acetic anhydride, then there is an essentially infinite reservoir of acetic anhydride to replace reacted molecules in the cell wall. However, if the impregnated wood is heated in an oven, then it is possible that the anhydride will be used up before complete reaction with the cell wall has occurred.

Microwave Reactions

Whereas the majority of reactions of acetic anhydride with wood are thermally assisted, there has been some interest in using other methods for delivering energy. Larsson Brelid (2002), Larsson Brelid and Simonson (1999) and Larsson Brelid *et al.* (1999) studied the use of microwave heating to acetylate wood in order to reduce reaction times, improve the distribution of bonded reagent within the wood and achieve more efficient removal of process chemicals and by-products.

3.2.4 Sample Clean-up

For work where it is critical to examine the properties of the chemically modified wood with the certainty that there are no by-products or solvents present, it is advisable not to rely solely on an oven-drying stage, or on vacuum-heating. More effective extraction can be achieved with the use of Soxhlet apparatus followed by oven-drying. However, the efficiency of this process is very dependent upon sample size and the scale at which the work is being performed. To ensure total removal of residual chemicals, it is usually a good idea

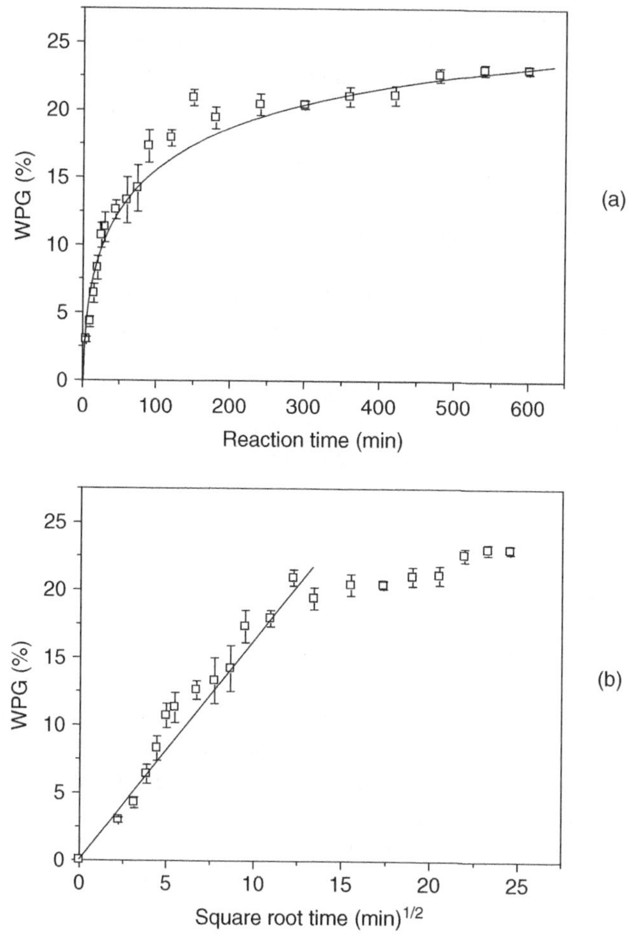

Figure 3.4 *The reaction kinetics of acetylation plotted as WPG against time (a) and in terms of diffusion (WPG against square root time) (b).*

to replace the solvent several times during the extraction process. On a laboratory scale, this can be accomplished relatively easily, but on an industrial or pilot scale ensuring that the wood is free of contamination can present major problems and be a significant cost in the process. One method for dealing with residual acetic acid is to react the wood in a post-treatment with aniline (Singh *et al.*, 1992).

Reagent and by-product recovery can be accomplished by applying a vacuum on the sample whilst maintaining a temperature in the region of 120–130 °C (Goldstein *et al.*, 1961). Often, a vacuum-heating stage is then followed by a conventional heating period in an oven under ambient pressure and atmospheric conditions. The efficiency of such processes depends critically upon the size and dimensions of the sample. Goldstein *et al.* (1961) reported that it required 10 days' heating in a forced convection oven at 105 °C to remove the residual acetic anhydride and acid from a piece of ponderosa pine of 5×10×25 cm. In order to speed up the recovery process, azeotropic distillation at around 100 °C with ligroin was investigated, which resulted in 90 % recovery of residual chemicals in 70 hours, although total removal was said to be difficult. Increase of the distillation temperature to about 150 °C, by using mineral spirits, resulted in a large increase in the rate of removal, requiring 14 hours to remove all the residual chemicals. Organic vapour stripping with xylene was also studied. Rapid removal of residual chemicals was found to result in checks forming in the wood samples. Rowell *et al.* (1990) used the application of a heating vacuum procedure to remove residual chemicals after acetylating wood chips. A pilot-scale reaction developed by Beckers and Militz (1994) has used a clean-up procedure that involves application of a vacuum with heating, and rinsing in water to convert anhydride to acid, followed by oven-drying for 10 days whilst increasing the temperature from 40 °C to 100 °C. More recently, steam post-treatments have been used to remove residual chemicals (Bongers and Beckers, 2003).

The use of microwave energy combined with applied vacuum to remove residual chemicals has been studied (Larsson Brelid and Simonson, 1999). Larsson Brelid (2002) also studied the effect of post-reaction treatment upon the final WPG obtained by the wood samples.

3.3 Cell Wall Reactivity

3.3.1 Cell Wall Accessibility

As noted in the previous section, access to the cell wall interior requires that the reagent molecules can penetrate the cell wall micropore network. The generally accepted maximum diameter (as determined by solute exclusion studies) of the cell wall micropores in fully swollen wood is of the order of 2–4 nm, whereas an acetic anhydride molecule has been estimated as having a diameter of the order of 0.7 nm (Hill *et al.*, 2004b). Clearly, when the cell wall is fully swollen, the interior is readily accessible to the reagent. However, when wood is oven-dried to zero moisture content, these micropores collapse and the interior of the cell wall would no longer be expected to be accessible to reagent. From the above, it is clear that for efficient reaction and reagent penetration, the wood is required to be in a swollen state; and that if dried from a water-saturated condition, it is desirable to maintain an open structure by some means. For this reason, solvent exchange drying has been investigated as a means of ensuring good cell wall access by reagent (Hill *et al.*, 2004b). However, this is not a feasible proposition for large-scale reactions, especially in an industrial environment.

3.3.2 The Reactivity of Cell Wall Polymers

It has been stated that the reactivity of the wood cell wall polymers to acetic anhydride decreases in the order lignin > hemicelluloses > cellulose, both within the wood cell wall (Rowell, 1982) and with the isolated polymers (Callow, 1951; Rowell *et al.*, 1994; Efanov, 2001). A comprehensive series of NMR studies has been performed investigating the substitution of the cell wall polymeric OH groups at various WPGs (Ohkoshi and Kato, 1992, 1993, 1997a,b). These have shown that not all of the lignin OH groups are substituted even at a WPG of 20%, and that the hemicelluloses are more reactive. Above 20% WPG, the crystalline cellulose begins to react. X-ray diffraction studies have indicated that the degree of crystallinity of wood being acetylated only changes above WPGs of 25% (Zhao *et al.*, 1987). However, if the wood is pre-treated with ethylenediamine, crystalline regions show much greater reactivity (Shiraishi *et al.*, 1973). Rowell (1982) reported that lignin was completely substituted at a WPG of 20%. Boonstra *et al.* (1996) found high levels of acetylation of isolated alkali Norway spruce lignin (43% WPG) after reaction in acetic anhydride vapour at 120°C, compared with a WPG of 33% for wood fibres. However, in this study, it was found that the holocellulose was comparatively unreactive (5% WPG under the same conditions), with cotton cellulose exhibiting a similar low level of reactivity. This low level of holocellulose reactivity is unusual. UV microscopy of acetylated wood has shown that the lignin UV absorption of the compound middle lamella decreases as a result of acetylation, whereas that of the S_2 layer is not affected (Sander and Koch, 2001). This was attributed to a lignin-leaching phenomenon.

Studies of thermomechanically pulped (TMP) wood fibre showed that delignification of the fibre (to give holocellulose) increased the rate of reaction with acetic anhydride (Haque and Hill, 1998, 2000; see Figure 3.5a). However, delignification of wood flour reduces the reactivity (Figure 3.5b). Haque and Hill suggested that removal of the lignin from TMP fibres improved accessibility of the cell wall to reagent and hence increased the rate of reaction. The reactivity of TMP fibre is strongly influenced by the refining conditions. Thus, where a glassy lignin-rich surface is produced, reactivity decreases because the lignin coating acts as a barrier to penetration of reagent; whereas a polysaccharide rich, fibrillated, surface exhibits higher reactivity. Such results show that, in these cases at least, the reactivity of the various hydroxyl groups associated with the cell wall components is not a significant factor controlling the rate and extent of reaction, but that substrate morphology is more important.

The distribution of adduct within the cell wall is dependent upon the conditions under which reaction is performed. A more even distribution of acyl groups will result when the cell wall is pre-impregnated with reagent prior to reaction, but if samples are not pre-impregnated then the distribution would be expected to be less regular, at low WPGs. A study of the cell wall distribution of chloroacetic anhydride using scanning electron microscopy – electron diffuse X-ray analysis (SEM–EDXA) has shown that, at low levels of reaction, the anhydride adduct is located predominantly within the S_2 cell wall, but after longer reaction times the middle lamella exhibits higher concentrations (Rowell *et al.*, 1992). It should be borne in mind that the use of chloroacetic anhydride may not exactly reproduce the reaction conditions pertaining with the use of acetic anhydride, since the presence of the chlorine atom will reduce the electron density at the acyl carbonyl carbon and thus promote reactivity. This problem is circumvented by the use of radioactive isotope labelled acetic anhydride. A study of the reaction of wood with radioactively labelled acetic anhydride showed that the distribution of acyl groups in the cell wall was more uniform than that reported for chloroacetic anhydride (Rosenqvist, 2000, 2002). In Rosenqvist's study, samples were vacuum impregnated for 15 minutes prior to reaction, which would be expected to result in a more even distribution of adduct. In the study by Rowell *et al.* (1992), samples were treated using a dip-draining method for 4 minutes before reaction, which presumably did not result in significant cell wall penetration by the reagent prior to reaction.

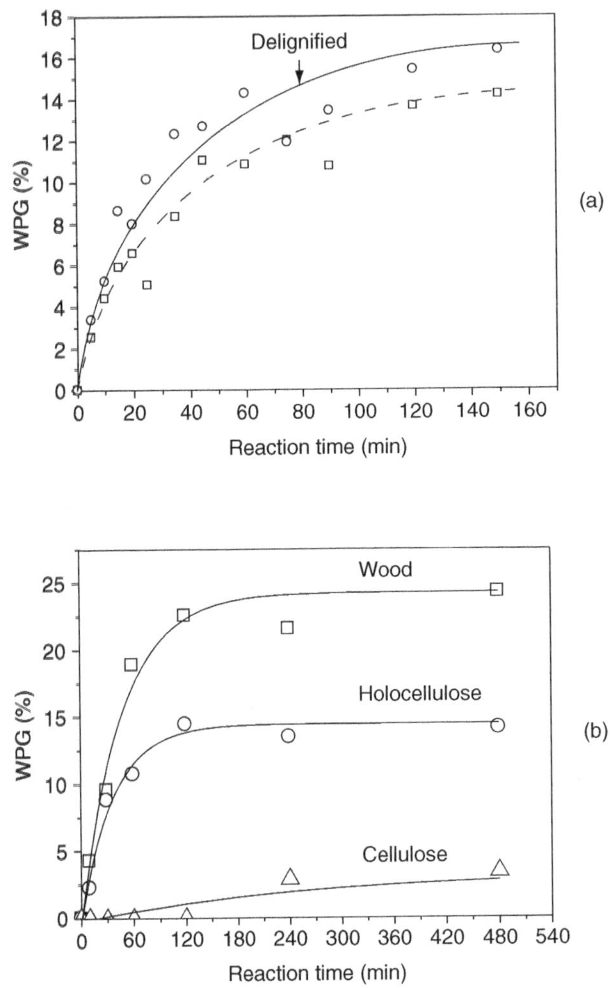

Figure 3.5 *The effect of lignin removal upon reactivity of acetic anhydride with MDF fibre (a) (from data of Haque and Hill, 2000) and wood flour (b) (from data of Çetin, 1999).*

There have been a number of studies of the reaction kinetics of anhydride reactions within the cell wall. Most studies have shown that the reaction kinetics of acetylation reactions are diffusion limited (Aboltins and Morozovs, 1998; Hill and Hillier, 1998; Hill *et al.*, 1998; Hill and Papadopoulos, 2002; Hill, 2003) (Figure 3.4b). Modelling of the reaction kinetics on a macroscopic scale (Aboltins and Morozovs, 1998; Morozovs and Aboltins, 1999; Morozovs *et al.*, 2003) and at a cell wall level (Hill and Hillier, 1999) has been the subject of a few studies. With diffusion-limited reactions, the different reactivities of the various cell wall OH groups do not influence the rate of reaction with anhydride molecules. The main factor controlling the reaction rate is the accessibility of the reactive site to reagent (i.e. the time taken for a reagent molecule to travel to a reactive site). Reaction with that site is essentially instantaneous compared with the time taken for the reagent molecule to travel to that site. This phenomenon has been studied using a computer model based upon percolation theory (Hill and Hillier, 1999). The effect of varying the rate of reaction of 'molecules' diffusing on a random network with reactive sites upon generated reaction profiles was studied and compared with experimental data. It was found that data for acetylation reactions was best described by a simple diffusion-limited kinetic model, but for longer chain length anhydrides, the reaction kinetics was more complex and required simultaneous consideration of both the rates of diffusion and reaction with active sites. This model was based upon data where anhydride molecules were present solely in the lumen when the reaction was initiated.

In general, the level of substitution of whole wood with acetic, or other linear chain, anhydrides rarely exceeds a WPG of 25 %, unless cell wall damage occurs. This maximum level of substitution corresponds to a degree of acetyl substitution of 5.9 millimoles of OH groups per gram of oven-dry wood. As noted in Chapter 2, theoretical calculations, based upon the composition of pine softwood, indicate a total OH content of the order of 14.9 mmoles g^{-1}, which corresponds to an accessible OH content of 8.6 mmoles g^{-1} when the crystalline cellulose OH content is subtracted (Hill and Jones, 1996b). It can be appreciated from the above argument that, at the highest levels of WPG achievable (about 25 %), there is a significant accessible OH content that remains unsubstituted. The factor that determines the extent of reaction is not the number of accessible OH groups but, rather, the inability of the cell wall to swell any further to accommodate additional adduct.

3.4 Analysis of Anhydride-modified Wood

The extent of reaction is generally determined by gravimetric methods for small samples on a laboratory scale, but with larger samples, where there is likely to be heterogeneous distribution of the bonded acyl groups, other methods for determination of the extent of reaction are required. The most commonly used reliable methods involve saponification of the bonded acyl groups followed by titration, or high-pressure liquid chromatography (HPLC) to determine the concentration of liberated acid. Analysis of acetylated wood is complicated by the fact that hemicelluloses are naturally acetylated to varying degrees and this needs to be taken account of when the acetyl content is calculated. Beckers *et al.* (2003) recently performed a study comparing various analytical techniques for the determination of the acetyl content in modified wood. This examination was undertaken to find a quick and reliable method for acetyl determination for quality control in a commercial context. The acetyl content was determined by titration using the Eberstadt method (Goldstein *et al.*, 1961), HPLC, Fourier transform infrared spectroscopy (FTIR), attenuated total internal reflection infrared spectroscopy (ATR), autofluorescence and time-of-flight secondary ion mass spectrometry. Absolute acetyl content as determined by titration was consistently higher than that measured using HPLC, which was attributed to wood components (e.g. phenolic hydroxyl, carboxylic groups associated with lignin and hemicelluloses) reacting with the sodium hydroxide, resulting in a higher titre of hydrochloric acid; however, acetyl gain as determined by the two methods was comparable. Results, as determined using HPLC, were rather variable. The acetyl content was determined by comparing the ratios of two absorptions, either 1740 cm^{-1} ($\nu_{C=O}$) to 1060 cm^{-1} ($\nu_{C\text{-}O}$), or 1740 cm^{-1} to 1510 cm^{-1} ($\nu_{C=C}$). Of the two methods, the 1740 cm^{-1}/1510 cm^{-1} ratio gave a better correlation to acetyl content (as determined by HPLC). This was considered to be due to the lack of substitution of the crystalline cellulose with acetyl, resulting in little change to the C-O stretch region of the spectrum; whereas changes in the C=C stretch region, attributable to lignin, could potentially occur due to reaction with acetyl. An analysis of wood particle size used to prepare KBr pellets for IR analysis showed that with softwoods, there was little effect upon the IR analysis for acetyl content, but this was not so for hardwoods, especially beech. Since ATR analysis does not require ground wood for sample preparation, it was thought that this might prove more satisfactory for acetyl analysis, but it was found that this technique could lead to errors. However, it was possible to analyse a greater number of samples in a given time, due to simpler sample preparation, and it was concluded that this method showed promise. Autofluorescence showed interesting results. There was a decrease in fluorescence in samples as levels of acetylation increased, even at low levels of acetylation. There were many factors affecting the accuracy of this technique and it was not considered reliable as a method for determining acetyl content. However, this method appears to offer some potential for studying the effect of substitution upon the lignin component of the cell wall. Very little work has been performed using ToF–SIMS, which appeared to show some promise, although it would be an expensive technique for quality control. It was noted that Goldstein *et al.* (1961) used a colouring agent that could be used to determine acetyl distribution in modified wood samples and also gave an approximate indication of acetyl content. However, this indicator (manufactured by Cyanamid) has not been available for some time.

3.5 Dimensional Stability

Anhydride modification is accompanied by a swelling of the wood cell wall due to the volume occupied by the bonded acyl adduct. Since the wood is swollen to some extent due to chemical modification, exposure to conditions of high RH result in a lower degree of swelling than would otherwise be the case if the wood was unmodified (Rowell, 1984a). The possibility that reduced swelling due to chemical modification may be related to blocking of OH groups has been raised by a number of workers (e.g. Popper and Bariska, 1975). However, it has been conclusively shown, by modifying with a variety of anhydride reagents, that dimensional stabilization is related solely to the WPG, or bulking of the cell wall, and that the extent of OH substitution is irrelevant (Stamm and Tarkow, 1947; Hill and Jones, 1996a; Li *et al.*, 2000b) (Figures 3.6 and 3.7).

Many studies have shown that the extent of swelling of the wood is directly proportional to the WPG due to reaction, but this is not always found to be the case. Wood swelling due to reaction is generally measured by determination of the external dimensions of a wood sample. Any conclusions regarding the swelling of the cell wall then rely on the assumption that the lumen volume remains unchanged as the wood swells. It has been shown that this assumption is invalid (Stamm, 1964). A recent study has shown that the volume change of the cell wall due to acetylation (determined by helium pycnometry) was proportional to the level of reaction, but that the apparent volume change, as determined by measurement of external dimensions, was not (Hill and Ormondroyd, 2004). This discrepancy was explained by considering the changes in lumen volume as modification

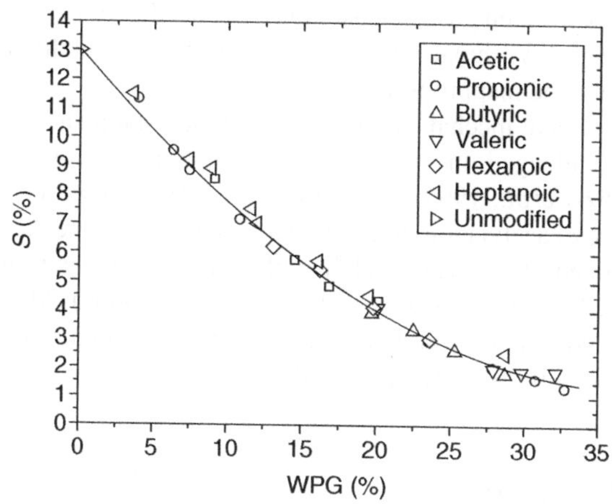

Figure 3.6 *The relationship between swelling coefficient (S %) and WPG for a variety of linear chain anhydrides (from data of Hill and Jones, 1996a).*

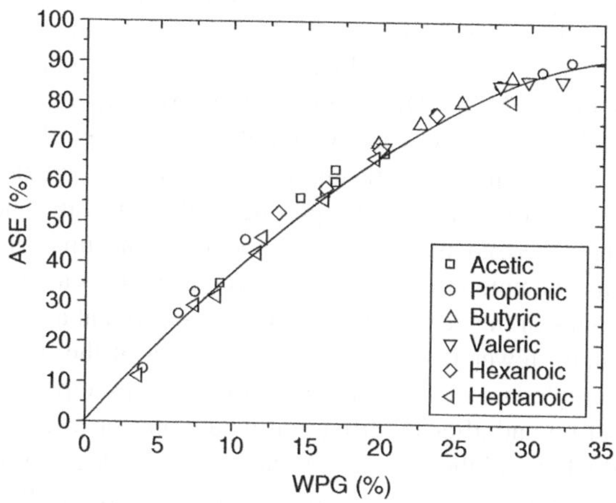

Figure 3.7 *The relationship between anti-swelling efficiency (ASE %) and WPG for a variety of linear chain anhydrides (from data of Hill and Jones, 1996a).*

proceeded. At low levels of modification, the lumen volume increased, but as WPG increased, this trend reversed and the lumen volume decreased, with the cell wall expanding into the lumen. It was concluded that it was not possible to determine cell wall volume changes from measurement of external sample dimensions. This means that the conclusions drawn by Hill and Jones (1999) regarding the cell wall void volume of anhydride-modified wood are incorrect, and that attempts to correlate volume increase with the volume of added chemical are also suspect.

SEM observations of acetylated wood have also shown that at WPGs in the region of 20 %, expansion of the cell wall into the lumen occurs (Evans *et al.*, 2000; Mohebby and Militz, 2002; Sander *et al.*, 2003). The decrease in lumen cross-sectional area may be related to the observation by Kumar *et al.* (1979) that the axial permeability of acetylated fir (*Abies pindrow*) to nitrogen gas is considerably reduced compared to control samples. However in Kumar's study, thioacetic acid was used for the modification reaction and direct comparisons may be inappropriate.

The generally accepted mechanism for dimensional stabilization due to cell wall bulking has been challenged in some recent work, and a cross-linking mechanism proposed instead (Pizzi *et al.*, 1994; Boonstra *et al.*, 1997). However, if cross-linking is responsible in the case of acetylation, it fails to explain why the same degree of dimensional stabilization is obtained at the same WPG with other linear chain anhydride reagents (Stamm and Tarkow, 1947; Hill and Jones, 1996a; Li *et al.*, 2000b). Although cross-linking cannot be totally ruled out, it is undoubtedly a very minor contributor to dimensional stabilization compared with bulking of the cell wall.

3.6 Mechanical Properties

3.6.1 Static Mechanical Properties

There are two competing effects that influence the mechanical properties of wood when it is acetylated. Acetylation of wood results in a decrease in equilibrium moisture content at a given relative humidity, which in turn yields a concomitant increase in tensile strength, MOR and MOE (Dinwoodie, 2000). Acetylation will also reduce the potential for forming H-bonding networks within the cell wall. Furthermore, degradation of the cell wall may also occur, due to the application of heat over extended periods and also due to the generation of acetic acid by-product in the cell wall. The extent of degradation will be strongly dependent upon the temperature and time of the modification reaction. It should also be noted that because the wood swells as a result of the presence of bonded acetyl groups in the cell wall, a given cross-sectional area of modified wood would therefore contain fewer fibres. If account is taken of the increase in the cross-sectional area of a sample as a result of modification, then a sample exhibiting the same strength/stiffness before and after modification would appear to exhibit a reduction in these properties, which could then be falsely interpreted as degradation due to modification. Alternatively, matched samples may be compared, where the increased cross-sectional area of the modified sample is not considered in the calculations of the mechanical properties, with results being presented as a percentage change. Another approach is that adopted by Dreher *et al.* (1964), in which all samples are machined to give the same cross-section. Unfortunately, in many reports it is unclear how the mechanical properties of the modified samples have been determined. As a consequence, Rowell (1996) noted that direct comparisons in mechanical properties between modified and unmodified wood have to be treated with caution.

A comprehensive study of the properties of acetylated wood was undertaken by the Forest Products Laboratory in Madison, Wisconsin, and included an investigation of the mechanical properties of acetylated wood (Tarkow *et al.*, 1946). The property changes were not significant, but there was some variation between species. For example, Sitka spruce and basswood exhibited increases in strength and MOE upon acetylation to about 20 % WPG, whereas yellow birch showed a decrease in these properties at 16 % WPG.

A study of the mechanical properties of wood samples (ponderosa pine, red oak and sugar maple) acetylated on a pilot scale was reported by Dreher *et al.* (1964). Samples were divided into matched pairs and one batch modified by drying in xylene vapour, then acetylating in a solution of acetic anhydride in xylene at 125–130 °C, followed by clean-up using xylene vapour stripping. Mechanical properties were determined from static bending or impact tests on pre-conditioned samples. All samples were prepared to the same dimensions, so that the additional bulked volume of the wood caused by modification was removed by planing. Fibre stress and work to proportional limit, compressive strength perpendicular to the grain and hardness increased for all species upon acetylation, but shear strength parallel to the grain decreased. Toughness increased for the pine and marginally for the oak specimens, but decreased slightly for maple.

In a study of the effect of acetylation upon Scots pine (*Pinus sylvestris*) and lime (*Tilia vulgaris*), microtomed strips (nominal thickness 100 μm) were reacted with acetic anhydride at 120 °C to a range of WPGs (Rowell and Banks, 1987). The tensile strength of pre-conditioned samples (expressed directly as force to failure) and the work to failure of acetylated samples were not found to be significantly different compared to unmodified controls, in view of the high standard deviations recorded. Birkinshaw and Hale (2002) also found that acetylation did not significantly affect the mechanical properties of the softwoods studied (pine, spruce and larch). Militz (1991a) reported a slight increase in MOR and a small decrease in MOE when beechwood was acetylated.

Larsson and Simonson (1994) studied the mechanical properties of acetylated *Pinus sylvestris* and *Picea abies*. The MOR and MOE decreased by about 6 % for pine, but increased by about 7 % with spruce samples after acetylation. Samples for this study were vacuum/pressure impregnated with acetic anhydride, excess anhydride was then drained off and samples were heated at 120 °C for 6 hours. The hardness of the acetylated wood samples was also found to increase, which was considered to result from the lower MC of the modified wood. Acetylated samples were also found to be less susceptible to deformation when subjected to varying RH.

There have been reports in which acetylation has a detrimental affect upon mechanical properties. Reaction of Scots pine in an acetic anhydride/xylene solution for 4 hours at 145 °C resulted in a 50 % decrease in the tensile modulus (Ramsden *et al.*, 1997). Spruce modified in acetic anhydride at 100 °C was found to exhibit a reduction in toughness of about 20 %, compared to unmodified wood (Reiterer and Sinn, 2002).

Bongers and Beckers (2003) performed a comprehensive analysis of the mechanical properties of wood, obtained from samples up to 4 m in length that had been acetylated in a pilot-scale reactor. Samples for the determination of mechanical properties were then machined from these larger pieces. The species studied were beech (*Fagus sylvatica*), poplar (*Populus* sp.), Scots pine (*Pinus sylvestris*) and radiata pine (*Pinus radiata*). MOR, MOE, impact resistance, Janka hardness, shear strength, resistance to axial withdrawing of screws and compression parallel to grain were determined. The EMC and density of samples were also determined. Variable results were obtained, with some species showing increases in some properties and others decreases, but the results were consistent within a species. For example, Scots pine either showed an increase or no change in MOR (eight batches), whereas radiata pine showed either a decrease or no change in this property (six batches).

3.6.2 Dynamic Mechanical Properties

The dynamic viscoelastic properties of acetylated wood have been determined and compared with other wood treatments in a number of studies. Both the specific dynamic Young's modulus (E'/γ) and *tan* δ are lower in acetylated wood compared with unmodified wood (Akitsu *et al.*, 1991, 1992, 1993a,b; Korai and Suzuki, 1995; Chang *et al.*, 2000). Acetylation also reduces mechanosorptive creep deformation of the modified wood (Norimoto *et al.*, 1992; Yano *et al.*, 1993). In a study of the dynamic mechanical properties of acetylated wood under conditions of varying humidity, it was concluded that the rate of diffusion of moisture into the wood samples was not affected by acetylation (Ebrahimzadeh, 1998).

The acoustic properties of wood are changed by acetylation and there have been several reports of the utilization of acetylated wood in musical instruments (Yano *et al.*, 1986a, 1988, 1993; Obataya, 1999). Both the sound velocity and sound absorption decrease as the WPG of wood is increased.

3.7 Microbiological Degradation

It is very well established that acetylation is able to impart a significant degree of decay resistance to wood, although considerably less is known about the mechanism(s) by which such resistance is imparted. Studies have been performed using laboratory pure culture tests (summarized in Tables 3.1 and 3.2), sterile and unsterile laboratory soil burial tests, and in long-term outdoor exposure tests. Results are often reported in terms of a decay threshold, the WPG above which no microbial degradation (as determined by some criterion, usually weight loss) of the wood occurs. It is difficult to draw comparisons between different studies, since different test protocols, fungal strains, wood species and so on are used. The data given in Tables 3.1 and 3.2 is an attempt to draw together the results from these studies, where it is possible to determine a threshold. There has also been an attempt to indicate the effectiveness of the fungal attack by reporting virulence, which is the mass lost by unmodified wood in the decay experiment.

3.7.1 Pure Culture Tests

Stamm and Baechler (1960) studied the decay resistance of Sitka spruce modified using a variety of techniques (including acetylation), against *Gloeophyllum trabeum* (3 months' exposure). Weight loss due to decay was correlated to the dimensional stability of the modified wood rather than the WPG, to allow direct comparisons between the various treatments. With acetylated wood, weight loss due to decay reduced to zero at about 75 % ASE. The authors were of the opinion that decay resistance was afforded by the blocking of OH groups by acetyl, which prevented enzymatic attack.

Goldstein *et al.* (1961) exposed acetylated ponderosa pine for 12 weeks to six basidiomycete fungi, according to ASTM D1413–56T, with a WPG of 18 % reported to be sufficient to provide decay resistance. Peterson and Thomas (1978) acetylated loblolly pine (*Pinus taeda*), green ash (*Fraxinus americana*) and yellow poplar (*Liriodendron tulipfera*), also

Table 3.1 *Table summarizing experiments on decay resistance of acetylated wood exposed to brown rot fungi*

Fungus (brown rot)	Wood species	Threshold [WPG (%)]	Virulence (%)	Reference
Coniophora puteana (Mad 515)	Ponderosa pine	17	37	Goldstein *et al*. (1961)
Coniophora puteana (BAM 15)	Beech (*Fagus sylvatica*)	17	27	Beckers *et al*. (1994)
Coniophora puteana (BAM 15)	Poplar (*Populus* spp.)	17	40	Beckers *et al*. (1994)
Coniophora puteana (BAM 15)	Scots pine (*Pinus sylvestris*)	>20	25	Beckers *et al*. (1994)
Coniophora puteana (FPRL 11E)	Corsican pine (*Pinus nigra*)	24	58	Forster *et al*. (1997)
Coniophora puteana (FPRL 11E)	Corsican pine (*Pinus nigra*)	18	62	Hill *et al*. (2003)
Coniophora puteana	Radiata pine (*Pinus radiata*)	20 to >20	22 to 57	Van Acker (2003)
Gloeophyllum trabeum (Mad 617)	Southern pine	20	54	Rowell (1982)
Gloeophyllum trabeum (*Lenzites trabea*) (Mad 617)	Ponderosa pine	17	35	Goldstein *et al*. (1961)
Gloeophyllum trabeum (BAM 109)	Beech (*Fagus sylvatica*)	17	32	Beckers *et al*. (1994)
Gloeophyllum trabeum (BAM 109)	Poplar (*Populus* spp.)	17	44	Beckers *et al*. (1994)
Gloeophyllum trabeum (BAM 109)	Scots pine (*Pinus sylvestris*)	20	39	Beckers *et al*. (1994)
Gloeophyllum trabeum (FPRL 108N)	Corsican pine (*Pinus nigra*)	23	58	Forster *et al*. (1997)
Gloeophyllum trabeum	Southern pine	14.8	68	Ibach and Rowell (2000)
Lentinus lepideus (Mad 534)	Ponderosa pine	17	25	Goldstein *et al*. (1961)
Lentinus lepideus (Mad 534)	Southern pine	20	62	Rowell (1982)
Meruliporia (*Poria*) *incrassata* (Mad 563)	Ponderosa pine	17	>40	Goldstein *et al*. (1961)
Oligoporus placenta (*Poria monticola*) (Mad 698)	Ponderosa pine	17	36	Goldstein *et al*. (1961)
Poria placenta (FPRL 280)	Scots pine (*Pinus sylvestris*)	>20	36	Beckers *et al*. (1994)
Poria placenta	Radiata pine (*Pinus radiata*)	17 to >20	18 to 45	Van Acker (2003)
Serpula lacrymans	Japanese cedar (*Cryptomeria japonica*)	15	53	Takahashi *et al*. (1989)
Serpula lacrymans	Japanese red pine (*Pinus densiflora*)	15	61	Takahashi *et al*. (1989)
Serpula lacrymans	Japanese beech (*Fagus crenata*)	>20	44	Takahashi *et al*. (1989)
Tyromyces palustris	Hondo spruce (*Picea jezoensis*)	18	45	Imamura and Nishimoto (1987a)
Tyromyces palustris	Japanese red pine (*Pinus densiflora*)	20	45	Takahashi *et al*. (1989)
Tyromyces palustris	Japanese cedar (*Cryptomeria japonica*)	22	57	Takahashi *et al*. (1989)
Tyromyces palustris	Albizzia (*Albizia falcata*)	20	29	Takahashi *et al*. (1989)
Tyromyces palustris	Albizzia (*Albizia falcata*)	20	17	Yusuf *et al*. (1989)
Tyromyces palustris (FFPRI 0507)	Makamba (*Betula maximowiczii*)	20	73	Ohkoshi *et al*. (1999)

Table 3.2 *Table summarizing experiments on decay resistance of acetylated wood exposed to white rot fungi*

Fungus (white rot)	Wood species	Threshold [WPG (%)]	Virulence (%)	Reference
Phanerochaete chrysosporium (FPRL S179)	Corsican pine (*Pinus nigra*)	10	46	Hill *et al.* (2003)
Pycnoporus sanguineus (DPF 44)	Corsican pine (*Pinus nigra*)	17	40	Forster *et al.* (1997)
Trametes (Polyporus) versicolor (Mad 697)	Ponderosa pine	17	25	Goldstein *et al.* (1961)
Trametes (Coriolus) versicolor	Hondo spruce (*Picea jezoensis*)	6	31	Imamura and Nishimoto (1987a)
Trametes (Coriolus) versicolor	Japanese beech (*Fagus crenata*)	16	72	Takahashi *et al.* (1989)
Trametes (Coriolus) versicolor	Japanese cedar (*Cryptomeria japonica*)	6	46	Takahashi *et al.* (1989)
Trametes (Coriolus) versicolor	Albizzia (*Albizia falcata*)	16	47	Takahashi *et al.* (1989)
Trametes (Coriolus) versicolor	Albizzia (*Albizia falcata*)	15	29	Yusuf *et al.* (1989)
Trametes (Coriolus) versicolor (CTB 863A)	Beech (*Fagus sylvatica*)	12	34	Beckers *et al.* (1994)
Trametes (Coriolus) versicolor (CTB 863A)	Poplar (*Populus* spp.)	12	38	Beckers *et al.* (1994)
Trametes versicolor (CTB 863A)	Corsican pine (*Pinus nigra*)	17	47	Forster *et al.* (1997)
Trametes (Coriolus) versicolor (FFPRI 1030)	Makamba (*Betula maximowiczii*)	12	22	Ohkoshi *et al.* (1999)
Trametes (Coriolus) versicolor (CTB 863A)	Corsican pine (*Pinus nigra*)	10	45	Hill *et al.* (2003)
Trametes versicolor (CTB 863A)	Beech (*Fagus sylvatica*)	20	18	Militz *et al.* (2003)

using acetic anhydride in xylene. The modified samples were exposed to the brown rot fungus *G. trabeum* and the white rot fungus *Coriolus versicolor* for 6 weeks. It was found that there was a large reduction in weight loss due to brown rot decay at WPGs in excess of 18 % with poplar and pine, but acetylated ash showed a relatively low decay weight loss even at 8 % WPG. However, ash was degraded by white rot, even at WPG levels in excess of 20 %, whereas pine and poplar showed much lower weight losses at all WPGs studied. Acetylation appeared to provide a high level of carbohydrate protection against both white and brown rot fungi, and it was thought that this unavailability of carbohydrate as an energy source may have also resulted in protection of the lignin component. Substitution of lignin OH groups was not considered to be the primary protection mechanism, due to the relative infrequency of such groups in the lignin. Microscopic examination showed that hyphal counts were substantially lower in acetylated wood samples and that there was no indication of cell wall degradation at higher WPGs. Examination of regions where hyphae were in direct contact with the cell wall revealed that there was very little evidence of cell wall degradation (and where this occurred, it was confined to the hyphal tips). Penetration of hyphal tips was completely prevented at WPGs in excess of 15 %.

The effect of the level of acetylation on the decay resistance of Japanese red pine, beech, cedar and albizzia was studied by Takahashi *et al.* (1989) and Takahashi (1996). Modified samples were exposed for 12 weeks in pure culture tests according to JWPA Standard 3–1979. Acetylated Japanese cedar was also exposed to Fenton's reagent, where it was found that weight loss due to Fenton's attack was reduced as WPG increased. Amongst other findings, the experiments indicated that acetylated softwoods were protected at much lower WPG levels when exposed to white rot, compared to hardwoods. Ohkoshi *et al.* (1999) studied the decay of acetylated makamba when exposed to *T. palustris* or *C. versicolor*. Protection against the brown rot fungus required a WPG of about 20 %, whereas protection against the white rot fungus was achieved at a WPG of around 12 %.

Beckers *et al.* (1994) examined the decay resistance of acetylated beech and poplar against *C. versicolor*, finding that the decay protection threshold was around 12 %, whereas acetylated pine required a WPG of 20 % to prevent attack by *P. placenta*. Attack by *C. puteana* and *G. trabeum* was prevented at about 17 % WPG, for acetylated beech and poplar, with pine requiring a slightly higher WPG of 20 %.

Forster *et al.* (1997) exposed acetylated Corsican pine sapwood blocks to *C. puteana*, *G. trabeum*, *T. versicolor* or *Pycnoporus sanguineus* in vermiculite overlay (high MC) tests and in conventional soil block tests without the vermiculite overlay. The purpose of the vermiculite overlay test was to ensure that the MC of the modified wood was high enough to allow for decay in the modified wood. The results from a number of other modifications were also obtained. Thresholds for zero weight loss at end of test were determined, and were found to be slightly lower for the white rot compared to the brown rot fungi. It was found that there were differences in the results obtained with the brown rot fungi, with *G. trabeum* producing greater weight losses in the high-MC test. It was also noted that some of the blocks exhibited pockets of decay in the latewood region of the interior of the samples, which was attributed to an uneven distribution of acetyl groups in the wood.

Suttie *et al.* (1997, 1998) tested wood samples modified with acetic, propionic or butyric anhydride in EN113 tests, and in vermiculate overlay tests based upon BS 1982.

In the EN113 tests, the highest decay weight losses were found for exposure to *G. trabeum* and *C. puteana*, with *T. versicolor* and *P. placenta* showing little sign of attack. In the vermiculite overlay tests, only *C. puteana* attack resulted in significant weight losses, contrasting with the findings of Forster *et al.* (1997). There was some concern that residual pyridine in the wood may have inhibited fungal attack, particularly in the case of *P. placenta*.

Hill *et al.* (2003) studied the decay resistance of wood modified with a range of linear chain anhydride reagents ($R=CH_3$ to C_5H_{11}) to *C. puteana*. The same linear relationship between weight loss due to decay and WPG was found, irrespective of anhydride, showing that protection arose due to cell wall bulking and not the extent of OH substitution (Figure 3.8). This work has been extended to show that the decay protection mechanism arises due to a reduction in cell wall moisture content, or due to blocking of the cell wall micropores (Hill *et al.*, 2005).

3.7.2 Unsterile Soil Tests

A number of studies have also been performed examining the performance of modified wood in unsterile soil burial (fungal cellar) tests under laboratory conditions. Takahashi *et al.* (1989) exposed acetylated blocks to soft rot attack in unsterile soil for 6 months. Japanese cedar had a decay protection threshold of 17 % WPG, with unmodified samples exhibiting weight losses of 37 %, red pine a threshold of 17 % WPG (control 13.2 % weight loss) and Japanese beech a threshold of 10 % WPG (control weight loss 35.2 %). Militz (1991b) reported that a 20 % WPG was sufficient to protect acetylated beech against attack by brown rot and white rot fungi. Whilst Militz *et al.* (2003) found that acetylated beech was more resistant to white rot attack compared to soft rot. Beckers *et al.* (1994) reported that a WPG of around 10 % was sufficient to protect wood against soft rot decay, as determined by weight loss. Relatively low thresholds for protection against

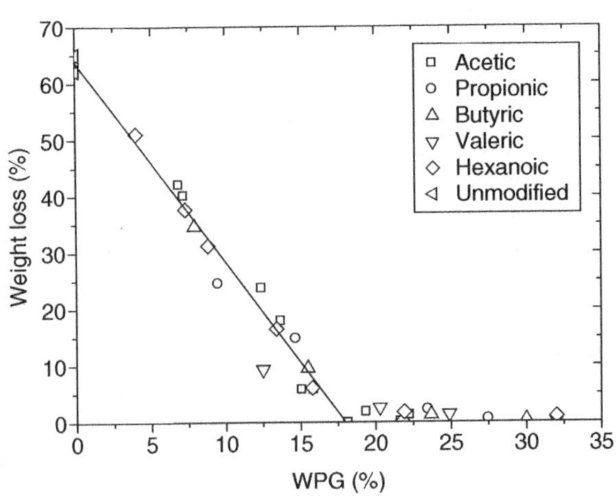

Figure 3.8 *The relationship between WPG and weight loss due to decay by Coniophora puteana for Corsican pine sapwood modified with a variety of linear chain anhydrides (from data of Hill* et al.*, 2003).*

soft rot were also found in a study by Beckers *et al.* (1995), who reported that protection was afforded at a WPG of 10.7 % for pine, 14.4 % for poplar and 12.8 % for beech, using a modified version of ENV807. In this later study, the test was conducted in vermiculite medium that was inoculated with a mixture of pure cultures of soft rot fungi. In this case, progress of decay was determined by using a three-point bending test every 8 weeks over a period of 3 months. Forster *et al.* (1998) exposed acetylated Corsican pine sapwood stakes in an unsterile soil burial test based upon ENV807 methodology, but using a higher soil MC and much longer exposure period. Stakes were removed after 13 and 20 months and decay was determined by weight loss. No decay protection threshold was determined (the stakes were modified to a maximum WPG of 15 % only), but a projection of the data suggested that this might be around 20 % WPG, with a weight loss for the unmodified controls of around 35 %, after 20 months' exposure.

Suttie *et al.* (1998) modified Scots pine with acetic, propionic, butyric or hexanoic anhydrides and determined resistance to soft rot attack by weight loss using a modified ENV807 stake test in unsterile soil. It was found that a threshold of *c.* 23 % was required to ensure full protection against soft rot attack, regardless of the anhydride used, although relatively low weight losses (11 %) were found with the unmodified Scots pine control samples after 52 weeks' exposure (30 % weight loss with beech controls). The progress of decay over time was determined by removing sets of stakes at intervals during the test. This showed that the threshold did not shift as the level of decay increased. Larsson Brelid *et al.* (2000) exposed Scots pine reacted to three acetyl contents (15.1, 18.5 and 20.9 %) in three different unsterile soils in a laboratory test. One soil had a dominant white rot decay, another brown rot dominance and the third was dominated by soft rot and tunnelling bacteria. Progress of decay was determined by weight loss after 6 months' exposure. The dominant forms of attack were dependent upon the soil type used, with weight loss due to attack being considerably lower with acetylated wood. At the highest WPG, only the wood exposed in the compost soil showed any visual evidence of attack (by tunnelling bacteria). Stakes modified to 15.1 % acetyl content exposed in compost or forest soil often showed soft rot attack that was more severe in the radial walls of the summerwood than the tangential walls.

In a study of the soft rot decay of acetylated beech and Scots pine, it was found that there was 'significant prevention' of fungal growth with acetylation at WPG levels of 8 and 10 %, respectively (Mohebby and Militz, 2002). Both weight loss and dynamic MOE were used to determine the degradation of wood exposed in fungal cellar tests, at 30-day intervals extending up to 300 days. A threshold of 17–18 % WPG was sufficient to prevent decay after 300 days' exposure. Beckers *et al.* (1995) also examined the decay resistance of acetylated wood following soft rot attack, by determining changes in strength properties. It was found that unmodified hardwoods were more susceptible to attack, with strength losses being more than twice those recorded for treated softwoods. Decay protection thresholds of around 10 % WPG were found for both hardwoods and softwoods. The loss of cell wall polymeric components from acetylated wood after 300 days' exposure was also studied. There appeared to be loss of lignin from the modified beech, even at the highest WPG studied, a result that appears to contradict the observation that beech was protected against soft rot at WPGs in excess of 18 %. Lignin protection was achieved with Scots pine at a WPG of around 16 %. Whereas cellulose was protected from attack at around 18 % WPG, hemicellulose protection required WPGs in excess of 20 %. Microbial activity of acetylated wood in soil burial tests has been found to be essentially suppressed at WPGs in excess of 17 %, as determined by ergosterol assay and microcalorimetry (Mohebby *et al.*, 2003).

3.7.3 Resistance to Moulds and Staining Fungi

A few studies have been performed examining the resistance of acetylated wood to mould and stain fungi in a laboratory environment (Wakeling *et al.*, 1991, 1992). These studies found that acetylation reduced the rate of mould growth, but at the end of the test there was no difference in the extent of colonization. Furthermore, samples modified to a WPG of 10 % exhibited a higher rate of colonization for reasons that were not clear. Samples acetylated up to a WPG of 20 % did not exhibit any improvement in performance in terms of mould growth, compared to unmodified samples, when exposed in a field trial.

3.7.4 Decay Resistance of Wood-based Composites

Particleboards made from acetylated wood have also been tested for resistance to biodegradation. Imamura *et al.* (1986) formed low-density particleboard from acetylated seraya (*Shorea* sp.) chips (17 % WPG) bonded with an isocyanate (IS) resin. The composites were formed from 0 %, 50 % or 100 % composition of modified chips, and examined using a standard weight loss test according to JWPA 3-1979 with *T. palustris* and *T. versicolor*, and in a bending creep test with *T. palustris*. Composites formed from 100 % acetylated chips were resistant to decay, as determined by weight loss and in bending creep tests (100 days' duration). The bending creep test has been used in a series of studies of the decay of acetylated wood (Takahashi, 1981; Imamura and Nishimoto, 1987b; Imamura *et al.*, 1987; Imamura *et al.*, 1988, 1989; Rowell *et al.*, 1988, 1989; Yusuf *et al.*, 1989).

Phenol formaldehyde (PF) bonded flakeboards, made from acetylated southern pine or aspen, required 2 weeks' leaching in distilled water to remove residual toxic chemicals (from the PF resin) before a pure culture test (Rowell *et al.*, 1987b). When challenged with *G. trabeum*, a WPG of about 15 % was required to ensure zero weight loss after 12 weeks' exposure in an ASTM D 1413 test. In a fungal cellar test, a 15 % WPG was also found to be adequate to ensure protection after 6 months' exposure. Initial attack in the soil burial test was by tunnelling bacteria, followed by fungal colonization. Chow *et al.* (1994) reported improved decay resistance for PF bonded particleboards made from acetylated aspen, or southern pine chips, when tested against *P. placenta*, *G. trabeum* or *T. versicolor*.

Okino *et al.* (2003) tested composites manufactured from acetylated rubberwood bonded with UF resin against *G. trabeum* and *T. versicolor*. In the 12-week exposure test, weight losses were found to considerably less than exhibited by control samples.

3.7.5 Field Tests

Compared with the relatively large number of laboratory-based studies, there is rather less data in the literature regarding anhydride-modified wood exposed in outdoor situations. In ground contact stake tests of acetylated pine samples performed at two Scandanavian test sites, it was found that an acetyl content of 20 % prevented attack by brown, white and soft rot fungi (Larsson Brelid *et al.*, 1997, 2000). Tests were run over a period of 8 years and the efficacy of acetylation treatment determined using a decay index method. Acetylation to a WPG of 22 % was able to provide durability that was superior to reference samples treated with CCA at retentions of 10.3 kg m^{-3}. One stake that had failed prematurely during the outdoor exposure test was analysed for acetyl content, which revealed that the acetyl content below ground level was lower than in the above-ground part of the stake. It was not known whether this difference was due to uneven acetylation, or due to hydrolysis/enzymatic attack of bound acetyl. Laboratory unsterile soil tests on acetylated mini-stakes of pine showed that an acetyl content of 18.5 % was able to provide significant protection against fungal attack. However, acetyl contents above 20.9 % were required to prevent attack by tunnelling bacteria. A comprehensive study of the decay resistance of a number of wood modifications, including acetylation, at three outdoor Scandinavian test sites has been reported recently (Westin *et al.*, 2004b). Acetylation to 23 % WPG was sufficient to prevent fungal attack at all test sites, with 21 % WPG virtually eliminating attack in laboratory fungal cellar tests.

Composite board stakes made from wood (aspen, spruce or rubberwood) acetylated to a low (*c.* 10 %) and high (*c.* 20 %) WPG were exposed at seven outdoor sites (in the USA, New Zealand, Sweden and Indonesia) in a ground contact test (Rowell *et al.*, 1997). All composites (MDF in the case of aspen and spruce, and rubberwood particleboard) were bonded with PF resin and a decay rating used to determine efficacy of the modification. After 3 years' exposure, most of the high WPG samples were found to be quite sound, with the exception of the two exposure sites in Indonesia, where the all the stakes were found to be entirely destroyed. Acetylated rubberwood samples did show some sign of decay at the USA and New Zealand sites. Although it was thought that the destruction of the samples at the Indonesian sites may have been due to termite or fungal attack, no termite attack was found at the USA site (after 32 months' exposure).

3.7.6 Mechanisms of Decay Protection

In general, most studies of acetylated wood have shown that a WPG of the order of 20 % is required to protect the modified wood from attack by brown rot fungi. With attack by white rot fungi, many studies indicate that a WPG of less than 10 % appears to be adequate to protect softwoods, whereas a WPG of the order of 20 % is necessary to protect hardwoods. These decay protection thresholds are determined from standard tests, which are designed to determine the efficacy of biocidal treatments. As a consequence, they cannot be relied upon to provide adequate information regarding the mechanisms by which decay protection occurs. For example, the concept of a decay protection threshold is undoubtedly false, since this merely refers to a WPG threshold at which a zero weight loss was recorded at the end of test. If a test is run for a longer time, then the threshold will probably shift. There is little information in the literature regarding the effect of time of exposure upon the 'threshold'.

In a round robin test conducted as a work package within the European Thematic Network for Wood Modification, considerable differences in decay resistance were found when acetylated wood samples were tested at eight different European laboratories (Van Acker, 2003). These results indicate that it is very unwise to make service predictions on the basis of only a few tests of acetylated wood. It has also been shown that thresholds derived from EN113 tests are highly dependent upon the virulence of the cultures used (Leithoff *et al.*, 1999).

Various hypotheses have been advanced to explain the mechanism by which acetylation imparts decay resistance to wood. Acetylation changes the chemical nature of the substrate, in that hydrophilic OH groups are replaced by acetyl moieties, and it has been hypothesized that, as a consequence, various specific enzymes are no longer capable of recognizing the substrate and hence cannot break down the cell wall polymers. As noted in Chapter 2, it is known that such enzymes are not capable of penetrating the cell wall of sound wood and that, as a consequence, if such a mechanism is operative, then this would presumably only be important for erosion attack from the lumen inwards. It is also important to note that many wood-decaying microorganisms are capable of expressing acetyl esterase enzymes, which can de-acetylate modified wood (Tsujiyama and Nakano, 1996). Furthermore, cellulose acetate is degraded by microorganisms even at degrees of substitution in excess of 2.5 (Buchanan *et al.*, 1993; Komarek *et al.*, 1993). The rate of decay of cellulose acetate is also strongly influenced by the morphology of the substrate, with higher surface area samples degrading much more rapidly (Zee *et al.*, 1998b). These facts further support the contention that decay protection is not related to the biochemical blocking of enzyme activity. It may, however, be possible that the action of LMWDAs is chemically, or sterically, prevented by substitution of the cell wall polymer OH groups.

It is well established that acetylation of wood results in a decrease in the EMC of the wood at a given RH, which is discussed in more detail elsewhere in this chapter. It is also well known that fungal attack of wood will only occur if the wood exceeds a minimum moisture content (Eaton and Hale, 1993). For example, Viitanen and Paajanen (1988) determined a minimum MC of 25 % for the growth of *C. puteana* on wood. Acetylation may thus provide decay protection by simply lowering the cell wall MC below the decay threshold. Finally, acetylation may act to physically block the ingress of degradative agents into the cell wall. In practice, it is difficult to deconvolute these two mechanisms, but establishing the importance of OH substitution as opposed to a physical bulking (or blocking) mechanism can be achieved by modification of wood using anhydride reagents of different molecular weights.

At present, there are clearly many areas in which further work needs to be performed in investigating the possible mechanisms by which microbial resistance is imparted to acetylated wood. Most studies have been concerned with service life predictions rather than investigating the decay protection mechanism(s) in any depth. It is apparent that with brown rot fungi (or at least those studied so far), full decay protection is not realized until a WPG of around 20 %. The study of the resistance of acetylated wood to white rot fungi is particularly limited, and there is clearly a need for considerably more work to be performed in this area. There does appear to be a significant difference between the WPG levels required to provide protection against brown rot and white rot fungi, at least when conventional test protocols are applied. Most studies indicate that WPG levels of around 10 % are sufficient to provide protection of softwoods to white rot, according to the protocols used. There also appears to be an important difference in performance between acetylated hardwoods and softwoods when exposed to *T. versicolor*, although there is some conflicting data. With *C. puteana*, and possibly for *T. versicolor* and *P. chrysosporium*, it would seem that the extent of cell wall bulking rather than degree of OH substitution is the most important factor in determining decay resistance. It is also apparent that thresholds are not fixed, but are dependent upon factors other than simply WPG. It may be that the generally lower levels of acetyl substitution required to provide protection against white rot fungi is a time-delay effect rather than an absolute threshold. There are also differences in performance between the various unsterile soil tests that have been reported, and clearly more work is required to investigate the behaviour of acetylated wood in such a complex test environment.

3.8 Biological Degradation by Insects and Marine Organisms

Imamura *et al.* (1986) exposed particleboard made from acetylated wood to the termite species *Coptotermes formosanus* and *Reticulitermes speratus*. A forced feeding test according to the JWPA standard 11–1981, where the untreated or acetylated wood was the only food source, and a choice feeding test (where wood specimens were randomly placed on a termite breeding colony for 30 days) were used. With *C. formosanus*, there was limited attack of the fully acetylated boards, with about 50 % termite mortality after 3 weeks in forced feeding tests, whereas with *R. sparatus* there was virtually no attack and 100 % mortality.

Acetylated wood veneers of Hondo spruce (*Picea jezoensis*), larch (*Larix leptolepsis*) and Douglas fir (*Pseudotsuga menziesii*) were formed into laminated veneer lumber samples (resorcinol bonded) and tested against the termite species *C. formosanus* and *R. speratus* (Imamura and Nishimoto, 1986). A forced feeding and choice feeding test was used. Acetylation of wood reduced the weight loss due to termite attack and increased termite mortality in proportion to WPG. In the case of *C. formosanus*, nearly all termites were dead after 6 weeks and 100 % of *R. speratus* after 2 weeks in a forced feeding test. The mortality rates exceeded those found in starvation tests of termites, an effect attributable to the indigestibility of the acetylated substrate resulting in no energy gain, coupled with energy expenditure by the termites in gnawing the wood. When only acetylated wood was available as a food source, there was a sharp decrease in the number and variety of protozoa found in the intestines of the termites; a change very similar to that observed with starving termites.

Rowell *et al.* (1987b) produced PF-bonded flakeboard from acetylated southern pine (21.6 % WPG) or aspen (17.6 % WPG) flakes. This was not completely resistant to attack by termites (*Reticulitermes flavipes*) in a 4-week test. It was thought that acetylation was less effective in preventing termite attack than other chemical modifications because cellulose decomposition in the intestines of termites leads to acetic acid formation in any case.

Yusuf *et al.* (1989) exposed IS- or PF-bonded particleboard made from acetylated albizzia to *C. formosanus* according to JWPA standard 12–1981, and in a field test (3 months' exposure next to a nest of *Coptotermes* sp.). In both tests, weight losses were around 50 %, indicating a lack of resistance to attack by this species.

Hadi *et al.* (1995) examined the resistance of PF-bonded flakeboards of acetylated rubberwood to attack by dry wood (*Cryptotermes cynocephalus*) and subterranean termites (*Macrotermes gilvus*). Flakeboards of acetylated damar (*Agathis loranthifolia*), tusam (*Pinus merkusii*), sengon (*Paraserianthes falcataria*) and meranti (*Shorea pinanga*) have also been tested (Hadi 1992). Moderate protection was exhibited by acetylated wood composites against *C. cynocephalus* in the 30-day test. Significant improvements in performance were found in the subterranean termite test, apart from with acetylated sengon.

Pine sapwood acetylated to 22 % WPG exposed for 1 year at an outdoor test site in Indonesia showed a weight loss of only 2 % due to termite attack, whereas unmodified controls lost 93 % in weight in the same period (Westin *et al.*, 2004b).

The resistance of Scots pine modified with acetic, propionic or butyric anhydride to attack by house longhorn beetle (*Hylotrupes bajulus*) was studied by Suttie *et al.* (1999) using EN46 and EN47 protocols. Unfortunately, the results were compromised due to the presence of residual pyridine (used with all of the modification reactions), which was shown to have a toxic effect upon the larvae. Nonetheless, at the highest WPGs studied, the wood modification did influence the performance, with no significant difference between the different anhydrides.

Johnson and Rowell (1988) found that some protection against marine borers was obtained by acetylating wood, but this was lower than that obtained with CCA or creosote treatment. Larsson Brelid *et al.* (2000) also found that acetylation provided only slight protection against attack in a marine environment. However, Westin *et al.* (2004b) exposed five acetylated pine samples (21 % WPG) at a marine test site, and found that four of these were free from attack after 4 years' exposure. Untreated pine sapwood samples had failed after 1 year due to heavy attack by *Teredo navalis*.

3.9 Moisture Relationships of Anhydride-modified Wood

Although reductions in EMC for acetylated wood were reported by Narayanamurti and Handa (1953) and Risi and Arseneau (1957a), the first analysis of the sorption–desorption properties of acetylated wood is that reported by Spalt (1958), who studied white spruce modified with acetic anhydride to a 32 % acetyl content. At this level of substitution, cell wall damage will have occurred. Analysis of the results was made using the Hailwood–Horrobin theory to divide the isotherm into a monolayer (surface sorption) and a multilayer (capillary condensation) component. This revealed that esterification resulted in a 67 % reduction in monolayer and a 62 % reduction in multilayer sorption at saturation (FSP), with 76 % of the reduction in total water sorbed attributed to cell wall bulking. As noted in Chapter 2, a fraction of this reduction is, in fact, attributable to an increase in the weight of the wood due to the presence of bonded chemical. The appropriateness of the use of the H–H model in such studies will be discussed towards the end of this section.

Popper and Bariska (1972) studied the moisture sorption properties of wood chemically modified with acetic (or phthalic) anhydride and analysed the results using Brunauer–Emmett–Teller (BET) theory and the H–H model. Acetylation was found to reduce the number of sorption sites, whereas little effect was noted with phthaloylation. By dividing the sorption isotherm into a monolayer component and a multilayer component using the H–H model, it was shown that there was a large reduction in the monolayer sorption capacity with acetylated wood, but little change with phthaloylated wood. Reductions in polymolecular (capillary) water occurred as a result of modification, due to bulking of the cell wall. The FSP was calculated from the H–H data, being 29 % for the earlywood (EW) and 30 % for the latewood (LW) with the unmodified wood, and reducing to 14 % (EW) and 18 % (LW) upon acetylation. The reduction in EMC with acetylated wood was explained as a combination of reduction in OH content, which reduced monolayer sorption and cell wall bulking (micropore blocking), reducing polymolecular sorption. Heat of sorption of the monolayer (calculated using BET theory) was reduced for acetylated wood, showing that the binding energy was reduced between acetylated wood and water. This would be expected, since there is a reduced potential for hydrogen bonding in acetylated wood. Acetylation reduced the EMC of the EW to a greater extent than the LW, which is presumably related to differing levels of reactivity and hence ultimate levels of substitution, but acetyl contents were not quoted. The quoted reaction conditions of acetylation in a mixture of acetic anhydride and pyridine at 90 °C for 40 hours would suggest that WPG was in excess of 20 %. Popper and Bariska (1973) also studied the moisture sorption thermodynamics of the modified wood, reporting on the enthalpy (δH) and entropy (δS) changes due to sorption. The enthalpy change was calculated by obtaining sorption isotherms at 20 °C and 25 °C and the use of the Clausius–Clapeyron equation. The data presented indicated that δH increased until monolayer coverage was achieved and decreased thereafter. For latewood, there appeared to be little difference in δH between unmodified and acetylated wood, but with earlywood, δH was lower above monolayer coverage.

Yasuda *et al.* (1995) also used the Hailwood–Horrobin (H–H) model to interpret sorption data over a range of five temperatures with acetylated wood (21.3 % WPG) and reported findings consistent with those of Spalt (1958). The enthalpy, entropy and free energy (δG) change of sorption were also determined. Acetylation resulted in a reduction of δH and δS of sorption up to about 10 % moisture content, compared to unmodified wood. With unmodified wood, δH and δS of sorption exhibited a maximum at 3 % MC and decreased thereafter, but with acetylated wood, these two parameters decreased from 0 % MC onwards. These results differ significantly from those reported earlier by Popper and Bariska (1973). The results were also compared with formaldehyde, glyoxal, glutaraldehyde and dimethylol dihydroxy ethylene urea (DMDHEU) treated wood.

Rowell and Rowell (1989) acetylated Scandinavian spruce (*Picea abies*) wood chips, then subsequently reduced these to fibres in a laboratory disc refiner. Fibre production did not result in loss of acetyl content, but it was found that new water sorption sites were produced as a consequence of the refining process. In addition, these workers modified a variety of lignocellulosic materials and found that all of the materials studied exhibited the same reduction in EMC at comparable WPGs.

In measurements of the dielectric relaxation of water adsorbed on acetylated wood, a large change in the activation enthalpy and entropy of dielectric relaxation was found to occur at 6 % moisture content (Zhao *et al.*, 1994), this presumably being attributable to the onset of formation of capillary water in the cell wall.

A recent study of wood modified with a variety of anhydride reagents has questioned the validity of the use of the H–H model when interpreting the moisture sorption isotherms of modified wood (Papadopoulos and Hill, 2003). This has shown that although modification with anhydride reagents leads to a reduction in both monolayer and multilayer sorption components, this decrease is related to the WPG and hence the degree of bulking of the cell wall, rather than to the extent of OH substitution. Similarly, the increase in the molecular weight of polymer per sorption site (OH group) is proportional to the WPG and not to the extent of OH substitution. Finally, it was noted that the number of sorption sites at FSP as predicted by H–H theory from the monolayer sorption curve falls far short of the value determined using alternative methods (substitution by acetyl in anhydride reactions, deuterium exchange etc.). This last conclusion relies upon the assumption that each water molecule in the monolayer is associated with one primary sorption site. Chauhan *et al.* (2001) also noted a reduction in monolayer water sorption for acetylated wood when analysed using the H–H model.

The ester linkage of anhydride-modified wood can be hydrolysed by water to yield the corresponding carboxylic acid and regenerate the wood cell wall OH group. This reaction is catalysed by low and high pH; thus any residual acid within the cell wall matrix will accelerate the rate of de-acetylation. It has been reported that a half-life for de-acetylation of 30 years is observed for acetylated wood when it is saturated with a buffer solution at pH 6 (Rowell *et al.*, 1992, 1993a). In this work, the stability of the ester bond in acetylated aspen and pine to hydrolysis was the subject of a very comprehensive study at a number of pHs and temperatures. Hydrolysis was found to be most rapid at pH 8 over all the temperatures studied, with stability to de-acetylation at other pHs being much greater at 24 °C and 50 °C. Hydrolysis was much more rapid at all pHs at 75 °C. In long-term cyclic and constant humidity exposure tests (5 and 6 years, respectively), less than 2 % of acetyl content was lost. It was concluded from these studies that acetylated wood would exhibit satisfactory stability under normal service conditions.

3.10 Composites Utilizing Acetic Anhydride Modified Wood

There has been considerable research directed at using acetylated wood in composites to produce boards with improved dimensional stability and water resistance. Klinga and Tarkow (1966) subjected Asplund and Masonite hardboards to vapour phase acetylation using acetic anhydride. The boards expanded as a result of the acetylation reaction to an extent about 20 % greater than calculated on the basis of the specific volume of the acetyl groups in the cell walls. This additional expansion was attributed to a release of stresses induced in the board during the manufacturing process. The acetylated boards exhibited improved dimensional stability and reduced EMC, with the modified Masonite boards being superior in performance. Modified boards exhibited improved mechanical properties compared to unmodified, which was in part attributed to the reduced EMC of the fibres.

Research on resin-bonded boards has used wood chips, or fibres, that are acetylated prior to composite manufacture. In almost all cases, acetylation has been achieved by using uncatalysed acetic anhydride. Significant improvements in dimensional stability proportional to the WPG of the wood particles have been reported when acetylated wood is used, and there is also a reduction in irreversible swelling (e.g. Bekere *et al.*, 1978; Arora *et al.*, 1981; Rowell *et al.*, 1986a–e, 1989, 1990, 1991, 1995; Yoshida *et al.*, 1986; Youngquist *et al.*, 1986a,b, 1988; Larsson and Tillman, 1989; Subiyanto *et al.*, 1989; Vick *et al.*, 1991; Clemons *et al.*, 1992; Kajita and Imamura, 1993; Sasaki and Kawai,

1994; Chow *et al.*, 1996a,b; Papadopoulos and Traboulay, 2002). Creep deformation of particleboards is also reduced by acetylation (Takino *et al.*, 1989). The extent of improvement in dimensional stability in relation to WPG is species dependent (Rowell *et al.*, 1986a,b) and also depends upon the resin system used. Acetylation is able to reduce inherent variability in dimensional stability arising from differences in location of the wood in the tree. A study of flakeboards made from acetylated heartwood and sapwood of *Pinus radiata* exhibited very similar swelling behaviour in water-soak tests, whereas flakeboards made from the unmodified wood were less dimensionally stable when made from sapwood (Rowell and Plackett, 1988). When wood flakes were acetylated to low WPGs using acetic anhydride vapour, it was found that composites made from the modified wood were less dimensionally stable than controls (Rowell *et al.*, 1986d). Houts *et al.* (2003) produced particleboard containing acetylated wood in targeted layers within the composite. This was done to produce composites that had improved properties, but at a lower cost than would result from the use of acetylated material in the whole of the composite. Improvements in dimensional stability were found with acceptable internal bond strength (IBS).

Acetylated wood particles have been found to be less compressible than unmodified wood, resulting in consolidated boards exhibiting larger springback and containing more voids than control boards. Higher press temperatures are required to achieve the target board thickness, and boards made from acetylated wood have slightly lower face densities and slightly higher core densities than control boards (Youngquist *et al.*, 1986b). Pressing boards to higher target densities results in greater springback and hence reduced dimensional stability (Vick *et al.*, 1991)

3.10.1 The Influence of Resin Type on Mechanical Properties

Although, in almost all cases, acetylated wood-based composites have been produced using resin binders, a binderless board was produced from acetylated wood by Kiguchi and Suzuki (1985). However, in this case acetylation was achieved by using acetic anhydride in the presence of sulphuric acid, which resulted in degradation of the wood. Where a resin is used to bond the wood particles, the strength of the interaction between the wood surface and the resin is an important factor determining the mechanical properties of the composite. Reaction with acetic anhydride reduces the surface energy of the wood, since polar hydroxyl groups are replaced with less polar acyl groups, which therefore alters the wettability of the surface with water-based resin systems. In addition, where the strength of interaction between the resin and the wood relies upon H-bonding, or direct covalent bonding to OH groups, a reduced bond strength will be obtained as a result of acetylation. For this reason, polyvinyl acetate adhesive shows poor performance when used on acetylated wood (Larsson *et al.*, 1992).

An early study by Rudkin (1950) showed that substitution of OH groups with acetyl resulted in a significant decrease in bond strength between the wood and a UF resin in lap-joint tests. Vick and Rowell (1990) studied the adhesive bonding of acetylated yellow poplar, with 18 different thermoplastic and thermosetting adhesives. The effectiveness of the adhesives was examined by determination of bond shear strength (and wood failure) of 6 mm thick, bonded wood strips after conditioning at 27 °C and

65 % RH and after water soaking. Various classes of adhesives were investigated (polyurethane and isocyanate, polyvinyl acetate, rubber-based contact, casein, epoxy-polyamide, amino resin, resorcinol and phenol–resorcinol, and phenolic). Generally, adhesion was reduced by acetylation of the wood relative to WPG, with only a neoprene contact adhesive and a polyurethane hot melt unequivocally showing improved behaviour with the acetylated wood. Thermosetting adhesives produced the strongest bonds in both wet and 'dry' conditions. Under wet testing conditions, all resorcinol-based resins, a cold-set polyvinyl acetate, a MF resin and an acid-catalysed PF yielded improved strength values with acetylated wood.

There is some inconsistency in the reported data for the effect of acetylation on mechanical performance of composites with similar, or even identical, resin systems. Arora *et al.* (1981) reported a consistent increase in both MOE and MOR with (presumably resol) PF-bonded particleboard made from acetylated wood particles from five species. However, in another study, when boards were produced from acetylated aspen flakes using an aqueous-delivered resol PF, it was found that resin dispersion on and penetration into the chips was poor, resulting in inferior mechanical properties (Youngquist *et al.*, 1986b; Rowell *et al.*, 1987a). Youngquist *et al.* (1986a) also reported that acetylation did not markedly affect IBS, but this was based upon the study of a limited number of samples. Chow *et al.* (1996b), found a decrease in MOR, MOE and IBS when acetylated hemlock fibres were used in composites bonded with an aqueous PF resin. Papadopoulos and Traboulay (2002) found a decrease in IBS with oriented stand boards formed from resol-bonded acetylated fir strands. Fibreboards made from acetylated aspen fibres bonded with different levels of an aqueous PF resin did not show any significant difference in MOR or MOE, when compared with boards made from nonacetylated fibres (Rowell *et al.*, 1991).

Some of the variability in the reported data may be related to the species used, since Rowell *et al.* (1986e) found that IBS decreased with Douglas fir, but increased with aspen and southern pine acetylated flakes for boards bonded using an aqueous PF resin. Evidence of the difference between species was also seen in the work of Chow *et al.* (1996b), who acetylated aspen and southern pine fibres to a WPG of 23 % and formed fibreboards from these, using PF resin with and without wax treatments. In general, a reduction in MOR, MOE and IBS was found when acetylated fibres were used, particularly with the southern pine fibres. The effect was less marked with the hemlock, and when wax was used with the acetylated hemlock fibres, increases in MOR and MOE were observed. Fuwape and Oyagade (1999, 2000) acetylated gmelina (*Gmelina arborea*) and Sitka spruce (*Picea sitchensis*) and formed boards from the modified flakes using PF, mangrove tannin PF or quebracho tannin PF resins. Acetylation resulted in a reduction in MOE values for all boards, but both MOR and IB values increased for acetylated gmelina, and decreased with spruce.

Youngquist *et al.* (1988) found improvements in both resin distribution and IB values when suitable emulsifiers were used in conjunction with waterborne resins, but considered it unlikely that the improvement in performance could be justified on cost grounds. It was postulated that acetylated wood interfered with the polymerization of the resin, so that it was not fully cured. It has also been suggested that acetic acid, which may be released during board pressing, could accelerate resin curing of resol type resins. In a study to determine whether this was so, acetylation of wood was found to slightly reduce resol PF resin curing rates and to result in a reduction of bond strength, as the WPG level increased (Chowdhury and Humphrey, 1999).

A study of the formation of composites from a variety of acetylated lignocellulosic materials bonded with powdered novalak resin exhibited slightly improved mechanical properties (MOR, MOE and IBS) compared with unmodified controls, with the exception of wheat straw (Gomez-Bueso *et al.*, 1999b, 2000). This data was incorrectly presented in an earlier publication (Gomez-Bueso *et al.*, 1999a). It was not known whether the improvement reflected improved compatibility between the resin and fibre surface, or was due to the use of a less pH dependent resin compared with other studies. The use of a powdered novolak PF resin was also found to slightly improve the MOR and IBS of boards made from acetylated wood fibres, but MOE decreased somewhat (Mahlberg *et al.*, 2001).

The generally poorer mechanical properties exhibited by acetylated lignocellulosic material in composites bonded using aqueous resin systems was considered by Korai *et al.* (2001). Fibres of yellow cedar (*Chamaecyparis nootkatensis*) were acetylated to a WPG of 24.8 % and then ozonated to different extents to increase the hydrophilicity of the fibre surface. Boards were fabricated from the fibres using an aqueous MF resin. Ozonation improved IBS of boards fabricated from acetylated fibres, proportional to level of ozone charge, and resulted in IBS values comparable to those of nonacetylated controls at higher levels of ozonation. However, although ozonation also improved MOR, the values obtained for acetylated fibres were always less than those obtained with unmodified fibres.

Korai (2001) also considered the importance of density profile of composites made from acetylated fibres in determining mechanical properties. Fibres of yellow cedar were acetylated with vapour-phase acetic anhydride and fibreboards were made from these, bonded with melamine formaldehyde resin. The results from this study indicated that bonding between fibres was the most important property determining mechanical properties.

Okino and co-workers produced composites from acetylated rubberwood (Okino *et al.*, 2001) and acetylated cypress (Okino *et al.*, 2004), bonded using a UF resin. The composites exhibited inferior mechanical properties compared to those produced from unmodified wood. The reduction in mechanical properties is undoubtedly due to poor wetting and weak interaction of the surface of the modified wood with the UF resin due to a reduction of H-bonding sites on the acetylated wood surface.

Composites of acetylated wood bonded using isocyanate (IS) adhesives have been reported to exhibit mechanical properties that are comparable with unmodified control specimens (Youngquist and Rowell, 1990). Although other studies have shown that there is a reduction in internal bond (IB) strength by about 35 % when IS resins are used with acetylated particles, with little change in MOE and MOR (Rowell *et al.*, 1989). Since acetylation blocks OH groups, the potential for IS resins to chemically bond to the wood surface is presumably reduced, whereas with a PF resin the main factor affecting bond strength is compatibility with the surface. Yoshida *et al.* (1986) reported a slight decrease in MOR for boards made from acetylated seraya (*Shorea* sp.) bonded with a polyurethane resin, but no change in IBS. Subiyanto *et al.* (1989) also found that increasing the WPG of acetylated albizzia reduced the IBS, MOR and MOE of IS-bonded boards, whereas with PF resin, there was little effect up to about 12 % WPG and a decrease thereafter. When boards were tested after 2 hours of boiling followed by 1 hour of water immersion, the MOR of the IS-bonded board was reduced when acetylated wood was used at all but the highest WPG, whereas with PF-bonded boards an improvement was recorded; MOE was improved with both resins. After 2 hours of boiling, the IS-bonded composites made from acetylated wood at 5 and 12 % WPG swelled much more than boards made from unmodified wood.

Vick *et al.* (1991) studied the effect of chip type (disc- or ring-cut), board density and resin content upon the mechanical properties of IS-bonded acetylated quaking aspen (*Populus tremuloides*) before and after aging tests (ASTM D1037 and AWPA D-5). Low-density boards with 3 % resin loading formed from acetylated ring-cut flakes exhibited markedly reduced breaking strength, MOR and MOE, compared to boards made from nonacetylated disc-cut material, both before and after accelerated aging tests. The longer disc-cut acetylated flakes performed much better in these tests, giving values much closer to those obtained with the control samples, and generally rather better values after aging. The IBS of acetylated samples was slightly reduced prior to aging but superior to control boards after aging, with both the disc- and ring-cut flakes. Properties of boards made from acetylated material were improved by increasing the resin content and by increasing the board density.

3.11 Conclusions

Modification of wood with acetic anhydride has been the subject of intensive study for over 50 years. It would appear that the changes in many properties are a function of the extent of bulking of the cell wall rather than the level of OH substitution. This has been shown conclusively for dimensional stability, and for at least one example of decay resistance, and is likely to be the explanation for the reduction in MC observed in water vapour isotherm studies. The general consensus appears to be that the strength properties of wood are not significantly affected by acetylation, provided that the acetylation temperature is not too high (120–140 °C). Work will no doubt continue in this area: in particular, it would appear to be necessary to establish appropriate testing protocols at an international level. The present standards have been developed for unmodified timber and require changes in order to reflect the different properties of acetylated wood.

4

Chemical Modification of Wood (II): Reaction with Other Chemicals

4.1 Introduction

A very large number of reagents have been studied for their potential for the chemical modification of wood. In almost all cases, these involve reaction with the cell wall polymeric hydroxyl groups. In the main, reactions described in this chapter are unlikely to be suitable for scale-up to industrial processes. In many cases, the cost of reagent and/or difficulties in handling large quantities of potentially toxic reagent remain considerable barriers to exploitation.

4.2 Reaction of Wood with Other Noncyclic Anhydrides

Although acetylation with acetic anhydride has been widely studied, work with other linear (or branched) chain anhydrides has been relatively limited. Longer chain anhydrides show much lower reactivity with wood, which decreases as the molecular weight of the anhydride increases (Stamm and Tarkow, 1947; Risi and Arseneau, 1957b,c; Hill and Jones, 1996a,b; Hill *et al.*, 1998; Dawson *et al.*, 1999; Hill and Hillier, 1999; Li *et al.*, 2000b; Chang and Chang, 2002). Rate of reaction is however, markedly improved if pyridine is used as a swelling solvent/catalyst (Hill and Jones, 1996b). A combination of N_2O_4 with DMF has been used to catalyse the reaction of wood with a range of anhydrides to impart thermoplasticity to the substrate (Shiraishi *et al.*, 1979). Microwave heating has recently been used for butyrylation reactions (Chang and Chang, 2003).

Li *et al.* (2000b) investigated the reaction of wood with acetic, propionic, butyric, isobutyric or hexanoic anhydrides. The pre-extracted samples were vacuum impregnated

Wood Modification: Chemical, Thermal and Other Processes C. Hill

with the anhydrides without any co-solvent or catalyst. Following this, the samples were wrapped in polyvinylidene chloride film and then aluminium foil and heated for various time periods and temperatures. At 90 °C, only the acetylated samples exhibited any significant weight gain due to reaction, with the propionylated samples showing a slight weight gain. As reaction temperature increased, reaction rates with other anhydrides also began to increase, with all anhydrides exhibiting significant WPGs after 24 hours' reaction at 145 °C. At the highest reaction temperature, acetylated samples exhibited WPGs of the order of 40 % (30 % with propionic anhydride), indicating that cell wall damage had occurred. Isobutyric anhydride exhibited lower reactivity than butyric anhydride, presumably due to steric factors. Due to the potential for wood degradation at high reaction temperatures, Li *et al.* (2001a) studied the use of a number of potassium and sodium salts upon the rate of propionylation at 90 °C and 110 °C, finding some improvement with acetates, sodium propionate and sodium thiosulphate.

Hill *et al.* (2004b) studied the effect of swelling of the cell wall of Scots and Corsican pine samples upon the rate of reaction with acetic or propionic anhydride under scrupulously dry conditions. Oven-dried samples of both species showed no reaction with propionic anhydride in xylene. However, when the wood was taken from a water-saturated state, through a solvent series of 50 % ethanol, 95 % ethanol, dry ethanol, dry acetone and finally dry xylene, the rate of reaction was markedly improved. Oven-dried samples were also vacuum impregnated with propionic acid, before reaction in a solution of propionic anhydride in propionic acid, which also markedly improved the reactivity. With Corsican pine, oven-dried samples showed good reactivity with acetic anhydride in xylene, although swollen samples reacted faster. With Scots pine, oven-dried samples were found not to react with acetic anhydride in xylene. This latter result is anomalous, being explained by the authors as being due to a difference in cell wall microporosity distribution between the two species.

Li *et al.* (2000a, 2001b) studied the flammability of wood modified with acetic or propionic anhydrides, finding that this parameter, as determined by limiting oxygen index (LOI), was the same as that found for unmodified wood. Samples were also prepared which were impregnated with sodium silicate prior to anhydride modification. The latter treatment resulted in a decrease in flammability, as shown by an increase in the LOI of the wood.

As noted previously in Chapter 3, the improved dimensional stability of wood as a result of anhydride modification has been found to be a function of WPG only, irrespective of the anhydride used for modification (Stamm and Tarkow, 1947; Hill and Jones, 1996b; Li *et al.*, 2000b). This shows that improved dimensional stability arises due to a bulking phenomenon, caused by the volume occupied by the bonded acyl adducts in the cell wall. The dimensional stabilization of wood modified with crotonic anhydride has also been reported (Çetin and Ozmen, 2001). Reductions in the EMC have also been attributed to a bulking phenomenon (Papadopoulos and Hill, 2003), and in at least one example of decay protection (Papadopoulos and Hill, 2002). Further research is needed to determine if this is the case for resistance to other decay organisms. Only with the phenomenon of surface wettability does the relationship between a physical property of chemically modified wood and WPG seem to be broken (Hill and Jones, 1996c) (Table 4.1).

The efficacy of anhydride modification to provide envelope protection to wood was examined using an ENV839 test against *G. trabeum* or *C. versicolor*. Wood samples were

Table 4.1 The effect of chemical modification on water contact angles

Anhydride	WPG (%)	Contact angle (degrees)
Unmodified	0	45
Acetic	23.9	98
Propionic	3.7	108
Propionic	8.3	117
Propionic	10.6	123

modified with propionic, butyric or hexanoic anhydride. Although modification reduced the rate at which fungi reached the bait sticks in the unmodified core of the sample, penetration was not prevented, in contrast with samples using a tributyl tin oxide (TBTO) envelope treatment. The importance of achieving adequate levels of modification to provide decay protection in the core of wood was emphasized (Suttie *et al.*, 1997).

4.3 Reaction of Wood with Cyclic Anhydrides

Reactions of wood with cyclic anhydrides do not yield a by-product, leaving the modified wood polymers with a covalently bonded carboxylic group: an example is reaction with succinic anhydride (SA) (Figure 4.1a). With reactions above 100 °C there is some formation of diester observed (Figure 4.1b), resulting in cross-linking within the cell wall matrix (Matsuda, 1987).

Apart from the alkenyl succinic anhydrides, other cyclic anhydrides are solids, requiring the use of a delivery solvent, or a solid-state reaction. Apart from succinic anhydride, phthalic (PA) (Figure 4.2a) (Risi and Arseneau, 1958) and maleic anhydride (MA) have also been extensively studied (Figure 4.2b) (Clemons *et al.*, 1992). Goethals and Stevens (1994) have reported reaction of wood with glutaric (Figure 4.2c) or cyclohexane-1,2-dicarboxylic anhydride (Figure 4.2d) and subsequent reaction with epichlorohydrin.

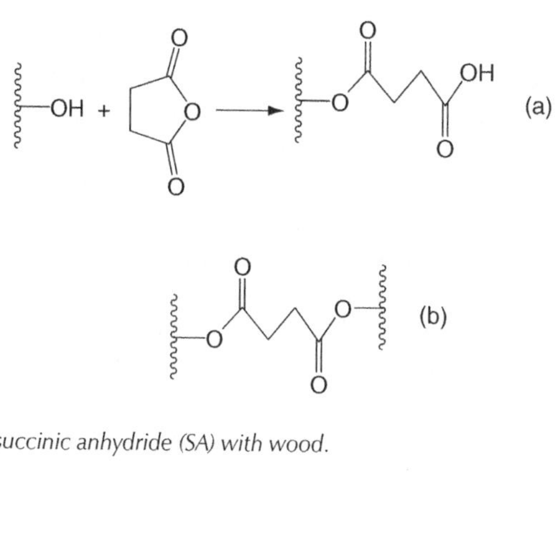

Figure 4.1 Reaction of succinic anhydride (SA) with wood.

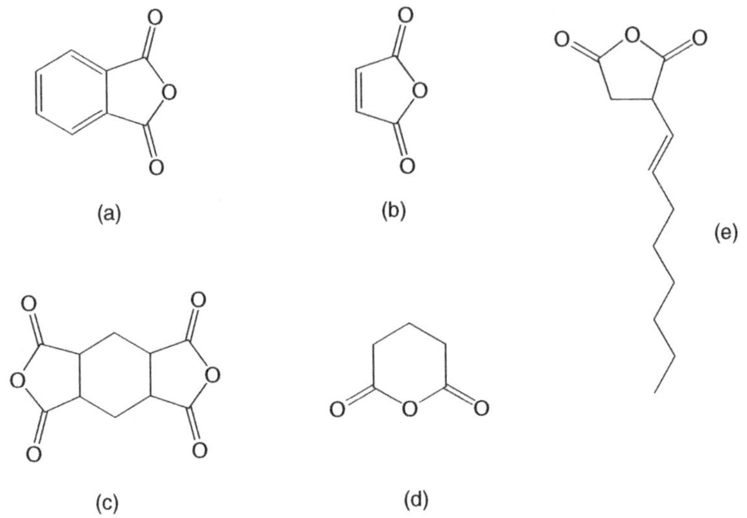

Figure 4.2 Structures of phthalic (a), maleic (b), glutaric (c), cyclohexane-1,2-dicarboxylic anhydride (d) and octenyl succinic anhydride (e).

Reaction results in a pendant carboxylic moiety attached to the wood via a covalent ester bond, providing a site upon which further reactive chemistry is possible. For example, reaction of wood with SA, MA or PA, and subsequent reaction with a variety of epoxides has been reported in an extensive series of publications (Matsuda *et al.*, 1984a, b, 1987, 1988a–f; Matsuda and Ueda, 1985a–f; Matsuda, 1987, 1993; Murakami and Matsuda, 1989a,b, 1990a,b; Ueda *et al.*, 1994) (Figure 4.3). It was found that esterification imparted thermoplasticity to the wood, with this property decreasing in the order SA > MA > PA. Softening points were in the range 55–77 °C. Values of anti-shrink efficiency (ASE) were reported for MA- and PA-modified wood of *c.* 38 % and 45 % at WPGs of 25.7 % and 29.3 %, respectively. Dimensional stabilization decreased with subsequent water-soak (WS)/oven-dry (OD) cycles, indicating loss of reagent with water soaking, due to hydrolysis of the ester bond. MA-treated wood was more susceptible to hydrolysis compared to PA-treated wood.

Stability was much improved by oligoesterification with epoxide. Composites were prepared by hot pressing of the esterified wood, or by using oligoesterification as a method of coupling the fibres. It was found that thermoplasticity decreased as the diester content increased. Increasing quantities of diester were formed above 120 °C. This work has been further developed by Timar *et al.* (1997a,b, 1998, 1999a,b, 2000a,b, 2004).

Loss of reagent following water soaking had been previously noted for PA-treated wood; where an initially impressive (and unusual) ASE of 100 % at 60 % WPG was reduced to 60 % by the sixth water-soak/oven-dry (WS/OD) cycle (Risi and Arseneau, 1958). Dimensional stabilization of MA-treated wood has also been studied, with ASE values of 50 % at 20 % WPG being quoted (Nishino *et al.*, 1991), much lower than the values obtained with linear chain anhydride modified wood. Modification with cyclic anhydrides results in a slight reduction in hygroscopicity of the wood, due to a bulking

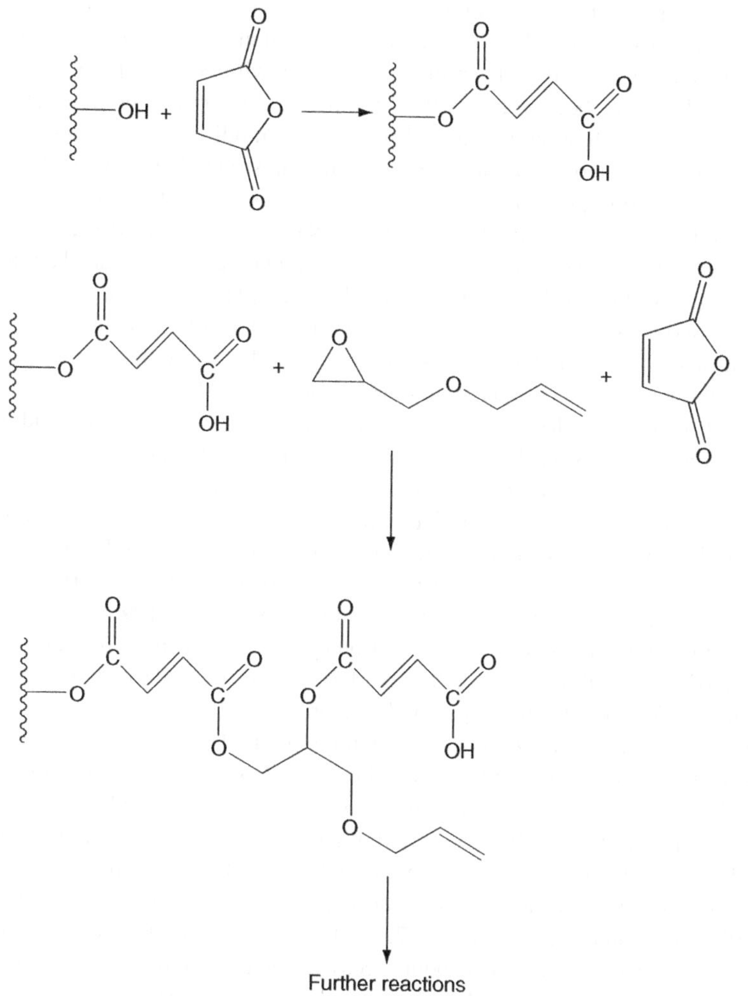

Figure 4.3 *Reaction of wood with maleic anhydride, followed by oligomerization of the activated wood with epoxide and maleic anhydride.*

effect, although in solid wood, such studies are complicated by degradation of the wood material, particularly at higher WPGs (Popper and Bariska, 1972, 1973; Matsuda, 1987).

The production of moisture resistant particleboard by treatment with a maleic anhydride – glycerol mixture and using phenol formaldehyde as the matrix material has been investigated (Fujimoto *et al.*, 1987). Boards prepared from modified wood showed considerable improvements in modulus of elasticity and internal bond strength when compared to control boards. Composites made from aspen fibres modified with SA, MA or acetic anhydride using phenol-formaldehyde (PF) or polypropylene as binder have also been studied (Clemons *et al.*, 1992; Rowell *et al.*, 1993b). The reaction of wood with MA was found to proceed at a slower rate than with SA. The volume increase due to modification was found to be proportional to weight gain at WPGs in excess of 30 %, but was larger than theoretically expected below a WPG of 30 %. Scanning electron microscopy (SEM) studies revealed severe cell wall splitting occurring with SA-modified wood at 61 % WPG, which has also been found with epoxide modification (above *c.* 25 % WPG) (Rowell *et al.*, 1976). It is commonly observed that modification of solid wood with cyclic anhydrides leads to severe degradation of the substrate after prolonged reaction.

The stress relaxation properties of SA-modified wood in media of different pH have also been studied (Nakano *et al.*, 1990a,b, 1991; Nakano, 1993a,b). It was found that an increase in pH led to an increase in stress relaxation properties due to side chain ionization, with maximum swelling of the wood occurring between pH 5–8. In addition, cross-linking within the wood was achieved by treatment with bivalent or trivalent metal ions. The rheological properties of maleic acid esterified wood also show that thermoplasticity is imparted to the wood by such modification (Norimoto *et al.*, 1992).

The use of maleic anhydride as a compatiblizer between wood particles and bisphenol A-based polyesters resins has been investigated (Han *et al.*, 1991). In this study, the MA was added directly to the composition of woody matrix filler and resin rather than by pre-modification of the wood. It was found that composite properties were improved by addition of MA, probably due to esterification of the wood occurring during the kneading process. The modification of sawdust using maleic anhydride has been performed in order to provide a compatible filler for polyester resins (Marcovich *et al.*, 1996). Modification was performed at room temperature using a solution of maleic anhydride in acetone, in some cases catalysed with sulphuric acid. It was claimed that bonding occurred under these mild conditions from IR spectroscopic evidence only.

Modification of Scots pine wood veneers with either maleic, succinic or phthalic anhydrides in DMF resulted in strength losses in the wood. No protection was afforded to such samples when exposed to UV irradiation (Evans, 1998). Hill and Mallon (1998) found that succinic anhydride modified wood exhibited poor dimensional stability properties compared to acetylated wood at comparable WPG values, and that there was also evidence for hydrolysis of the ester bond during water-soak/oven-drying cycles. Forster *et al.* (1997, 1998) and Suttie *et al.* (1997, 1998) have reported that succinic anhydride modified wood exhibits markedly poor decay resistance. Indeed, such modification appeared to encourage attack, rather than providing protection, in that higher weight losses were recorded in fungal cellar tests as the WPG was increased (Suttie *et al.*, 1998). It was hypothesized that this phenomenon was due to fungal metabolism of the hydrolysed succinic acid molecules, although direct hydrolysis would also explain the observed results. Forster also found that both succinic and phthalic anhydride modified wood exhibited high weight losses in sterile control tests, attributed to hydrolysis of the ester linkages during decay experiments.

Alkenyl succinic anhydrides (Figure 4.2c) are well known in the pulp and paper industry, where they are used extensively as paper sizing agents. The use of these reagents as wood-modifying agents has been the subject of only a few investigations (Codd *et al.*, 1992; Hill and Mallon, 1998), although there has been some interest in their use as compatibilizers in cellulose–propylene composites (Caulfield *et al.*, 1993). Codd *et al.* (1992) studied the decay resistance of heptadecenylsuccinic anhydride modified Scots pine, in a study of wood modified with various chemical agents. Reaction with the anhydride was accomplished using dimethyl formamide as a catalyst/swelling agent, with WPGs as high as 37 % being achieved. Such high WPGs undoubtedly resulted in cell wall damage occurring, and this is confirmed by the large percentage volume increases noted (26 %). Modified samples were tested against *C. puteana* and *G. trabeum*, according to ASTM D-1413-76 (1986), with weight loss due to 12 weeks' exposure being zero at about 35 % WPG. Forster *et al.* (1997) modified Corsican pine with alkenyl succinic anhydride (average side chain length 17 carbon atoms), reporting thresholds of 38 % for *C. puteana* and *G. trabeum*, 36 % for *T. versicolor* and 40 % for *P. sanguineus*.

4.4 Acetylation Using Ketene Gas

Since reaction of wood with acetic anhydride leads to the formation of acetic acid by-product, which must be removed from the wood, there has been some interest in the use of ketene gas for acetylation (Figure 4.4a). Ketene, for reaction with wood, is produced by pyrolysis of diketene. Provided that the wood contains no moisture, no acetic acid by-product is produced. However, ketene presents handling problems: it is very toxic and explosive, and it also has a tendency to dimerize. A comprehensive series of studies of ketene-based acetylation has been performed in Latvia and this work has been reviewed by Morozovs *et al.* (2003). Hardwoods have been found to be more reactive to ketene than softwoods and the optimal temperature for reaction has been determined as 47 °C. Application of vacuum and treatment of wood with ammonia solution has been used to remove the excess ketene. The reaction of wood with liquid diketene was also studied, with a WPG of 35 % being obtained after reaction for 3 hours at 52 °C.

Rowell *et al.* (1986f) reacted oven-dry aspen and southern pine wood flakes with gaseous ketene at 50–60 °C and formed flakeboards from these using aqueous PF resin. It required reaction times of 10–15 hours to achieve WPGs of the order of 20 %, with southern pine reacting more slowly and to lower ultimate WPGs (17 %) than aspen (23–25 %). Reaction at temperatures above 65 °C resulted in a marked discolouration of the wood flakes. Acetyl analysis was performed on wood flakes reacted with ketene to various WPGs. The acetyl analysis agreed closely with the WPG values up to about 12 % WPG, but were much lower at higher WPGs (12 % acetyl at 20 % WPG), which was

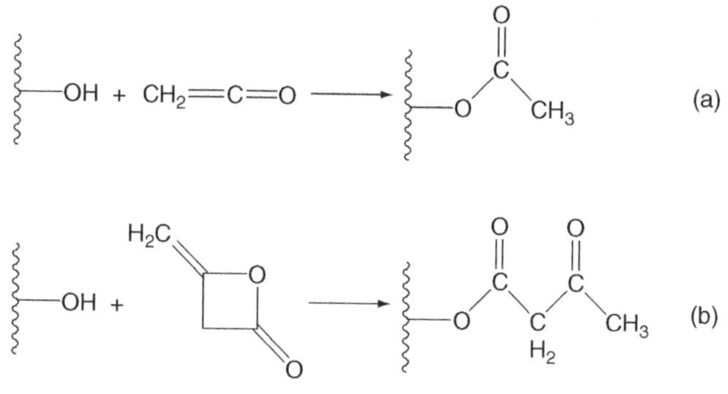

Figure 4.4 Acetylation of wood with ketene gas (a) and acetoacetylation of wood with diketene (b).

attributed to ketene undergoing polymerization reactions at the wood surface. Reaction of diketene with wood was also studied, with acetoacetylated wood being produced as a result (Figure 4.4b). Ketene modification was found to improve the dimensional stability of aspen boards more than southern pine, but was not found to be as effective at improving this property as acetic anhydride modification.

Four different methods (vapour-phase acetylation using acetic anhydride, acetylation using ketene gas, liquid phase acetylation using acetic anhydride/xylene, or neat acetic anhydride) were used to acetylate pine wood chips to a variety of WPGs for the production of MUF-bonded particleboards (Nilsson *et al.*, 1988). Composite boards were exposed to unsterile soil in fungal cellar tests. Boards made from ketene acetylated chips were not found to be resistant to decay at the maximum WPG level achieved (17 %): with a liquid acetic anhydride modification, no decay was recorded at a WPG level of *c.* 18 % after 12 months' exposure, whereas with a vapour-phase treatment at the same WPG, evidence for decay was found.

4.5 Carboxylic Acid Modification

The reaction of carboxylic acids with wood is of the same generic type as with the anhydride reactions (nucleophilic substitution at acyl carbon), but due to the lower positive charge on the acyl carbon in carboxylic acids compared to anhydrides, the reaction proceeds exceedingly slowly (Bender, 1960). For this reason, an acid catalyst is often used to promote the reaction (Figure 4.5). Thus, in an early study of the reaction of carboxylic acids with wood, trifluoroacetic acid (TFA) or trifluoroacetic anhydride were used as catalysts (Arni *et al.*, 1961a,b). The use of trifluroacetic anhydride to catalyse the reaction of unsaturated acids with wood has been extensively studied (Nakagami, 1978; Nakagami and Yokota, 1978a,b).

Reaction of wood with a mixture of acetic anhydride and crotonic acid in pyridine was reported by Risi and Arseneau (1957b), who were of the opinion that crotonylation of the wood had not occurred. Wood has been reacted with methacrylic acid by using the TFA impelling method, in order to activate the surface for grafting or covalent bonding between wood particles (Nakagami and Yokota, 1981a,b). Nakano (1994) and Nakano and Nakamura (1986a–c, 1987) also used the TFA impelling method to react wood with propionic, *n*-valeric, *n*-caproic, *n*-capric, lauric and palmitic acids in order to produce thermoplasticized wood. This work has been discussed in detail by Nakano (1996).

Kumar and Agarwal (1983) modified wood with thioacetic acid in the vapour phase, which results in hydrogen sulphide being generated as the by-product. The decay and termite resistance of the wood was determined. The thioacetic acid modified wood exhibited

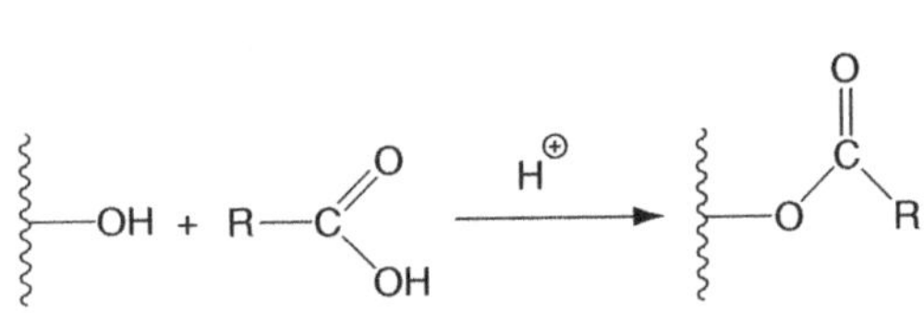

Figure 4.5 Acid-catalysed reaction of wood with a carboxylic acid.

good resistance to *O. placenta* (*Poria monticola*) at a WPG of 18 %. Thioacetic acid modified wood was also found to provide superior dimensional stability compared to an acetyl chloride treatment (Kumar *et al.*, 1991). A disadvantage with using this reagent is residual smell due to trapped H_2S after the reaction, although water leaching followed by vacuum drying appears to remove this (Singh *et al.*, 1979).

4.6 Acid Chloride Modification

Wood can be esterified using acid chlorides, but hydrogen chloride is released, which leads to degradation of the substrate (Figure 4.6). Wood samples have been acetylated using acetyl chloride, using lead acetate as a catalyst (Singh *et al.*, 1981; Kumar *et al.*, 1991). Such treatments are not viable for scaling-up for solid wood.

It has been reported that thermoplastic properties can be imparted to wood by modification of wood particles with fatty acid chlorides in a dinitrogen tetroxide – dimethylformamide – pyridine mixture (Funakoshi *et al.*, 1979; Shiraishi *et al.*, 1979, 1983). A method has also been developed for the modification of wood sawdust without the addition of organic solvents (Thiebaud and Borredon, 1995), and the thermal properties of such modified wood determined (Thiebaud *et al.*, 1997).

4.7 Isocyanate Modification

The reaction of isocyanate functionalities with the cell wall OH groups of wood involves the formation of a carbamate ester bond (Figure 4.7). No by-product is formed during the reaction, and the chemical bond between the adduct and the cell wall polymers is hydrolytically stable. However, these advantages are outweighed by the difficulty of handling many isocyanate reagents, which has precluded commercial exploitation and relegated studies to the laboratory.

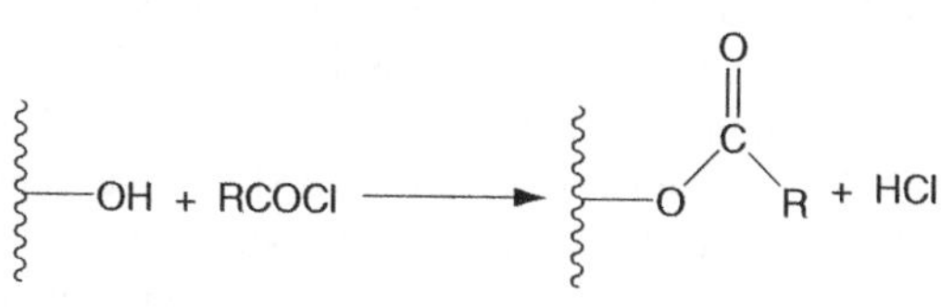

Figure 4.6 Reaction of wood with an acid chloride.

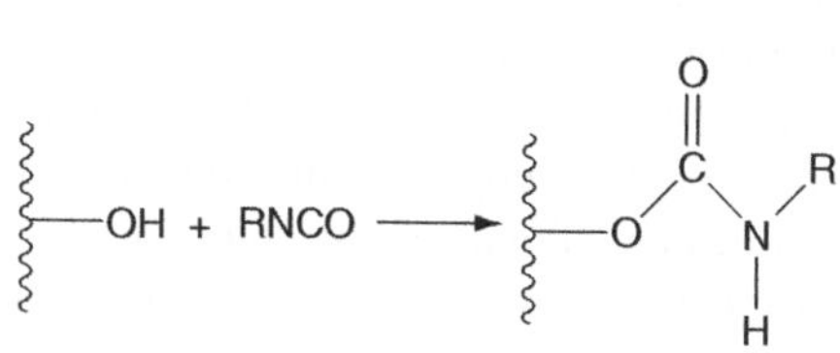

Figure 4.7 Reaction of wood with an isocyanate.

The isocyanate reaction involves addition to the $-C{=}N-$ bond, and in reactions involving compounds with an active hydrogen (such as OH groups), this hydrogen bonds to the isocyanate nitrogen. Most compounds that contain a hydrogen atom bonded to oxygen react with isocyanates, and in most of these reactions, aromatic isocyanates are more reactive than aliphatic isocyanates. Substitution of electronegative groups on the aromatic ring enhances reactivity and steric hindrance on the isocyanate or the alcohol retards the reaction. All these reactions are catalysed by acids, but much more strongly by bases. Reaction with alcohols decreases in the order primary >> secondary > tertiary. Phenols, being much more acidic than aliphatic alcohols, react more slowly. Conventional isocyanates react rapidly with water to yield a di-substituted urea. For this reason, it is vital that moisture is rigorously excluded when reaction between isocyanates and wood is attempted. The reaction between phenols and isocyanates is reversible at around 150 °C, which is exploited commercially as a method to form free isocyanates *in situ*. These phenyl isocyanates are called blocked isocyanates, or splitters, and do not readily react with water. Although reaction of blocked isocyanates with wood is possible, the high temperatures required would result in cell wall degradation. Tertiary amine catalysts are often used to catalyse the reaction of isocyanates with OH groups, and the effectiveness of these is usually in direct proportion to their base strengths.

Isocyanates are capable of co-reacting to form dimers, oligomers and polymers. For example, aromatic isocyanates will readily dimerize when heated, although the presence of a substituent ortho to the –NCO group reduces this tendency. For example, toluene diisocyanate (TDI) is less susceptible to dimer formation than diphenylmethane diisocyanate (MDI). The dimerization reaction is reversible, with dissociation being complete above 200 °C. It is unusual for aliphatic isocyanates to form dimers, but they will readily form trimers, as do aromatic isocyanates. The polymerization of aromatic isocyanates is known, but requires the use of metallic sodium in DMF.

It should be noted that isocyanates react with DMF to form bicyclic ureas or isocyanurates (Chadwick and Cleveland, 1981). DMF is an unsuitable solvent for wood isocyanate reactions because of these side reactions.

The first literature report of a reaction of an isocyanate with wood is that due to Clermont and Bender (1957). In this study, DMF impregnated wood samples 1/8 in thick were suspended above phenylisocyanate liquid in a vessel heated at temperatures from 100 °C to 125 °C for various time intervals. Treated samples were washed with DMF, then water, then acetone, and dried in an oven at 105 °C. ASEs in the range of 60–80 % were reported for these samples. In view of the reactivity of DMF with isocyanates, the lack of an efficient clean-up procedure and the fact that ASE values were calculated from the first water-soak cycle only, this study is of limited value.

The next study of wood modification was that reported by Baird (1969), who performed vapour-phase reactions of spruce with ethyl, *n*-butyl, *t*-butyl, allyl and phenyl isocyanate (PhNCO). Unfortunately, DMF was used as a catalyst for the reactions, which resulted in polymerization of PhNCO in the cell wall of the wood, leading to unpredictable results. No evidence was presented in support of the contention that polymerization had occurred, and since this requires an anionic catalyst initiator, this is considered unlikely. However, the presence of side reactions when DMF is used in conjunction with isocyanates has already been mentioned. Greater success was reported when butyl isocyanate was reacted with wood (presumably a consequence of the lower reactivity of this isocyanate with DMF). ASE was found to increase with increasing WPG of the isocyanate. The rate of weight increase was found to fall rapidly once a WPG of 50 % had been achieved. Increasing the loading of DMF in the wood increased the weight gain due to reaction, as did increasing the temperature of reaction. However, increasing the moisture content of the wood was found to lead to a reduction in weight increase, due to reaction of the isocyanate with water rather than the wood cell wall polymers. Little data was given for the other isocyanates.

One particular problem with the work of Baird (1969) was the lack of proof that the isocyanates had actually reacted with the wood cell wall OH groups. This deficiency was noted by Rowell and Ellis (1979), who additionally cited the work of Wakita *et al.* (1977) and Lutomski (1975), stating that 'It is not clear in any of the reported literature on the modification of wood with isocyanates, whether or not a reaction took place between the isocyanates and the wood cell wall component.' Rowell and Ellis (1979) reacted oven-dry southern pine blocks with methyl isocyanate at 10 atmospheres pressure (1 MPa) and 120 °C. Samples with WPGs from 14 % to 52 % were prepared by varying the reaction time. No solvent clean-up was considered necessary, since the boiling point of methyl isocyanate is so low. It was found that the volumetric swelling of wood due to modification equalled the volume of add-on chemical (calculated from weight gain divided by density of reagent), indicating that addition took place within the cell wall. ASEs were determined using 7-day soak cycles, over a total of four cycles. It was found that some weight loss occurred with modified samples over the four cycles, which was attributed to the leaching of bulking but nonbonded chemical. Since extractives were not removed prior or subsequent to vapour treatment with the isocyanate, the loss of extractives (or modified extractives) may be the reason for this behaviour. SEM studies showed that cell wall splitting occurred at high WPGs.

Southern pine was reacted with a range of isocyanates (ethyl, *n*-propyl, isopropyl, *n*-butyl, *p*-tolyl, phenyl, 1,6-diisocyanatohexane (HDI), toluene-2,4-diisocyanate (TDI) and isophorone diisocyanate: Ellis and Rowell, 1984). Triethylamine, DMF or DMSO were used as catalysts/swelling agents. Treatments were performed at 10 atmospheres pressure (1 MPa) and 120 °C. Specimens were extracted for 2 hours with benzene/methanol, post treatment. In the absence of swelling agents, it was found that higher molecular weight isocyanates did not penetrate the cell wall. Rowell and Ellis (1984) also investigated the effects of wood MC upon the reaction of epoxides or isocyanates with wood. It was found that such reactions were very sensitive to even low levels of moisture in the wood.

Spruce meal and beech blocks were modified by using a mixture of perfluoroalkyl ethanol and MDI. Improved dimensional stability and water repellency were reported (Engonga *et al.*, 1999). In order to avoid the handling problems associated with the use of most isocyanates, the generation of isocyanates within the wood by thermal rearrangement of acyl azides has been studied (Gérardin *et al.*, 1995) (Figure 4.8). This is potentially a

Figure 4.8 *Thermolysis of an acyl azide to produce an isocyanate.*

very interesting method for the preparation of wood carbamates, but is unlikely to have any commercial potential.

A study of the reaction kinetics of the reaction of butyl isocyanate with wood has been performed (West and Banks, 1986; West and Banks, 1987). Reactions were performed without catalyst and using pyridine, triethylamine, 1,4-diazobicyclo[2,2,2-octane] or di-butyl-tin-diacetate as catalyst. The data showed that no catalyst was effective without the presence of a swelling solvent. Kinetic profiles were obtained, which were deconvoluted to yield two component reaction curves. It was considered that these two curves represented reaction with lignin and the holocellulose component of the cell wall.

Martins and Banks (1991) modified pine and wood samples with phenyl or butyl isocyanate and investigated the water vapour sorption properties of the modified wood. Modification resulted in a reduction in EMC at a given RH up to about 35 % WPG with butyl isocyanate, although reaction with phenyl isocyanate above 25 % WPG resulted in an increase in sorption due to cell wall damage caused by the reaction. The butyl derivative was more effective at reducing hygroscopicity over all the WPGs studied.

4.7.1 Decay Resistance

Methyl isocyanate modified southern pine was found to be resistant to attack by *G. trabeum* at a WPG of 19 % in an ASTM D1413 soil block test (Rowell, 1980). Rowell also determined the partitioning of reacted chemical between the various cell wall polymeric components by using various isolation techniques. The lignin fraction was isolated by sulphuric acid digestion and the holocellulose component by chlorite treatment. The amount of nitrogen in each sample was determined by the Kjeldahl method. As WPG increased, it became harder to separate the fractions, and extrapolation was used to determine the actual extent of substitution in each component. This work showed that the degree of substitution (DS) of OHs in the lignin component was higher than with the holocellulose at all WPGs. Evidence for reaction of the isocyanate with wood was obtained using IR spectroscopy. It was found that at the level of substitution where resistance to attack by *G. trabeum* occurred, only 59 % of lignin and 11.7 % of holocellulose OHs were substituted. This might indicate that OH blocking is not the primary mechanism responsible for decay resistance.

Wood blocks were modified by methyl isocyanate, allyl isocyanate, allyl isothiocyanate and phenylhydrazine, by refluxing in pyridine (no reaction when benzene or methylene chloride were used as solvents; Kalnins, 1982). The wood was not extracted before reaction, but was extracted for 8 hours with benzene after the reaction. Allyl isothiocyanate formed what appeared to be a polymer within the wood cell wall, whilst the other reagents were thought to react with the cell wall OH groups. Decay resistance was determined against *G. trabeum, L. lepideus* and *C. versicolor*. Nearly complete protection against decay was found when WPGs exceeded 10 %, with the modified wood being less resistant to attack by *G. trabeum*. Reaction of the isocyanates with OH groups was confirmed using IR spectroscopy.

Chlorinated phenyl isocyanates have also been reacted with wood and found to reduce attack by *G. trabeum*. The mechanism of protection was attributed primarily to substrate blocking, rather than toxicity of the reacted chemical (Rowell, 1984b). Chen *et al.* (1990)

compared the fungal resistance of southern pine sapwood blocks impregnated with methyl fluorophenyl carbamates compared with wood reacted with fluorophenyl isocyanates. Although reaction with the isocyanates did provide decay resistance, a much higher molar concentration of reacted chemical was required to provide the same level of protection. However, it was found that the chemically bonded reagent was more stable to hydrolysis. The fungal resistance of the fluorophenyl carbamates increased as the degree of fluorine substitution increased.

The fungal resistance (*G. trabeum*) of loblolly pine reacted with *p*-toluene sulphonyl chloride (PTSC) or isocyanate was determined (Chen, 1992a). Reactions were performed on pyridine-impregnated blocks, which were soaked in acetone for 1 day then Soxhlet extracted with acetone for 12 hours subsequent to reaction. Both modifications were found to provide some decay resistance, although the PTSC modified wood was not stable to hydrolysis. Pyridine impregnation was also used for the reaction of loblolly pine sapwood with methyl, or phenyl isothiocyanate by heating at 85 °C for various periods (Chen, 1992b). After reaction, blocks were soaked in acetone for 1 day, and then Soxhlet extracted using acetone for 12 hours. Proof of bonding was obtained using IR spectroscopy. Decay resistance to *G. trabeum* was determined. It was found that wood reacted with the phenyl derivative was more resistant to decay, based upon the number of micromoles of reagent reacted with the wood. The permanence of the thiocarbamyl ester group was determined by measuring the sulfur content of samples following the 12-week decay experiment. It was found that the phenyl thiocarbamate ester was more stable.

Loblolly pine modified by 1,6-diisocyanatohexane (HDI) was found to be resistant to attack by *G. trabeum* at a WPG of 26 % (Chen, 1992c). At 26 % WPG, 6 % of bonded chemical was lost during a 12-week soil decay test. When moist wood was used for reaction, the HDI reacted mainly to form ureas and biurets. It was stated that the decay resistance of HDI modified wood was probably due to the inability of the modified cell wall to absorb sufficient amounts of water to support decay. Although wood reacted with chlorosulphonyl isocyanate lost only 1.3 % mass when exposed to *G. trabeum* in a decay test, it was reported that 50 % of the bonded chemical was lost in this test.

Forster *et al.* (1997) found that butyl isocyanate modification was more effective in enhancing decay resistance compared to the anhydrides studied in the same experiment. Decay protection thresholds were of the order of 15–20 % WPG, with the fungi studied.

Suttie *et al.* (1998) exposed butyl isocyanate modified (WPG ranging from 9 % to 23 %) Scots pine sapwood to soft rot decay in soil beds, removing samples at 8-week intervals up to a total exposure time of 52 weeks. The weight lost in all experiments was low (no greater than 3.5 %) and there was no significant change in the decay protection threshold (29 % WPG) over the duration of the experiment. Unmodified wood showed a steady increase in weight loss during the experiment. However, Forster *et al.* (1998), obtained a threshold of about 12 % WPG for butyl isocyanate modified samples exposed in unsterile soil for 20 months. This difference between the two experiments can be attributed to the WPG ranges used for the former set of experiments, which were effectively higher than the threshold. Forster found weight losses of the order of 10 % at 6–8 % WPG, and of the order of 20 % at 2–4 % WPG.

Cardias Williams and Hale (1999, 2003) studied the effect of the carbon chain length of the isocyanate upon decay resistance of the modified wood. Wood was reacted with butyl or hexyl isocyanate, or HDI in a pyridine/acetone solvent mix. Modified wood was exposed to one of four fungi (*C. puteana, G. trabeum, C. versicolor* or *P. sanguineus*) in an ASTM D1413 test, but for 16 weeks. The isocyanate modified wood showed the least resistance to *C. puteana*, with butyl isocyanate exhibiting a threshold for decay protection of around 15 % WPG with all the test fungi. There appeared to be no significant difference in performance related to the carbon chain length of the isocyanate. The decay of wood blocks modified to sub-threshold levels revealed that there was extensive decay in the interior of the samples, particularly in the latewood. This showed that there was envelope protection, even in blocks of dimension 10 mm in the longitudinal direction.

4.8 Epoxide Modification

The reaction of an epoxide reagent with wood is thought to lead to the formation of an ether linkage with the cell wall polymers, due to reaction of the cell wall hydroxyl groups with the highly strained three-membered ring of the epoxide group. Such a reaction would result in the formation of a product to which is attached a new OH group, that may then be involved in reaction with another epoxide (Figure 4.9) (Kirk, 1973; Rowell, 1983a; Kumar, 1994). This results in the formation of oligomers attached to the original OH site.

Epoxides can react with alcohols via acidic or basic catalysed reaction mechanisms. However, since both strong acids and bases will degrade the cell wall polymers of wood, the reaction is usually catalysed via the use of amines, which are more strongly nucleophilic than the OH group. For example, whereas the production of epoxy-phenolic resins requires temperatures in the region of 180–205 °C, reaction between epoxides and primary or secondary amines takes place at 15 °C (Turner, 1967). Reaction of epoxides with wood often involves the use of tertiary amines as catalysts (Sherman *et al.*, 1980). The sapwood is more reactive towards epoxides than heartwood (Ahmad and Harun, 1992).

McKelvey *et al.* (1959) investigated the reaction of epoxides with cellulose in alkaline conditions, reporting that alkaline cellulose reacted readily once the concentration of sodium hydroxide was sufficiently high. However, no evidence was found of reaction between cotton yarn and cellulose with a range of epoxides under a variety of reaction conditions. It was concluded that the apparent reactivity of cellulose with epoxides was primarily due to alkaline swelling of the cellulose, self-polymerization of the epoxide monomers then occurring within the interior structure of the fibres. It was also noted that the reactivity with phenol OH groups was very low (e.g. only 1 % conversion of ethylene oxide with various phenols).

In the absence of catalyst, ring opening of the epoxide requires the presence of alkoxide, or phenoxide ions in the cell wall matrix, which are only present in exceedingly

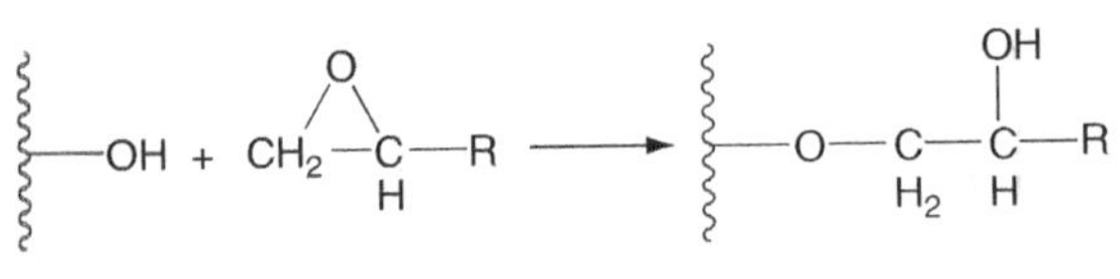

Figure 4.9 *Reaction of wood with an epoxide.*

low concentrations under normal reaction conditions. Thus, it would seem that the reaction of epoxides with cell wall OH groups is less likely than oligomerization of epoxide monomers within the cell wall matrix under the reaction conditions commonly employed, and especially in the presence of amine catalysts.

The first reported study of a reaction of wood with an epoxide appears to be that of McMillan (1963). This involved the use of gaseous ethylene oxide (Figure 4.9, R = H) at a temperature of 93 °C and a pressure of 3 atmospheres (0.3 MPa). In some cases, the wood was diffusion pre-treated with trimethylamine vapour as the catalyst. A 65 % ASE at 20 % WPG was obtained, attributed to a bulking effect due to *in situ* polymerization of the epoxide. There was no effect on the static bending strength of samples, and the modified wood became distinctly brown at higher levels of treatment.

In another study using ethylene oxide (Barnes *et al.*, 1969), the wood was pre-treated with trimethylamine and then with ethylene oxide vapour. It was found that oscillating pressure during treatment, rather than constant pressure, gave better results. An ASE value of 42 % was obtained at a WPG of 11.4 %. Rowell and Gutzmer (1975) performed a series of experiments in order to identify the optimum conditions for the epoxidation of wood. From these studies, a reaction pressure of 150 psi (1 MPa), a temperature of 120 °C, a 0.5–1.0 hour reaction time, and an epoxide to catalyst (amine) ratio of 95/5 were selected. The reactions of propylene oxide, butylene oxide (Figure 4.9, $R = C_2H_5$), epichlorohydrin and styrene oxide in the vapour phase were studied. It was found that above a WPG of 30 %, there was a reduction in ASE. This was thought to be due to space occupied by the formation of oligomer in the cell wall, causing damage to the cell wall microstructure, thereby leading to previously inaccessible OH sites being exposed.

The reaction of propylene and butylene oxide was further studied by Rowell *et al.* (1976), using the same reaction conditions (Rowell and Gutzmer, 1975). WPGs from 4 % to 45 % were obtained. Above 33 % WPG, there was a reduction in ASE. This was due to damage occurring in the cell wall, predominantly in the earlywood. Rowell (1977b) studied organic vapour drying of southern pine before treatment with butylene oxide, propylene oxide or propylene oxide with epichlorohydrin. It was hoped that this drying method would improve accessibility of reagent into the wood samples by limiting pit aspiration, but no improvement in reactivity was observed. In a study of the reaction of propylene and butylene oxide with wood (Rowell, 1978), it was found that there was a reduction in ASE during a sequence of water-soak/oven-drying cycles. This was particularly apparent when moving from cycle 1 to cycle 2. This was thought to be due to loss of bulking, but nonbonded, chemical. A reaction mixture of propylene oxide with epichlorohydrin using triethylamine as catalyst was used to determine the distribution of reagent using SEM–EDXA. The earlywood and sapwood both showed higher levels of reaction than the latewood or heartwood. Although the heartwood showed higher weight gains initially, there was a greater weight loss with subsequent solvent extraction than from the sapwood.

Sugar maple was reacted with propylene and butylene oxide (Rowell *et al.*, 1982). The modulus of elasticity (MOE) and modulus of rupture (MOR), fibre stress at proportional limit, and maximum crushing strength all exhibited a reduction, compared to unmodified samples. Nilsson and Rowell (1983) reacted ponderosa pine with butylene oxide and exposed the wood in an unsterile soil decay test. At low WPGs, severe surface decay due to soft rot and tunnelling bacteria was observed. Such attack was reduced at 15 % WPG, and was suppressed at 24 % WPG. The modified wood was also exposed to the brown rot fungus *Fomitopsis pinicola*. Attack was evident at the two lower WPGs. At 24 % WPG, apparently undecayed wood samples exhibited microcracks in the ML regions and cell wall corners of latewood tracheids. The fungus had gained entry into the ML via these microcracks.

Razzaque and Banks (1982) reacted Corsican pine wafers with propylene oxide at 125 °C and 8–10 bar (0.8–1.0 MPa). Triethylamine was used as catalyst. The modified samples showed a lower volumetric swelling, but higher water uptake than control samples. This was thought to be due to the rigid oligomeric chains causing the cell wall to be in a swollen but porous state. This then allowed the water to enter the cell wall without causing swelling. Such behaviour would be expected if the oligomers occupied space within the cell wall, but without substitution of the cell wall OH groups. Investigations of the dynamic mechanical properties of propylene and butylene oxide treated wood showed that the epoxide oligomer bulked the cell wall and was hydrophilic (Akitsu *et al.*, 1993a).

Chen (1994) reacted wood with epicholorohydrin, using triethylamine as a catalyst. Weight loss due to decay by *G. trabeum* in a 12-week exposure test was less than 3 % for a WPG of 11 %. Some of this weight loss was found to be due to loss of epicholorohydrin. IR and chemical analysis data was presented, which was interpreted as indicating that cross-linking of cell wall polymers had occurred, with reference to other work where this had been found with polysaccharides. However, it is not clear from the evidence presented that such a cross-linking reaction had indeed occurred.

Wood wafers were treated with mixtures of propylene oxide and oligomeric isocyanate (Guevera and Moslemi, 1983). The best treatment was found to be a mixture of 9:1 propylene oxide to isocyanate. In another study, Guevera and Moslemi (1984) studied the swelling properties of wood modified with propylene or butylene oxide and compared the data with modifications using a furan resin, or vinylpyrrolidinone. The best results were obtained by the use of alkylene oxides in combination with a cross-linking agent (trimethylol propane trimethacrylate).

Scots pine wood blocks and chips were modified with ethylene, propylene or butylene oxide, with trimethylamine as catalyst (Tillman, 1988). Particleboard was produced from the chips. Alkoxylated solid wood samples exhibited superior resistance to *P. placenta*, compared with unmodified samples. TMP fibres were reacted with glycidyl methacrylate (GMA), using pyridine as catalyst (Rozman *et al.*, 1994a,b). Modified fibres were then reacted with vinyl monomer in the presence of benzoyl peroxide (BPO) in a hot-press and composites produced. These exhibited a significant reduction in thickness swelling in water-soak tests, compared to boards produced using unmodified wood with phenol–formaldehyde resin. Çetin and Hill (1999) also studied the reaction of GMA and allyl glycidyl ether (AGE) with wood, following the procedure used by Rozman and co-workers. The results from this study indicated that AGE did not react with the OH groups of wood and that GMA probably did not react with the wood OH groups.

The resistance of propylene and butylene epoxide modified wood to subterranean termites was studied (Rowell *et al.*, 1979). A threshold of 34 % WPG was found to give good protection. Giant ipil ipil (*Leucaena leucocephala*) was modified with acetic anhydride, maleic anhydride – glycerol or propylene oxide (Mallari *et al.*, 1988, 1990). All modified samples showed good decay resistance to *C. versicolor* and *Tyromyces palustris*.

Ibach and Rowell (2000) reacted wood with propylene oxide and butylene oxide, and determined the decay resistance of the modified wood. Propylene oxide modification was found to be ineffective in preventing decay by *G. trabeum*, whereas butylene oxide modification proved to be effective at 23 % WPG. Reaction of epicholorohydrin with wood was found to provide decay protection at 31 % WPG against *G. trabeum*, although such modification did not lower the EMC significantly (Ibach and Lee, 2002). It was concluded that decay protection was due to substrate modification.

4.9 Alkyl Halide Modification

In virtually all studies of the reaction of wood with an alkyl halide, pre-treatment of the wood with a strong base (such as aqueous NaOH) is used. Kenaga and Sproull (1951) and Kenaga *et al.* (1950) attempted to react wood with allyl chloride in pyridine, but no reaction occurred. Instead, such improvements in ASE that were observed were due to bulking by water-leachable allyl-pyridinium chloride polymers (Risi and Arseneau, 1957d). Many studies of the reaction of alkyl halides have been concerned with the modification of wood surfaces for bonding purposes, and are fully discussed in Chapter 6. Surfaces of wood chips and wood blocks have been treated with aqueous NaOH and reacted with allyl bromide for surface bonding (Kiguchi, 1990a,b; Ohkoshi, 1990). Reaction of wood with crotonyl chloride (Risi and Arseneau, 1957b), methyl iodide (Narayanamurti and Handa, 1953) and butyl chloride (Risi and Arseneau, 1957c) in pyridine has been studied, with only temporary improvements in ASE being reported.

Benzylation has been performed on wood in order to impart thermoplastic properties to the substrate (Hon and Ou, 1989). Wood is pre-treated with aqueous NaOH solution, then with benzyl chloride. Benzylation of the surfaces of wood blocks and chips for self-bonding of wood surfaces has also been reported (Kiguchi, 1990a,b; Kiguchi and Yamamoto, 1992). A vapour-phase benzylation method has also been developed (Kiguchi, 1993). Carboxymethylation of NaOH-treated wood using various solvent systems has been studied (Shiraishi and Kishi, 1986; Honma and Nakano, 1991). Wood modified in this way has been used to make wood–phenolic adhesives (Kishi and Shiraishi, 1986).

4.10 Aldehyde Modification

4.10.1 Formaldehyde Modification

As for reaction of wood with carboxylic acids, the reaction of aldehydes also requires the use of an acid catalyst, which leads to degradation of the substrate. The most studied reaction in this category is that of formaldehyde with wood (Figure 4.10).

As can be seen from Figure 4.10, formaldehyde has the potential to react with two OH sites within the cell wall, resulting in cross-linking. The first stage of the reaction leads to the formation of a hemi-acetal, then an acetal. In both cases, the bond with the wood cell wall components is susceptible to hydrolysis. Because of the property of cross-linking, high values of ASE at low WPGs are observed (Stamm, 1959a–c). Wood has also been treated with trioxane and tetroxane (the cyclic trimer and tetramer of formaldehyde,

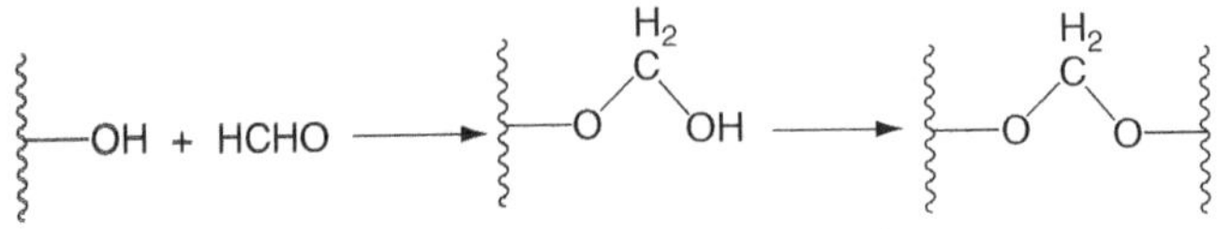

Figure 4.10 Cross-linking of wood cell wall polymers with formaldehyde.

respectively) in the presence of SO_2, which was found to be a very effective treatment (Minato *et al.*, 1990a,b; Yasuda *et al.*, 1993).

Wood does not react with formaldehyde in the absence of a catalyst, with Tarkow and Stamm (1953) finding that reaction temperatures as high as 160 °C had little effect on the dimensional stabilization of wood specimens. However, in the presence of small amounts of mineral acid catalyst, high ASEs were readily obtained. Acids used to catalyse the reaction of formaldehyde with wood include hydrochloric acid (Tarkow and Stamm, 1953; Ueyama *et al.*, 1961; Suzuki *et al.*, 1963a,b; Stevens and Schalck, 1978), nitric acid (Tarkow and Stamm, 1953), sulphur dioxide (Dewispelaere *et al.*, 1977; Stevens *et al.*, 1979), *p*-toluene sulphonic acid and zinc chloride (Stamm, 1959a; Ueyama *et al.*, 1961). The use of acid catalysts will result in degradation of the wood substance, with (for example) sulphur dioxide and hydrochloric acid reported to cause degradation of the fibre cell walls and loss of polysaccharide (Stevens and Parameswaran, 1981a,b). Similarly, the use of Lewis acid catalysts also causes a reduction in the mechanical properties of formaldehyde-treated wood (Vihavainen *et al.*, 1980). It has, however, been reported that wood treated with formaldehyde vapour in the presence of sulphur dioxide does not exhibit such a severe loss of strength as with other treatments (Vihavainen *et al.*, 1980; Minato and Yano, 1990). Because of the degradation associated with the use of acid catalysts, Burmester (1967) investigated the modification of wood with formaldehyde gas followed by gamma irradiation.

The kinetics of the reaction between formaldehyde vapour and wood using HCl as a catalyst have been determined (Minato and Mizukami, 1982). Two activation energies were determined, $69\,kJ\,mol^{-1}$ from the initial rates of reaction and $20\,kJ\,mol^{-1}$ from the diffusion part of the rate curve.

The modification of beech and pine wood with formaldehyde and tannin in a combined treatment to improve dimensional stability has been studied (Burmester, 1971a). Reaction of oak heartwood with gaseous formaldehyde did not improve dimensional stabilization, unless the heartwood was activated by pre-steaming (Burmester, 1971b).

The mechanical properties of formaldehyde-treated wood are poor, with severe embrittlement being a common observation (Burmester, 1967). This is due to the short chains in the cross-link being rigid, as well as acid degradation of the cell wall polymers. Evidence of such cross-linking has been reported from studies of the dynamic mechanical properties (Sugiyama and Norimoto, 1996). This is also found by examining the relationship between dielectric properties and dimensional stability, where it has been determined that greater stability is attained by formalization compared with acetylation at a given level of substitution (Kameyama *et al.*, 1994). This is because dimensional stabilization arising from formalization is due to cross-linking of the cell wall polymers, whereas stabilization by acetylation is due to bulking of the cell wall. Formaldehyde treatment is reported to improve the acoustic properties of wood (Yano *et al.*, 1986b; Akitsu *et al.*, 1991; Yano and Minato, 1992, 1993). Yusuf *et al.* (1995) tested the mechanical properties of formaldehyde-treated IS-bonded wood composites. Formaldehyde treatement of the boards resulted in significant reductions in thickness swelling at WPGs in the region of 3 %, a slight increase in IB and a decrease in MOR. Both decay and termite resistance were markedly improved.

Even low levels of formaldehyde modification result in significant reductions in the EMC of wood (Minato and Norimoto, 1985; Minato, 1993). For example, Yasuda *et al.* (1995) found that a WPG level of only 3.5 % resulted in a 50 % reduction of the EMC compared to unmodified wood. This is attributable to the formation of rigid cross-links within the cell wall matrix preventing swelling of the cell wall in the presence of atmospheric moisture.

In decay studies, it was found that formaldehyde-treated wood showed protection against *L. trabea* at only 2 % WPG of formaldehyde (Stamm and Baechler, 1960). This exceptional decay resistance at such low levels of treatment has been related to the extensive cell wall cross-linking preventing expansion of the cell wall and hence limiting the cell wall moisture content. Vihavainen *et al.* (1980) determined the decay resistance of SO_2-catalysed formaldehyde-treated wood. Sulphur dioxide treatment alone was found to improve decay resistance to *C. puteana, P. placenta* and *G. trabeum*, with formaldehyde/SO_2 showing a further improvement. Very good resistance to white rot fungus (*C. versicolor*) has been reported for formaldehyde treated wood, but resistance to brown rot (*Tyromyces palustris*) was poor (Minato *et al.*, 1992). Dewispelaere *et al.* (1977) found that WPGs of only 6 % were sufficient to provide good decay resistance. Schmidt (1983) modified wood wafers with formaldehyde catalysed with SO_2. Boards were made from the modified wood using PF resin. The boards were not resistant to attack by brown rot fungi (*G. trabeum* or *P. placenta*), but were resistant to decay by the white rot fungus *C. versicolor*.

Formaldehyde vapour treatment has been applied to improve the dimensional stability of medium-density fibreboard (MDF) (Minato *et al.*, 1992) and albizzia waferboard (Yusuf *et al.*, 1995c).

4.10.2 Modification with Other Aldehydes

Wood has been reacted with acetaldehyde (Tarkow and Stamm, 1953) and benzaldehyde (Tarkow and Stamm, 1953; Weaver *et al.*, 1960), using either nitric acid or zinc chloride as catalyst. Weaver *et al.* (1960) investigated the reactions of 16 aldehydes with wood, including glyoxal, glutaraldehyde, trichloroacetaldehyde, phthaldehydic acid and α -hydroxyadipaldehyde, using zinc chloride, magnesium chloride, *p*-toluene sulphonic acid, phenyldimethylammonium chloride and pyridinium chloride as catalysts. It was concluded that none of the aldehydes studied formed stable cross-links in the cell wall. Minato and Yasuda (1992) reported the reaction of wood with glyoxal and glutaraldehyde. The reaction of glyoxal and glyoxal/glycol with Japanese ash has been investigated (Nakano, 1993c). Whilst glyoxal showed little reactivity, the glyoxal/glycol mixture showed high weight increases following treatment. Evidence for bond formation between glyoxal and wood components using FTIR and NMR has been presented (Yasuda and Minato, 1995). The moisture sorption and creep properties of glyoxal and glutaraldehyde treated wood have been studied (Yasuda *et al.*, 1994).

4.11 Cyanoethylation

Cyanoethylation of wood involves the reaction of acrylonitrile with wood in the presence of an alkaline catalyst (Figure 4.11). With NaOH-treated wood, WPGs up to 30 % have been reported, giving ASEs in the region of 60 %. At 25 % WPG, samples were reported to be resistant to attack by *Poria monticola, C. puteana, G. trabeum* and *Lentinus lepideus* (Goldstein *et al.*, 1959). Treated wood with a nitrogen content of 8.5 % was found to be resistant to attack by *Poria vaporaria* (Fuse and Nishimoto 1961). With a nitrogen content of 1 %, treated wood was resistant to attack by *Lentinus lepideus, P. monticola, G. trabeum* and *C. versicolor* (Baechler, 1959; Stamm and Baechler, 1960). Cyanoethylated stakes were found to have an average lifetime of 7.8 years in ground contact tests, compared to 3.9 years for untreated stakes (Gjovik and Davidson, 1973). The decay protection mechanism is thought to be due to bulking, rather than toxicity of the treated wood (Baechler, 1959).

Cyanoethylation of wood meal has been used to impart thermoplasticity to wood (Morita and Sakata, 1986; Hon and San Luis, 1989). Treatment of cyanoethylated wood with a chlorine solution has also been found to improve the solubility of the treated wood in various organic solvents (Morita *et al.*, 1987). Films have been prepared by casting from DMF solution on to glass (Yamawaki *et al.*, 1990).

4.12 Beta-Propiolactone

The reaction of β-propiolactone with wood can be catalysed by acid or base, to yield two different products (Figure 4.12).

Reaction of wood with β–propiolactone to 25 % WPG resulted in good decay resistance to *L. lepideus, G. trabeum, P. monticola* and *C. puteana*. At 30 % WPG, an ASE of 60 %

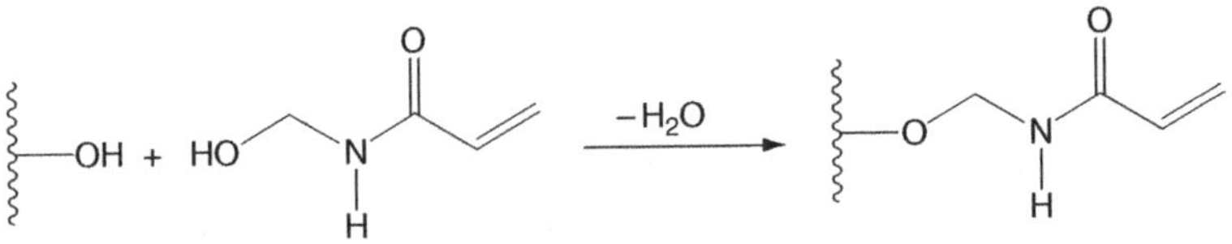

Figure 4.11 Reaction of acrylonitrile with wood.

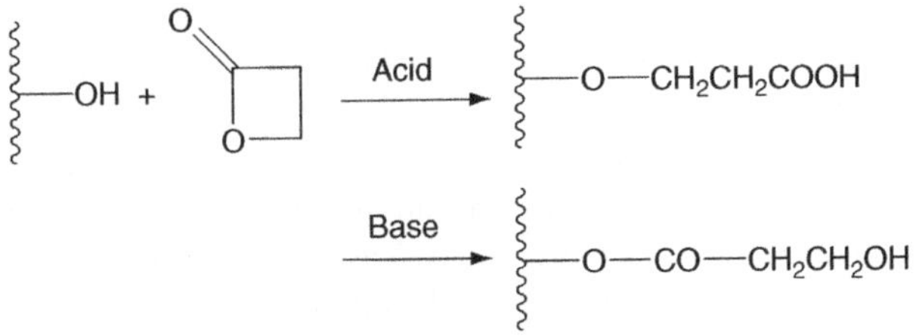

Figure 4.12 Reaction of β-propiolactone with wood.

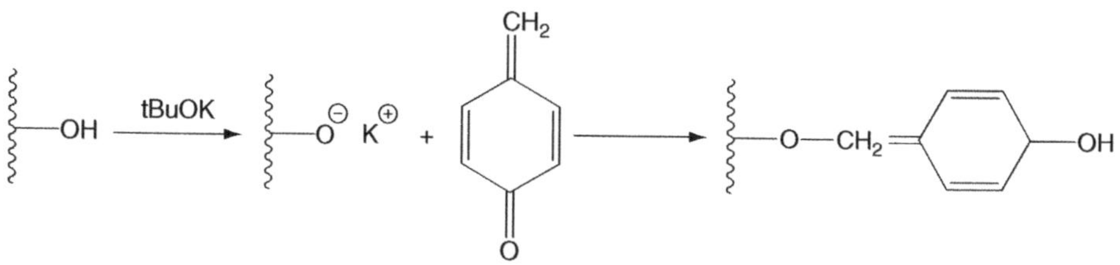

Figure 4.13 Reaction of quinone methides with wood.

was reported (Goldstein *et al.*, 1959). Due to the carcinogenic properties of this compound, no further research has been undertaken.

4.13 Quinone Methides

Solvent-extracted delignified beech wood pulp was treated with potassium tertiary butoxide, then with quinone methides at 25 °C or 45 °C (Figure 4.13). Following reaction, extensive solvent extraction was then performed, and the modified flour was pressed to form a disc, which was then exposed to *C. versicolor* or *G. trabeum*. No indication of biological attack was found (Loubinoux *et al.*, 1992).

4.14 Conclusions

A very large number of reagents have been studied for their potential to be used as chemical modifying agents for wood. Although these studies have been of great value in furthering research into wood modification, there has been little interest in them commercially. Most of the processes described in this chapter involve the use of reagents, solvents or catalysts that would present problems if handled in the quantities necessary for commercial operations. Although industrial exploitation is unlikely at present, there may be further developments that change this situation.

5

Thermal Modification of Wood

5.1 Introduction

This chapter is primarily concerned with laboratory-based research studies of thermal wood modification, with commercialization of thermal treatment processes being discussed in more detail in Chapter 8. Of all the various wood modification processes that have been studied, thermal modification is by far the most advanced commercially. The thermal modification of wood has long been recognized as a potentially useful method to improve the dimensional stabilization of wood and increase its decay resistance. Tiemann (1915) was one of the first workers to report on the effect of high-temperature treatment upon the physical properties of wood. He heated air-dry wood in superheated steam at 150 °C for 4 hours, which reduced the subsequent moisture sorption by 10–25 %, with relatively low reductions in strength found in most cases. Stamm and Hansen (1937) mention an unpublished report by the Forest Products Laboratory, dated 1916, which involved a study of the heating of black gum at 205 °C, resulting in a reduction in hygroscopicity.

Further studies of this technology led to the development of several commercial processes, such as 'Lignostone' and 'Lignifol' in Germany, and 'Staypak' and 'Staybwood' in the United States. More recently, several commercial processes have been introduced in Europe. These are the Thermowood process in Finland, the Plato process in Holland, and the Perdure process and Retification in France.

Thermal modification is invariably performed between the temperatures of 180 °C and 260 °C, with temperatures lower than 140 °C resulting in only slight changes in material properties and higher temperatures resulting in unacceptable degradation to the substrate. Studies of the thermal treatment of wood above 300 °C are of limited value, due to severe degradation of the material. There is also evidence to show that there is an abrupt change in the degradation kinetics close to this temperature (Elder, 1991). The literature discussed in this chapter is therefore limited to effects taking place below 300 °C. Modern thermal modification processes are limited to temperatures no higher than 260 °C.

Wood Modification: Chemical, Thermal and Other Processes C. Hill

As a result of thermally induced chemical changes to the macromolecular constituents, the physical and biological properties of the wood are altered. These changes include the following:

- improvements in dimensional stability, depending on the treatment conditions;
- reduced hygroscopicity (a decrease in EMC at a given RH and reduced wettability);
- improved resistance to microbiological attack;
- often, an increase in modulus during the initial stages of heating, with a reduction thereafter;
- a reduction in impact toughness, modulus of rupture and work to fracture;
- reduced abrasion resistance;
- a tendency for cracks and splits to form, knots to come loose and so on;
- a darkening of the material.

The properties of thermally modified wood are highly dependent upon the thermal treatment employed, and it is very important to take these into account when comparing the various treatment methods employed. This chapter examines the effect of the process variables upon the properties of thermally modified wood, and then considers the chemistry of thermal modification. Studies of physical changes are discussed, followed by an overview of the biological properties of thermally modified wood. A short description of some recent literature on the use of thermal treatment combined with compression and on hot oil treatments is also included.

5.2 Process Variables

There are a variety of thermal modification methods that can be applied to wood, and the exact method of treatment can have a significant effect upon the properties of the thermally modified wood. Important process variables include the following:

- time and temperature of treatment;
- treatment atmosphere;
- closed versus open systems;
- wood species;
- wet and dry systems;
- sample dimensions;
- use of catalysts.

5.2.1 The Effect of Temperature and the Duration of Treatment

As wood is heated, there is a decrease in weight of the material initially, due to the loss of bound water and volatile extractives, with less volatile extractives tending to migrate to the surface of the wood. As the temperature is increased, chemical changes to the macromolecular components of the cell wall occur, accompanied by further weight loss and colour changes. Several studies have shown that mass loss (as well as changes in other properties) obey a first order kinetic process, although other studies have shown that this depends upon treatment conditions and sample size.

5.2.2 The Effect of the Treatment Atmosphere

Treatment can be performed in air, in a vacuum or under an inert atmosphere. In some early studies, air was excluded by heating the wood under molten metal. Clearly, treatment in oxygen results in oxidative processes occurring, and there are significant differences in the chemistry of degradation as well as the material properties obtained. There has been some work in which heating of wood is performed in oil, which acts as a heat-transfer medium and additionally acts to exclude oxygen from the wood. This has resulted in the development of the Menzholz process in Germany. If water is present in sufficient amounts, this can also act as a blanket to prevent, or reduce the rate of oxidative processes.

5.2.3 Closed versus Open Systems

Heating of wood in a sealed reactor allows for the build-up of degradation products that can affect the chemical changes taking place to the wood; there will also be an increase in pressure in the reactor. The presence of acids produced from acetyl groups on hemicelluloses result in accelerated degradation of the polysaccharide components of the cell wall (Stamm, 1956). Thermal treatment using an open system allows for the removal of these products. If green wood is treated in a closed reactor, the process takes place in the presence of high-pressure steam, whereas the water escapes from an open system. In some processes a recirculating system is employed, in which condensable products (volatile extractives and water) and breakdown products (e.g. acetic acid) are removed before the treatment atmosphere is returned to the reactor.

5.2.4 Effect of Species

Differences have been noted between species in the way in which they respond to heat treatment, but most notably between hardwoods and softwoods. Thermal, hydrothermal or hygrothermal treatment of various woods results in weight losses that are generally found to be higher for hardwood compared to softwood species (MacLean, 1951; Zaman *et al.*, 2000; Militz, 2002).

5.2.5 Hydro- and Hygrothermal Treatments

The presence of water, or water vapour, affects the chemistry of thermal modification and heat transfer within the wood (Burmester, 1981). Under dry treatment conditions, the wood is dried prior to thermal modification, or water is removed by the use of an open system, or a recirculating system equipped with a condenser. In closed systems, water evaporated from the wood remains as high-pressure steam during the process. Steam can also be injected into the reactor to act as a heat-transfer medium, and can additionally act as an inert blanket to limit oxidative processes. Such steam treatment processes are referred to as *hygrothermal* treatments. Where the wood is heated in water, this is known as a *hydrothermal* process. Hydrothermal treatments have been extensively studied as a technology for the processing of biomass for the production of sugars and have been recently reviewed by Garrote *et al.* (1999).

5.2.6 Sample Dimensions

The inherent heterogeneity of the material leads to variations in the responses of wood to thermal modification. The rate of transfer of heat into the interior of the wood is of paramount importance in order to ensure that there is a constant temperature throughout the sample. The thermal conductivity of dry wood is low and the heating method employed must ensure that the treatment is as even as possible. Heat transfer into the interior may be improved by the use of steam-heating. Heat transfer is a very significant factor in the treatment of timber of larger dimensions.

5.2.7 Heating in the Presence of Catalysts

There has been some limited work done where the wood is heated in the presence of chemicals that are used to accelerate the degradation processes. This normally involves the use of catalysts that generate acids, thereby accelerating the degradation of the polysaccharide constituents. Catalysts are not used commercially. Stamm (1959c) studied the improvement of dimensional stability by applying heat treatment of wood in the presence of a range of catalysts. Catalysts reduced the time of heating required to produce a given increase in dimensional stability, but did not affect the correlation between mechanical properties and ASE, nor between weight loss due to heating and ASE. The use of acid catalysts combined with heat treatments increase the yields of furfural and furan derivatives (Fengel and Wegener, 1989).

5.3 Chemical Changes in Wood due to Thermal Modification

The heating of wood will lead to a variety of processes occurring, which depend upon the heating regimen employed. It is accepted that hemicelluloses are degraded to a greater extent than the other macromolecular components (Shafizadeh and Chin, 1977), but the relative stability of the cellulose and lignin is harder to determine. In general, loss of polysaccharide material becomes particularly significant at temperatures above 180 °C, but this depends very much upon the treatment conditions. The precise relative rates of degradation of the different components are dependent upon the experimental methods employed, and as a consequence the literature on this subject is complex and apparently sometimes contradictory. Much of the work in this area relies upon standard gravimetric methods to determine the relative proportions of the macromolecular components, and whilst such studies have an undoubted value, they need to be treated with some caution, since material loss is not the sole factor indicating that degradation has occurred.

There have been many studies that have attempted to elucidate the chemistry of thermolysis of wood by examining the thermal behaviour of isolated wood components, and much of the early work in this respect has been reviewed (Beall and Eickner, 1970). The use of thermal analysis techniques has shown that the results obtained are quite variable and heavily dependent upon the method of preparation of the materials as well as the experimental parameters, particularly heating rate and atmosphere. It is also undoubtedly the case that the chemistry of the thermal degradation of the isolated wood components is different from that taking place within the cell wall, where the various reactions can act synergistically within the wood material. There are also competing processes occurring during heating that can result in simultaneous endothermic and exothermic events, making it difficult to determine the onset temperature for the different processes.

As wood is progressively heated to higher temperatures, production of condensable fractions occurs, with loss of water and volatile extractives at temperatures below about 140 °C. Above this temperature, the production of cellular breakdown products derived from the more labile species attached to the cell wall polymers assumes greater significance. This results in the production particularly of acetic acid derived from the hemicelluloses; but additionally formic acid and methanol, as well the production of noncondensable gases (mainly CO_2) as the temperature is raised further. Above about 140 °C dehydration reactions also begin to occur, as so-called 'water of constitution' is lost, leading to a decrease in OH content, and these assume greater importance as the temperature is raised. As the temperature is further increased, CO and CO_2 are also increasingly detected in the gases produced (Bourgois and Guyonnet, 1988). At around 270 °C, there is a significant change in the reaction kinetics due to the onset of an exothermic reaction (Stamm *et al.*, 1946). Although the chemistry of thermal modification is complex and far from fully understood, it can be readily appreciated that there are nonetheless distinct changes in the nature of the reactions taking place as the temperature is increased. What is less certain is the exact points at which the different reactions become dominant.

As noted, the treatment atmosphere has a significant influence upon the reaction chemistry occurring. Heating under oxidizing conditions results in an increase in carbonyl-containing groups, whilst heating in a reducing or inert atmosphere leads to a loss of oxygen-containing species, although a decrease in OH groups may be associated with a small increase in carbonyl functionality. Heating of wood in the presence of oxygen has been shown to result in an initial decrease followed by an increase in carbonyl content (as determined from the IR spectrum) at extended heating times (Chow, 1971). The decrease was attributed to loss of ester-bonded and carboxylic-containing groups, with the subsequent increase being due to oxidative carboxylation. A decrease in carbonyl band intensity was found with heating in nitrogen, although a slight increase was noted for longer treatment times. Results were strongly influenced by the temperature of treatment and by the presence of wood extractives.

The presence of water also affects the reaction chemistry, although this depends upon the amount of water present and whether or not it is lost from the treatment system. The heating of wood in the presence of water or steam results in the accelerated formation of organic acids (primarily acetic acid) that catalyse the hydrolysis of hemicelluloses, and to a lesser extent the amorphous cellulose (Mitchell, 1988). The generation of these acids is also enhanced in the presence of air (wet oxidation). It should, however, be noted that the continued presence of water as steam during the treatment process can prevent oxidative processes from occurring. Hydrothermal processing results in the hydrolysis of polysaccharide due to the action of hydronium ions generated by autonionization of the water, although the formation of hydronium ions from acetic acid is more important (Garrote *et al.*, 1999). Temperature ranges from 150 °C to 230 °C are generally used, because hydrolysis is very slow at lower temperatures, whereas cellulose degradation begins to occur in the region 210–220 °C. Cellulose degradation becomes predominant at 270 °C.

5.3.1 Hemicelluloses

As noted above, when wood is heated, the most thermally labile of the wood polymeric components (hemicelluloses) begin to degrade, resulting in the production of methanol, acetic acid and various volatile heterocyclic compounds (furans, γ-valerolactone etc.). Degradation of hemicelluloses increases with temperature (Bourgois *et al.*, 1989) and time of heating exposure. Loss of hemicelluloses leads to an increase in the degree of crystallinity of wood samples, in addition to those changes related to degradation/rearrangement of the amorphous cellulose content. Stamm (1964) proposed that the initial process occurring in the thermal degradation of wood was a breakdown of hemicelluloses to form furfural polymers that are less hygroscopic than the hemicelluloses. Chemical analysis of wood components shows that there is good stability up to a temperature of 100 °C with 48 hours' heating (Fengel and Wegener, 1989). Above this temperature, the holocellulose content decreases, which is associated with loss of hemicelluloses, since the cellulose content remains unchanged up to 150 °C (changes in DP can occur, however). The loss of hemicelluloses is faster in closed systems (Mitchell *et al.*, 1953), due to the build-up of acidic vapours that catalyse sugar hydrolysis.

Although it is well established that hemicelluloses are less thermally stable than cellulose, there are variations in the literature regarding the exact temperature for the onset of degradation. The use of differential thermal analysis (DTA) of wood shows that there is a large endotherm extending from about 90 °C up to 150 °C, associated with the evaporation of cell wall bound water. Thereafter, a series of exothermal peaks are observed, starting at 200 °C, which are associated with degradation of the cell wall polymeric components. However, DTA thermograms of wood and isolated cell wall components differ (Fengel and Wegener, 1989). Beall (1969) studied the thermal degradation under nitrogen, or air, of nine isolated hemicelluloses by using thermogravimetric analysis (TGA). The temperature at which there was 10 % weight loss due to heating was found to be above 200 °C in all cases, and higher for the softwood compared with the hardwood hemicelluloses. Degradation temperatures were lower, and rates of degradation were higher in an oxygen atmosphere. The use of DTA under a nitrogen atmosphere showed the onset of an exotherm at 180 °C for a softwood xylan, but the glucomannan exhibited this first exotherm at a higher temperature. Comparison of the DTA and TGA results indicated that degradation was occurring before a significant exotherm was recorded. It was concluded that the degradation consisted of competing thermal events that were sufficiently balanced to produce little net heat of pyrolysis (Beall, 1971). Ramiah and Goring (1967) studied the thermal degradation of isolated birch xylan and a spruce glucomannan, by determining the rate of evolution of gases from the heated samples. These experiments showed that the hardwood xylan was less thermally stable than the softwood hemicellulose. Ramiah (1970) stated that the initial temperature of active pyrolysis of a glucomannan derived from spruce was 140 °C, although some weight loss from this hemicellulose was noted at temperatures as low as 100 °C.

A significant factor in the thermal degradation of hemicelluloses is the presence of acetyl groups that are thermally labile and lead to the formation of acetic acid, thereby causing acid-catalysed degradation of the polysaccharides. The removal of acetyl groups from larch galactoglucomannans has been shown to increase the thermal stability of the deacetylated hemicellulose (Shimizu *et al.*, 1972). Degradation of the *O*-acetylgalactoglucomannan was observed to initiate at 154 °C in air, whereas deacetylation resulted in an increase to 200 °C. Beall (1969, 1971) found that deacetylation increased the thermal stability of a hardwood xylan isolated from birch. Degradation and de-acetylation of the hemicellulose component accompanied by an increase in polysaccharide crystallinity was found in steam heat-treated pinewood, using ^{13}C CP-MAS NMR spectroscopy (Sivonen *et al.*, 2002). A slight increase in crystallinity was found even when samples were dried in saturated steam at 115 °C. Free-radical formation was also studied using ESR, which showed a large increase in radical concentration proportional to treatment temperature, with these radicals being highly stable and still observable several months after the treatment process. Loss of acetyl from hemicelluloses has also been confirmed in IR studies (Weiland and Guyonnet, 2003).

A probable pathway for the degradation of hemicelluloses via free-radical intermediates has been proposed by Fengel and Wegener (1989) and is shown in Figure 5.1. Hemicellulose polymers are depolymerized to form oligosaccharides and monosaccharides, which are dehydrated to form furfural (pentoses) and hydroxymethyl furfural (hexoses).

Hardwoods are less thermally stable than softwoods and this is attributable to differences in the hemicellulosic content and composition. Pentosans (which are found in higher proportions in hardwood hemicelluloses) are more susceptible to thermal degradation than hexosans (Fengel and Wegener, 1989). Additionally, hardwoods, in general, have a higher proportion of hemicellulose, and the hemicelluloses of hardwoods also have a higher acetyl content compared to softwoods.

5.3.2 Cellulose

It is generally accepted that cellulose degradation occurs at a higher temperature than that of hemicelluloses, although some of the evidence is contradictory. It is likely that minor thermal degradation does take place at relatively low temperatures, but at a much slower rate than with the hemicelluloses. The amorphous regions of cellulose are more susceptible to thermal degradation and these regions probably exhibit similar thermal properties to the hexose components of hemicelluloses. Crystalline cellulose degrades in the temperature range 300–340 °C (Kim *et al.*, 2001). The rate of cellulose degradation is reduced if water is present, which is assumed to be due to the enhanced ability of the amorphous regions to change structure to produce more thermally stable crystalline regions (Fengel and Wegener, 1989). With extended heating, chain scission of the cellulose occurs, producing alkaline-soluble oligosaccharides, with a concomitant decrease in the cellulose DP and the degree of crystallinity. It should be noted that this production of alkaline-soluble cellulose breakdown components can lead to errors in estimating the hemicellulose content. Reductions in cellulose DP are found with heating at temperatures of 150 °C, with a faster decrease in DP observed when cellulose is heated in air (Shafizadeh, 1984). Fengel and Wegener (1989) report that decreases in cellulose DP in thermally treated

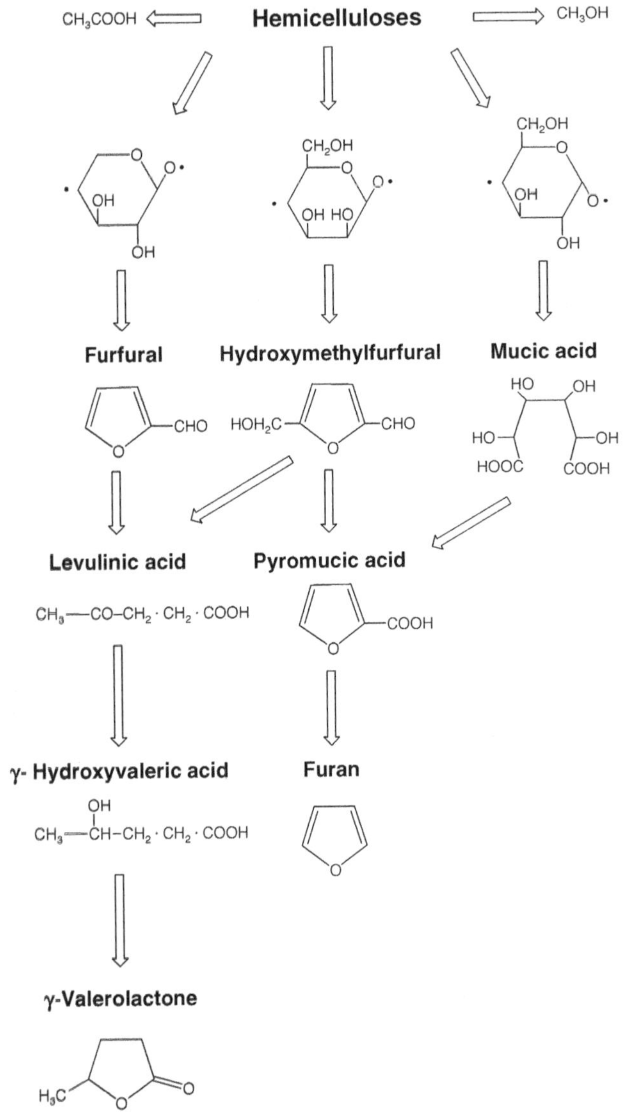

Figure 5.1 Probable thermal degradation pathways for hemicelluloses, according to Fengel and Wegener (1989).

spruce occur at temperatures above 120 °C, and in isolated cellulose at as low as 100 °C. Heating in air results in the production of carbonyl and carboxylic groups and short-lived hydroperoxide groups due to oxidation of the cellulosic OH groups. Heating for longer periods results in an increase in carbonyl groups at the expense of carboxylic moieties. A small increase in carbonyl groups is also found when cellulose is heated in nitrogen. The formation of such groups is associated with a yellowing of the cellulosic material. CO_2 and CO are produced when cellulose is heated at 170 °C, and the yields thereof are higher when heating is carried out in air compared to nitrogen (Shafizadeh, 1984). There is a change in reaction kinetics when pyrolysis is performed above 300 °C. The chemistry of thermal degradation is different when cellulose is located in the cell wall associated with the other macromolecular components. It is very likely that the breakdown products formed are capable of further reaction with other wood components, which is not possible with the isolated cellulose.

Levoglucosan has been identified as the primary breakdown product derived from the cellulose during thermal treatment (Sanderman and Augustin, 1964), but other anhydroglucoses, furan and furan derivatives are also produced (Fengel and Wegener, 1989) (Figure 5.2). The production of levoglucosan is most significant at temperatures in excess of 300 °C and appears to be associated with degradation of the crystalline cellulose regions, with only limited amounts of levoglucosan being found below 250 °C.

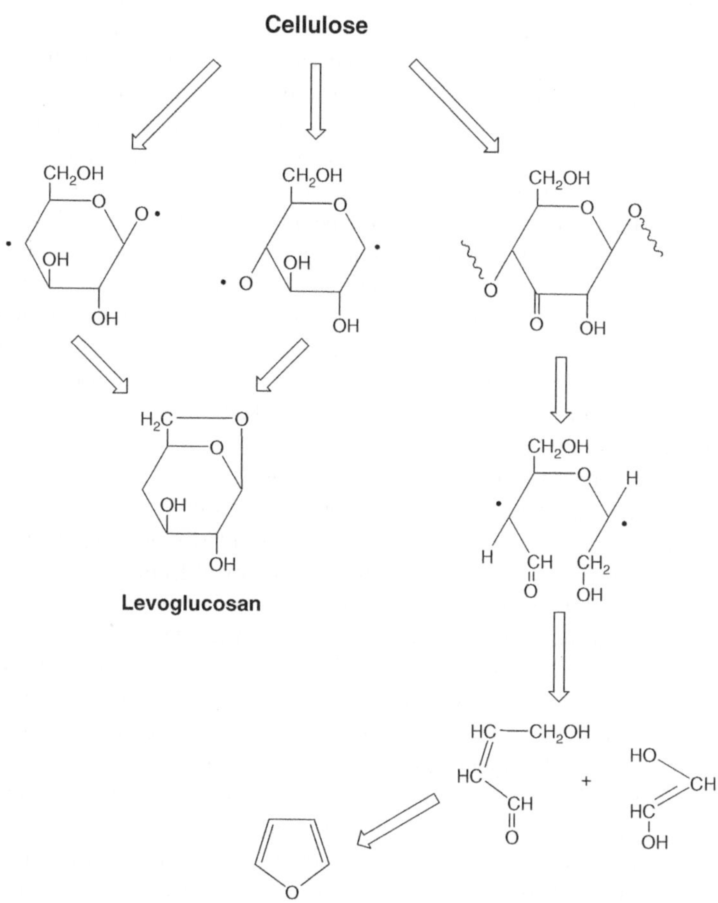

Figure 5.2 *Probable thermal degradation pathways for cellulose, according to Fengel and Wegener (1989).*

5.3.3 Lignin

The loss of polysaccharide material on heating leads to an increase in the lignin content of the wood. It is generally accepted that lignin is the most thermally stable component of the cell wall, but some thermal degradation of lignin can occur at relatively low temperatures, with the production of various phenolic breakdown products (Sandermann and Augustin, 1964). Examination of isolated lignin using DTA under helium gas has shown that the positions and sharpness of the main exotherms are strongly affected by the heating rates employed (Nassar and McKay, 1984). The presence of any residual polysaccharide material was also found to change the results obtained. The DTA thermograms showed that there were endotherm reactions extending from 50 °C to 200 °C, with a small exotherm occurring at 220 °C. Other, larger, exotherms were found above this temperature. The endotherm was not due to loss of moisture, but was thought to be related to molecular rearrangements associated with softening of the lignin. Significant degradation reactions were only found to occur at 280 °C and higher. Haw and Schultz (1985) investigated the thermal degradation of steam-exploded and HCl lignin under N_2 using DTA/TGA and analysis of the residues with CP-MAS NMR and FTIR. Amongst other conclusions, they reported that there was extensive cleavage of the aryl–alkyl ether linkages that initiated at 220 °C, and that methoxyl groups were not cleaved below a temperature of 335 °C.

Although results from DTA analysis indicate that lignin degradation will only occur at temperatures in excess of 200 °C, there is evidence to show that there are many changes occurring at lower temperatures. When oak chips were heated in air, it was found that guaiacyl moieties in oak were degraded at temperatures as low as 165 °C, with syringyl groups being less affected; this indicated that the methoxyl group attached to the aromatic ring confers a degree of thermal stability (Sarni *et al.*, 1990). An increase in the syringyl/guaiacyl ratio of oak lignin during heating was observed as the treatment temperature increased. In the temperature range 175–195 °C, significant amounts of monomeric lignin breakdown products were found in water/alcohol extracts of the thermally treated oak, but these decreased markedly above 195 °C. This may be because cross-linking reactions are now beginning to dominate.

Sudo *et al.* (1985) heat-treated beech (*Fagus crenata*) wood chips with pressurized steam at temperatures ranging from 183 °C to 230 °C for different times and examined the breakdown of the lignin. The changes to the lignin were strongly dependent upon the temperature and time of treatment. The lignin was removed from the steamed wood by dioxane extraction, and it was found that steaming resulted in an increase of the extractibility of the lignin and that this increased with treatment temperature. However, heating at 230 °C for longer periods resulted in a reduction in the amount of lignin extracted, indicating that cross-linking reactions were now occurring. Examination of the extracted lignin by gel permeation chromatography showed that there was an increase in the production of lower molecular weight fragments as the steaming temperature was increased. An increase in phenolic OH content was also found as a result of steaming, although this was reduced at higher temperatures. It was concluded that that steam treatment resulted in cleavage of the β-aryl ether linkages to produce phenolic OH groups. Lignin from wood steamed at temperatures below 215 °C was found to be richer in syringyl units and only slightly modified compared to treatments at higher temperatures, where loss of methoxyl groups was found to occur. It was also found that a softwood lignin (from red pine) was less susceptible to degradation, as determined by dioxane extractability. Although the above studies indicate the relative stability of syringyl groups, Shimuzu *et al.* (1989) reported that such groups were, in fact, less stable in hydrothermal reactions.

Nuopponen *et al.* (2004) found that lignin became partially acetone soluble when Scots pine was steam heat treated for 3 hours at temperatures of 180 °C and higher, with the amount of acetone extractible material increasing proportionally to temperature of treatment. Increased levels of phenolic OH groups were also detected in the lignin. This was attributed to cleavage of β-*O*-aryl linkages, although loss of methoxyl groups from syringyl and guaiacyl moieties in the presence of water (with the production of methanol) would also result in an increase in phenolic OH content. NMR spectroscopy of pine that was heat treated under steam has shown that there is a decrease in the intensity of signals attributable to the methoxyl content of the lignin (Sivonen *et al.*, 2002). A sharp increase in the free-radical content of the wood was also found when the wood was heated at temperatures above 200 °C. These changes were considered to allow for coupling reactions between aromatic rings, leading to more extensive cross-linking of the lignin network. Hydrothermal processing is shown to result in the production of soluble lignin fragments initially, but re-polymerization occurs as the reaction progresses, possibly involving furfural and other polysaccharide degradation products (Garrote *et al.*, 1999). Sanderman and Augustin (1964) also reported that extended heating resulted in cross-link formation within the lignin.

Tjeerdsma *et al.* (1998b) studied the molecular changes occurring in wood (when treated according to the two-step Plato protocol) using CP-MAS NMR. Evidence was presented showing lignin cleavage occurring at the Cα and O-4 positions, with condensation of lignin units through the Cα positions. Wood modification using the Plato protocol was found to lead to changes in the UV absorption spectrum associated with lignin degradation reactions (Sander and Koch, 2001). It was noted that changes in the UV spectrum, centred at 280 nm, are associated with the production of carbohydrate degradation products such as furfural, and that these could lead to erroneous conclusions regarding changes to the lignin, which also absorbs in this region. However, an increase in absorption at 330 nm was assigned to lignin degradation, which was thought to be due to the appearance of conjugated carbonyl groups and conjugated α–β double bonds. Although it seems likely that there are new cross-links formed in the lignin network structure as a result of thermal treatment, it is not clear what contribution this makes towards the improved dimensional stability and other property changes arising from thermal treatment of wood. As noted previously, Seborg *et al.* (1953) were of the opinion that this was not the mechanism responsible. However, in their study, the wood was treated at 300 °C and the results are unlikely to be applicable to thermal modifications performed at lower temperatures.

5.3.4 Extractives

Dry heat treatment of oak resulted in a decrease in the amount of ellagitannins and a simultaneous increase in ellagic acid: the lignan lyoniresinol was unaffected by heat treatment at low temperatures, but was degraded at 250 °C (Sarni *et al.*, 1990). Steam heat treatment of Scots pine planks (50 mm × 200 mm in cross-section and 1.5 m long) for 3 hours in the temperature range of 100–160 °C resulted in migration of fats and waxes along axial parenchyma cells to the surface of the wood (Nuopponen *et al.*, 2003). With treatment above 180 °C, fats and waxes disappeared from the wood surface. Resin acids could be found in the centre of the wood samples with heat treatments in the range 100–180 °C, but at a temperature of 200 °C resin acids were no longer found in the centre of the wood samples and had migrated towards the surface. With treatment above 200 °C, resin acids were no longer detectable.

The loss of volatile extractives during heat treatment is reflected in the volatile organic compound (VOC) emission profiles measured for heat-treated wood. In a determination of the VOC profiles of timber treated at 230 °C for 24 hours, there was only a very low emission of volatile terpenoids, but furancarboxaldehyde, acetic acid and 2-propanone originating from thermal breakdown of the wood components were identified (Manninen *et al.*, 2002). Although thermally treated wood exhibits considerably reduced emission levels of terpenes, rather higher levels of emission of acetic acid are found compared to unmodified wood (Mayes and Oksanen, 2002). Migration of extractives to the surface of wood during thermal treatment results in the formation of unsightly resin patches that need to be removed by planing.

5.4 Physical Changes in Wood due to Thermal Modification

5.4.1 Colour

The physical changes that occur to the wood depend very strongly upon the exact conditions of the treatment method. A darkening of the wood occurs, with the colour change being related to the temperature and time of treatment. Treatment in air results in a greater colour change compared with heating under nitrogen. Some attempts have been made to correlate colour change with the treatment conditions, in order to provide a quality control indicator to predict performance in service (Bourgois *et al.*, 1991; Bekhta and Niemz, 2003). Chow and Mukai (1972) reported that there was a relationship between colour change due to thermal treatment, degree of crystallinity, degree of polymerization and OH content. The colour stability of thermally modified wood is better than control samples in accelerated weathering tests (Ayadi *et al.*, 2003), but the colour will nonetheless fade if heat-treated samples are exposed in exterior situations (Syrjänen and Kangas, 2000). Heat treatment has been found to affect the performance of surface finishes. Feist and Sell (1987) found that semi-transparent and film-forming stains performed worse on heat-treated spruce, compared with controls, whereas performance on heat-treated beech was marginally improved.

Mitsui *et al.* (2003) studied the colour change of wood that had been previously irradiated with light and then subjected to a relatively mild heat treatment (no higher than 90 °C). The extent of colour change was dependent upon the wavelength of irradiation and was found to be considerably greater when the irradiated wood was heated under conditions of high humidity.

5.4.2 Mass Loss and Dimension Changes

Heating of wood results in a reduction in mass and a decrease in volume, the extent of which is dependent upon the treatment method, temperature and time of exposure (Seborg *et al.*, 1953; Stamm, 1956; Rusche, 1973b; Fung *et al.*, 1974). The effects of different treatment parameters upon mass loss are summarized in Table 5.1.

Heat treatment at lower temperatures results in low mass loss, mainly associated with loss of volatiles and bound water. Loss of macromolecular components can occur at temperatures above 100 °C, and this assumes greater significance as time and heating temperature are increased (Millett and Gerhards, 1972) (Figure 5.3).

Mass loss occurs at a greater rate in closed systems, but depends upon temperature and treatment atmosphere. Rusche (1973b) found that the mass loss due to thermal treatment in an open system was the same as in a closed system at a treatment temperature of 150 °C, but was higher in an open system at 175 °C in air. With treatment in nitrogen, or vacuum, mass loss was higher in a closed system at 175 °C. Mitchell *et al.* (1953) also found that mass losses were greater in a closed system. When wood was heated at 300 °C, mass losses were found to be slightly higher in an open system (Seborg *et al.*, 1953).

Table 5.1 *Treatment variables affecting weight loss*

Lower rate of mass loss	Higher rate of mass loss
• In inert atmospheres, anaerobic conditions, or under vacuum	• In air
• In open systems	• In closed systems
• In dry conditions	• In wet conditions
• Softwoods	• Hardwoods

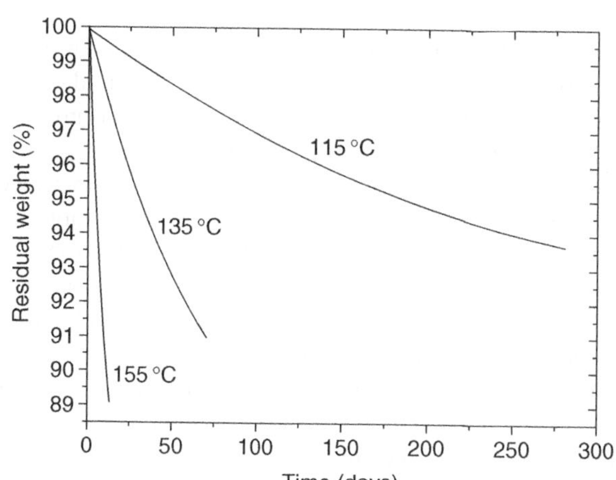

Figure 5.3 *The effect of temperature and time of treatment on thermal mass loss, according to data reported by Millett and Gerhards (1972).*

Furthermore, at this temperature the change in mass stabilized after about 30 minutes' heating in the closed system.

Loss of material from the cell wall results in dimensional changes of the wood, which can lead to distortion occurring with larger pieces if the process is not optimized. As a result of mass loss, the specific gravity of the wood decreases as the time and temperature of treatment increase, but the changes that occur depend upon the relationship between mass loss and volume change. Chang and Keith (1978) reported that the volumetric shrinkage of wood due to thermal treatment was found to correlate well with mass loss due to heating. A slight increase in longitudinal dimension was also found, which was attributed to a stress effect caused by radial and tangential shrinkage. Seborg *et al.* (1953) found that specimens subjected to heat treatment shrank more in the tangential compared to the radial direction.

During air heat treatment of Douglas fir at 170 °C, Lin (1969) found that the mass loss kinetics obeyed a first order rate law after 30 hours' heating, but not before. This was also found by Stamm (1956), who attributed the initial mass loss as being due to the evaporation of bound water. In Lin's study, samples (100 mm × 100 mm × 5 mm: $l \times r \times t$) were oven-dried at 105 °C for 24 hours prior to heat treatment: thus water loss would not be an explanation, although loss of volatile extractives remains a possibility. Changes in surface reflection did, however, obey first order kinetics, and Lin concluded that the thermal decomposition reaction was indeed first order, but was initially confined to the surface. As time of heating increased, the degradation reaction progressed into the interior of the wood, requiring 30 hours before equilibrium was established. Stamm (1956) collated a large amount of data on the mass loss of wood under different heating conditions, for times ranging from 1 minute to 2.4 years and temperatures ranging from 93.5 °C to 300 °C. When the logarithm of mass loss was plotted against time of heating, linear relationships were found in all cases, except for during the initial stages of heating (mentioned above). The rate constants obtained from these plots were found to obey the Arrhenius relationship. The mass loss due to thermal degradation of wood in moist conditions did not, however, obey first order kinetics (Mitchell, 1988).

Stamm (1956) found that the mass loss due to thermal degradation was much higher when heating was carried out in air. Rusche (1973a,b) also reported that the mass loss due to thermal degradation in air (temperature range 100–200 °C) was greater than that exhibited by wood samples heated in vacuum or under nitrogen.

When Seborg *et al.* (1953) heated wood specimens at 300 °C in sealed reactors under dry and moist conditions, it was found that mass loss was more rapid in the presence of water. However, the total mass loss stabilized at the same value (55 %) in both dry and moist conditions. Stamm (1956) found that thermal degradation was greater under steaming compared to dry heating conditions. Mass losses due to thermal treatment were found to be higher when a hydro- or hygrothermal process was used, compared to heating in air (MacLean, 1951). Additionally, the hydrothermal treatment was found to result in a higher rate of weight loss compared to heat steam treatment.

Hardwoods generally exhibit higher mass losses than softwoods when heated under identical conditions. Rusche (1973a,b) studied the thermal degradation of pine and beech wood, reporting that mass losses of beech were significantly higher at 200 °C. However, strength losses were related to mass loss only and were independent of species. Schneider and Rusche (1973) also found that spruce was more thermally stable than beech, as determined by mass loss during heating.

5.4.3 Crystallinity Changes

During the initial stages of heat treatment, there is an increase in the degree of crystallinity, but as the time of heating is extended, thermal decomposition of the cellulose occurs, accompanied by a decrease in crystallinity (Hirai *et al.*, 1972). Nakao *et al.* (1983) also found an increase in crystallinity over short heating periods. Kubojima *et al.* (1998) heated Sitka spruce at temperatures ranging from 120 °C to 200 °C in air or nitrogen. The width of the cellulose crystallite and crystallinity increased during the initial stage of heating in the range 120–160 °C and then remained constant. At higher temperatures, an initial increase in these two properties was followed by a gradual decrease with extended heating times. Bhuiyan *et al.* (2000) heated powdered samples of cellulose, spruce and buna (*Fagus crenata*) under oven-dry and moist conditions at 180 °C, 200 °C and 220 °C. Heating took place in closed microreactors, presumably under air. All samples exhibited an increase in crystallinity over short heating periods and a decrease with extended heating, with such changes taking place more rapidly at higher temperatures. Greater increases in crystallinity were found when the wood samples were heated under moist conditions, but this was not found to be the case with pure cellulose. The changes in crystallinity were explained as being due to molecular reorientation rather than the loss of amorphous material. It was noted that there was only a slight decrease in weight during the early stages of heating, but this became more significant for heating over longer periods. Thus, only when weight decreases were substantial was it considered that increases in crystallinity were due to degradation of amorphous regions. Major differences in behaviour between OD and moist heating were found when the relative change in crystallinity was plotted against the relative width of the crystal. Under moist conditions, increases in crystallinity were accompanied by an increase in the relative width of the crystal, which continued to increase slightly even when the degree of crystallinity decreased. The relative width of the crystal then began to decrease when the relative crystallinity fell below a value of 1.1. Only slight increases in the relative width of the crystal occurred with heating under dry conditions, followed by a rapid decrease when the relative crystallinity decreased. This behaviour is shown diagrammatically in Figure 5.4. Increases in the relative width of the crystal are apparently linked to the reorientation processes noted above, whereas decreases in crystallinity are a result of cellulose degradation. Bhuiyan *et al.* (2001) also studied the effect of a series of heat treatments under dry conditions at 200 °C in an oven for short time intervals compared with continuous heating for the same total period. Increases in crystallinity were recorded which were related to the time of heating, but when samples were heated over a number of cycles, there was no further change in crystallinity after the first heating stage. It was concluded that the cooling stage between heat treatments was responsible for a process that effectively locked the structure, thereby preventing further changes in crystallinity during subsequent heating cycles.

Bourgois *et al.* (1989) reported that there was no change in crystallinity following thermal modification of wood for 30 minutes at temperatures ranging from 240 °C to

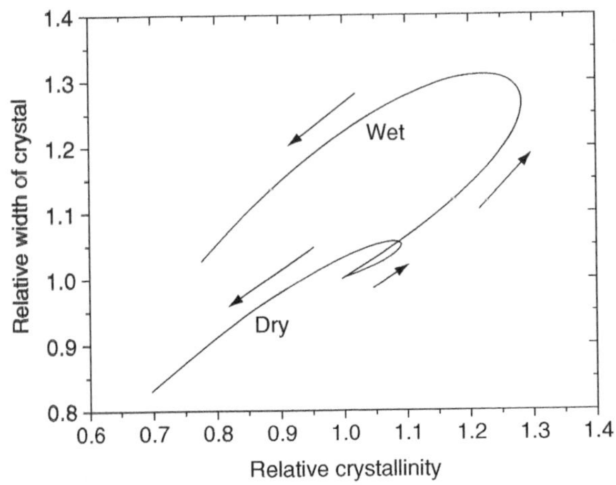

Figure 5.4 *The relationship between change in microfibril width and degree of crystallinity following thermal treatment in wet and dry conditions, according to the data of Bhuiyan* et al. *(2000).*

290 °C under nitrogen. In a study of the change in crystallinity due to heat treatment, it was found that steaming for 10 minutes, at a range of temperatures from 120 to 220 °C, resulted in an increase in crystallinity, whereas heating in air at the same temperatures for 20 hours resulted in a decrease in crystallinity (Dwianto *et al.*, 1996). Kim *et al.* (2001) studied the decomposition of cellulose crystallites in tension wood of cottonwood in the absence of moisture. No decomposition of cellulose crystallites was observed with 1 hour's heating at 300 °C. Decomposition was observed at 320 °C after 20 minutes' heating. A difference in the rate of reduction of crystallite size and relative intensity indicated that thermal decomposition of the crystallites was heterogeneous. The piezo-electric modulus has been found to increase for short heating times and decrease with more extended heating periods, with these changes being related to changes in the crystallinity of the samples (Hirai *et al.*, 1972; Bhuiyan *et al.*, 2002).

5.4.4 Mechanical Properties

Strength and Toughness

Treatment of wood at high temperatures results in a reduction in strength, toughness and abrasion resistance (Chang and Keith, 1978). Seborg *et al.* (1953) noted that commercial prospects for thermal modification would be limited unless some means was found of separating the improvement in dimensional stability from the embrittlement. The reduction in mechanical properties as a result of high-temperature treatments has led to the development of the two-step Plato process, in which it is claimed that strength reductions are minimized by performing the treatments at temperatures below 200 °C (Boonstra *et al.*, 1998). Loss in strength is strongly influenced by the type of heat treatment employed. Loss in strength is faster in closed systems compared to open systems, under hygrothermal compared to hydrothermal conditions, and in air compared to anaerobic conditions

(MacLean, 1954; Stamm, 1956). Hardwoods exhibit higher strength losses than softwoods when treated under the same conditions. For example, Chang and Keith (1978) found that elm and beech were more susceptible to thermal degradation compared to aspen and maple, as determined from toughness measurements of the thermally modified samples.

Hygrothermal treatment of Douglas fir, southern yellow pine and Sitka spruce resulted in significant reductions of work of fracture (MacLean, 1953). This was of the order of 50 % for a treatment of 4 hours at 177 °C, and 75 % after 8 hours' treatment. The percentage decrease in mechanical properties exhibited a linear relationship when plotted against the logarithm of the treatment temperature. By extrapolation of this data, it was concluded that a 32-hour treatment at a temperature of 95 °C would result in no loss of mechanical properties. Hardwoods were more susceptible to degradation by steam treatment and showed a greater tendency to shrink and collapse.

The change in mechanical properties of wood that had been thermally treated under molten metal (Stamm, 1956) was found to obey first order kinetics. Stamm *et al.* (1946) thermally modified wood in molten metal baths for different times and temperatures (160–320 °C) and determined the toughness of the treated specimens. At an ASE value of 50 %, the toughness had reduced by about half. Seborg *et al.* (1953) recorded large decreases in toughness when wood was heated at 300 °C, with a 50 % reduction at 13 % weight loss and a 90 % reduction at 30 % weight loss. No difference was found between samples heated under nitrogen or in air. The temperature used for these experiments is much higher than used for modern commercial thermal treatments. Rusche (1973a), in a study of the thermal treatment of beech and pine, found that strength decreases measured in compression were of the order of 20 % up to a mass loss of 1 %, but this had increased to 80 % when the thermal mass loss was 10 %. Strength losses as determined by tensile testing were not significant until 10 % thermal mass loss, but increased rapidly thereafter. Although relatively low strength losses were found for short heating periods, work to fracture at a thermal mass loss of only 1 % had declined to 60 % of that exhibited by unmodified wood.

A number of studies have been performed investigating the kinetics of the decrease in mechanical properties. Stamm (1956), in a study of the heat treatment of wood under a variety of conditions, plotted the logarithm of strength loss against linear treatment time and found linear relationships (as with weight loss), showing that the decrease in mechanical properties also obeyed first order rate laws.

Modulus of Elasticity

Many studies have shown that there is a slight increase in MOE when wood is thermally treated for short time periods. A reduction in MOE then occurs, with the rate and extent of reduction being dependent upon the treatment temperature and other conditions. In a study of the thermal degradation of beech and pine at various temperatures and time periods, it was found that although the decrease in strength and work to maximum load was directly proportionate to mass loss, irrespective of treatment and species, the decrease in MOE only became significant when the mass loss exceeded 8 % (Rusche, 1973a). Millett and Gerhards (1972) heated dry specimens of several wood species in air at temperatures ranging from 115 °C to 175 °C, finding that MOE increased initially. The subsequent rate of decrease in this property was related to the temperature of treatment, with higher temperatures leading to a greater rate and extent of decrease (Figure 5.5).

Kubojima *et al.* (2000a) heated wood samples in a sealed reactor at 160 °C under nitrogen or air for various time periods. The change in MOE was related to the treatment atmosphere. As with the above study, the MOE increased over short heating periods, and then remained relatively constant when heating was performed under nitrogen, but decreased when heating was performed in air. Bending strength also showed an increase initially and then decreased with extended heating, rather more so when heating was carried out in air. Kubojima *et al.* (2000b) also studied the vibrational properties of heat-treated wood and found that a mild heat treatment at 160 °C resulted in an increase in the dynamic Young's modulus and a decrease in the loss tangent. Although Bekhta and Niemz (2003) found a decrease in bending strength of around 50 % for spruce wood heated at 200 °C, there was little change in MOE. Chang and Keith (1978) found that the MOE of wood samples (elm, beech, aspen and maple) increased slightly after thermal modification, but noted that more severe treatments resulted in a reduction in MOE.

The reduction in MOE (and MOR) due to the thermal degradation of wood in moist conditions at 150 °C in a closed system was not found to obey first order kinetics (Mitchell, 1988). Dry treatments at this temperature generally resulted in a slight increase in MOE, except in the presence of oxygen for longer heating periods. When moisture was present, however, this led to reductions in MOE for all but the shortest heating periods (Figure 5.6). Kim *et al.* (1998) found that there was a slightly greater reduction in MOE when moist wood specimens were heated in air, compared to heating of oven-dry wood, but there was no indication of any increase at short heating times.

Most mechanical tests of thermally modified wood have been made with relatively small samples, where the effect of defects, such as knots, can be eliminated. Although such studies are valuable, they do have limitations when considering the use of thermally treated wood in real-life situations. Bengtsson *et al.* (2002) studied large spruce and pine beams (of dimensions 45 × 145 mm and length 4.5 m) hygrothermally treated at 220 °C and found a reduction of 50 % in the bending strength, accompanied by a slight decrease

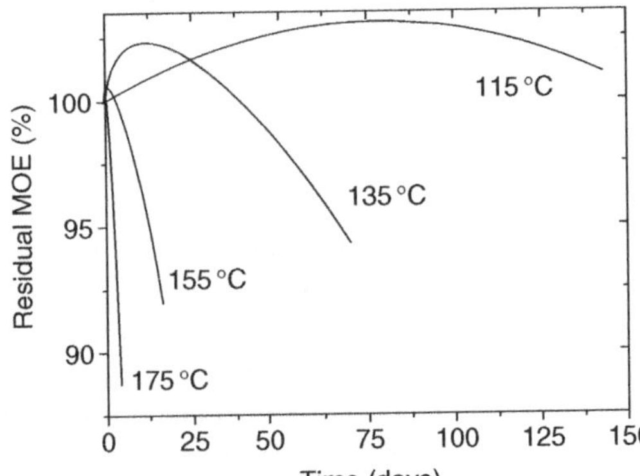

Figure 5.5 Changes in MOE as a result of thermal treatment at different temperatures and time periods, according to the data of Millett and Gerhards (1972).

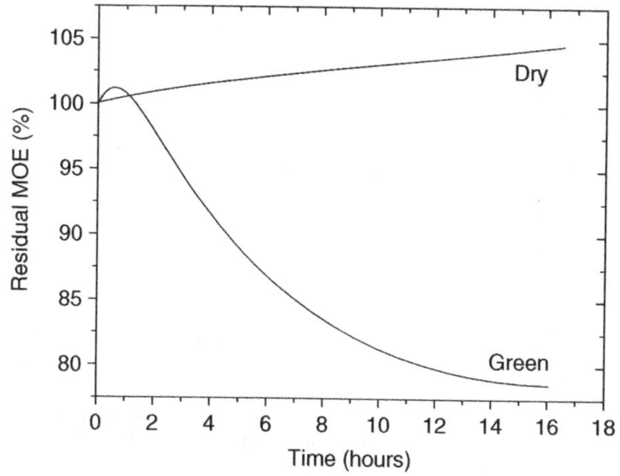

Figure 5.6 *Changes in MOE as a result of wet and dry thermal treatments, according to the data of Mitchell (1988).*

in MOE. There was an increase in the variability of results when compared to untreated wood and the failure mode of thermally modified wood was found to be brittle in all tests.

Modulus of Rupture

Unlike MOE, there does not appear to be any increase in MOR when wood is subjected to thermal treatment, even for short time periods (Figure 5.7). This is because MOR is actually the equivalent stress in fibres at the point of failure and, as such, is related to the ultimate bending strength of the wood (Dinwoodie, 2000).

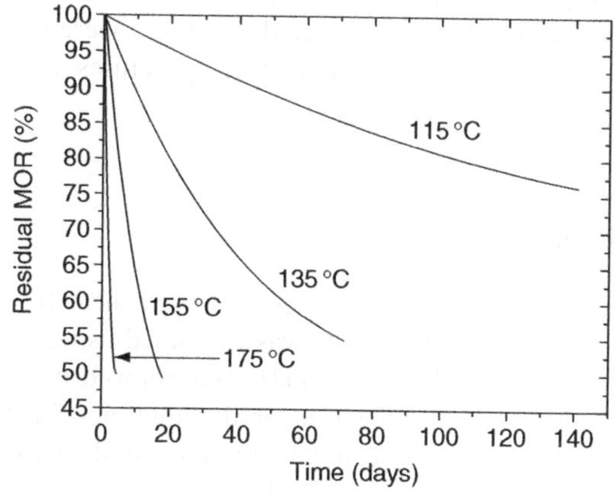

Figure 5.7 *Changes in MOR as a result of thermal treatments at different temperatures and times, according to the data of Millett and Gerhards (1972).*

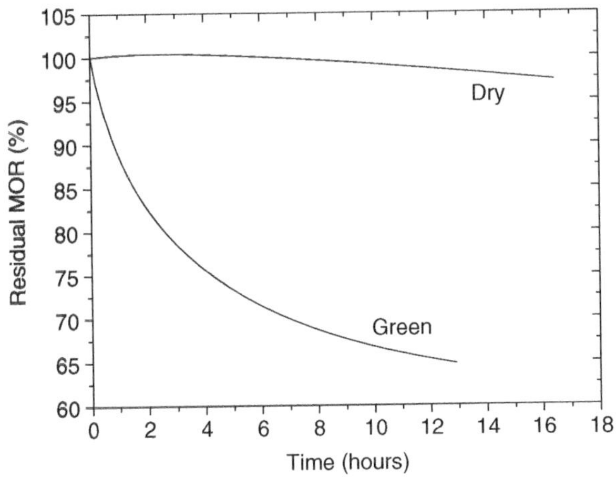

Figure 5.8 Changes in MOR as a result of wet and dry thermal treatments, according to the data of Mitchell (1988).

As with other physical properties of thermally modified wood, the decrease in MOR is a function of the treatment conditions. Stamm *et al.* (1946) found that reductions in MOR at a given ASE value were greater when wood samples were heated in air compared to in anaerobic conditions. Mitchell (1988) reported that dry treatments in air or nitrogen at 150 °C in a closed system resulted in negligible changes to MOR, although more substantial changes were found when treatment was performed in pure oxygen (Mitchell, 1988). However, when moisture was present, MOR decreased significantly, irrespective of the treatment atmosphere (Figure 5.8). The reduction in MOR due to the thermal degradation of wood in moist conditions did not obey first order kinetics, and a nonlinear equation was developed that more accurately described the changes in these properties. Kim *et al.* (1998) found that there was a greater reduction in MOR when moist wood specimens were heated in air, compared to heating of oven-dry wood. It was also found that there were significant reductions in MOR even when samples were heated at temperatures as low as 120 °C for extended periods.

5.4.5 Cell Wall Microporosity

The cell wall microporosity of wood samples that were heat treated using steam at temperatures ranging from 180 °C to 230 °C was examined using an NMR technique at different moisture contents (Hietala *et al.*, 2002). This showed that whilst mild thermal treatments (115 °C) did not change the size or distribution of cell wall micropores, temperatures in excess of 180 °C significantly increased both their size and broadened the microporosity distribution. The average size of these micropores was estimated to be tens of nanometres. The increase in the microporosity of the cell wall was thought to be due to component removal during the steam heating process.

5.4.6 Dimensional Stability

Increases in dimensional stability occur when wood is thermally modified, but the effect observed is dependent upon the heating protocol adopted. Stamm *et al.* (1946) reported that the ASE of softwood samples increased both with temperature and time of treatment. For example, an ASE of 20 % could be obtained by heating small wood samples either at 150 °C for 6 days or at 250 °C for 3 minutes. Three softwood species were studied and these showed very similar behaviour. The changes in ASE were found to obey first order kinetics. The increase in ASE as a result of heat treatment followed an asymptotic relationship when plotted against weight loss due to heating, reaching a maximum ASE value at about 20 % weight loss. However, the results obtained were influenced by the presence of air and lower ASE values were found for a given weight loss compared to anaerobic conditions. Stamm and Hansen (1937) found that heating of dry wood at temperatures ranging from 165 °C to 205 °C for up to 6 hours resulted in a substantial reduction in ASE, but there was no change when the wood was heated in the presence of water. It was considered that the presence of water suppressed those thermal reactions involving loss of water of constitution. Heating in air was found to result in greater reductions in ASE compared with heating in a reducing atmosphere.

Seborg *et al.* (1953) also found that the change in dimensional stability was highly dependent upon the treatment atmosphere employed. Wood samples were heated in air or nitrogen at 300 °C, and it was found that the dimensional stabilization reached a maximum at about 20 % weight loss. Beyond this weight loss, further increases in dimensional stability were only achieved when the wood was heated in a closed system. Dimensional stability did not change when thermal mass loss exceeded 20 %, when samples were heated in an open system in nitrogen and it deteriorated when samples were heated in an open system in air. It was also found that the swelling of thermally modified wood in concentrated aqueous sodium hydroxide, morpholine or pyridine was the same or greater than that exhibited by unmodified controls, with only the swelling in water reduced. From this data it was concluded that the increased dimensional stability was not due to the formation of ether cross-links during thermal modification. However, as noted elsewhere, the treatment methods employed in this study are severe when compared to more recent thermal modification methods, and the possibility of cross-link formation in the lignin particularly cannot be excluded as at least a contributory factor in the enhanced dimensional stability of thermally modified wood. Burmester (1975) was of the opinion that loss of hemicellulose was the primary contribution to the dimensional stabilization of heat-treated wood.

The effect of catalysts upon the change in dimensional stability as a result of heat treatment has also been investigated. Stamm and Baechler (1960) studied the dimensional stability imparted to wood as a result of heat treatment in an open system (180 °C) in the presence and absence of a catalyst ($ZnCl_2$ or NaCl). The relationship between the square of the fractional reduction in swelling against weight loss due to heating was plotted. This showed that with an uncatalysed thermal treatment, weight loss and dimensional stabilization were related until a weight loss of 16 %, and thereafter the dimensional stability began to decrease. In the case of treatment in the presence of either catalyst, the relationship broke down at 10 % weight loss.

Keith and Chang (1978) found that improvements in dimensional stability as a result of heat treatment were species dependent and more apparent in the radial direction, this being attributed to anatomical differences. Yun *et al.* (1999) studied the change in dimensional stability with heating duration and found that a linear correlation in property change was obtained when plotted against square root of heating time.

5.4.7 The Hygroscopicity of Heat-treated Wood

The hygroscopicity of the wood is markedly reduced as a result of thermal modification, the reduction being related to the time and temperature of the process. The treatment atmosphere used during the thermolysis process also affects the sorption behaviour of the thermally modified wood. Seborg *et al.* (1953) found that the EMC at 90 % RH of thermally modified pine heated in air at 300 °C decreased relative to untreated wood up to a treatment time of 1 hour, but then increased for more extended heating periods. When heating was performed under nitrogen, the sorptive capacity of the thermally treated wood decreased up to a heating time of 60 minutes and was constant thereafter. The sorption behaviour of wood thermally treated in the presence and absence of air was studied by Schneider and Rusche (1973), who found that the sorption capacity of wood specimens decreased as time and temperature of treatment increased. However, when samples were heat treated in air at 200 °C, the sorption capacity began to increase again with extended heating periods (24 hours for beech and 48 hours for spruce), corresponding to a thermal mass loss of about 20 %. When correlated against the relative loss of mass due to heat treatment, the decrease in sorption capacity was greater for heat treatments in the presence of air, with beech showing a greater reduction in sorption compared to spruce. Mitchell *et al.* (1953) heated Douglas fir sawdust at 300 °C for 15 hours under a nitrogen or an air flow, and determined the EMC of the samples at 90 % and 30 % RH over a total of 10 cycles. Although the EMCs recorded at the first cycle were reduced compared to untreated wood, these values increased with each cycle, and the EMCs of wood heated in air were actually higher that those found for untreated wood after the fifth cycle. This behaviour occurred to a much more limited extent with wood blocks. When Keith and Chang (1978) heated four hardwood species at 220 °C in air in a forced circulating oven for 2 hours, they found that the EMC was reduced by 50 % compared to unmodified control samples. When studying the effect of time of exposure, it was found that the effect of different treatment times upon the EMC was more apparent when samples were equilibrated at high RH values. There was no significant change in sorptive behaviour when samples were subjected to cyclic RH changes. Species-dependent differences in the sorptive behaviour were noted for heat-treated samples.

Rather less well attested is the change in the shape of the desorption isotherm, although some indications of this can be seen in earlier work, where it was stated that the differences in hygroscopicity due to heat treatment were more apparent during absorption compared with the desorption process (Keith and Chang, 1978). In most studies, the sorption–desorption behaviour has been determined using a restricted number of RH values and the change in shape of the desorption isotherm is not obvious. This becomes much more readily apparent when sorption–desorption data are recorded using a larger number of RH values, as can be seen in Figure 5.9. The shape of the desorption isotherm is much more linear

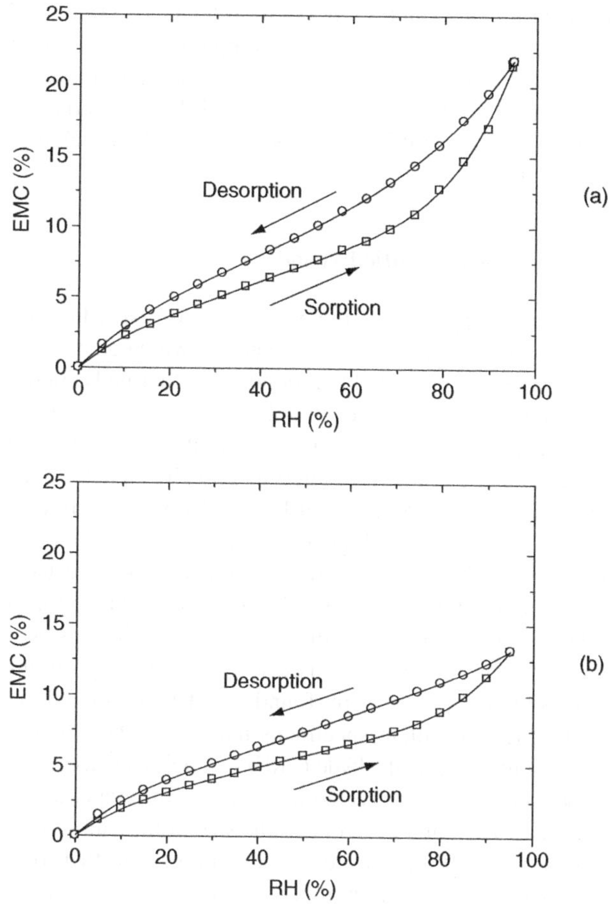

Figure 5.9 Sorption–desorption behaviour of unmodified wood (a) and wood modified by thermal treatment in N_2 at 250 °C for 2 hours (b).

compared with the classic sigmoidal shape of unmodified wood (Figure 5.9a). There is also a reduction in the hysteresis between the sorption and desorption curves.

Repellin and Guyonnet (2005) determined the reduction in FSP of wood subjected to heat treatment, using a differential scanning calorimeter. The reduction in FSP showed a good correlation with reduction in swelling. The results showed that the reduction in swelling was not just attributable to the loss of sorption sites associated with hemicellulose degradation.

The surface wettability of heat-treated wood decreases due to a reduction in the hydroxyl content of the modified wood (Pétrissans *et al.*, 2003). There is a reduction in the water-sorption capacity, which is related to a reduction in the number of primary sorption sites (OH groups) within the wood cell wall, largely as a result of the removal/degradation of the hemicellulosic component. As remarked upon earlier, hygroscopic properties are strongly influenced by the treatment method employed. Podgorski *et al.* (2000) heated wood samples for 1 hour at temperatures ranging from 60 °C to 200 °C, finding that contact angle increased markedly as the treatment temperature was increased (65° for unmodified wood to 145° for heating at 200 °C). Changes in wettability at lower temperatures were attributed to migration of lipophilic extractives to the surface, whereas higher temperatures result in macromolecular cell wall changes, leading to a reduction in OH content.

5.4.8 Bonding of Thermally Modified Wood

Changes to the chemical composition of the wood (primarily a reduction in OH content) lead to a reduction in the surface energy of the material, which changes the wettability of the material. Thus, it would be predicted that bonding of thermally modified wood using polar adhesives, or adhesives dispersed or dissolved in water, is likely to result in weak joints. Furthermore, the reduction in adhesion between fibres of the wood matrix will lead to premature failure of a bonded material even if the glue line itself remains sound. Chow (1971) studied the shear strength properties of PF-bonded veneers, finding that there was a reduction in bonding strength and a higher proportion of wood failure as temperature and treatment time increased. Chang and Keith (1978) bonded heat-treated wood specimens of aspen, beech, maple and elm with UF resin. There was a reduction in shear strength of the glue line as both temperature and time of treatment increased, although aspen performed rather better than the other woods tested. A high proportion of wood failure was also recorded. Similar results were found for PVAc-bonded heat-treated elm. Bengtsson *et al.* (2003) tested glulam specimens made from heat-treated wood bonded using either a phenol resorcinol formaldehyde (PRF) or a polyvinylacetate (PVAc) adhesive. Pine and spruce wood samples were heated at a maximum temperature of 220 °C for 5 hours, with a total process time of 4 days due to the size of the samples. The performance of specimens bonded with PVAc was extremely poor, but PRF-bonded specimens showed good results, with only a small number of glulam beams made from heat-treated pine failing in the delamination test.

5.4.9 Composites Made from Thermally Treated Wood

There have been some studies of the use of heat treatment to improve the dimensional stability of wood-based composites. Burmester (1973) found that heat treatment of wood particles reduced the water swelling of the composite by 50 %. Dry thermal treatment of wood strands under argon resulted in improved dimensional stability of oriented strand boards formed from the treated wood, but a reduction in the MOE and MOR of the boards (Goroyias and Hale, 2002). Improvements in dimensional stability were significant at treatment temperatures of 230 °C and above, but 230 °C was considered the maximum treatment temperature that should be used due to wood degradation and associated strength reductions. High-temperature dry post-treatments have also been used to improve the dimensional stability of wood-based composites (Hsu *et al.*, 1989). Obviously, such methods require the use of a heat-resistant adhesive to bond the wood particles. The use of short treatment times limits chemical degradation of the wood polymers, but requires sufficiently high temperatures (above the softening points of the polysaccharide and lignin) to allow for an annealing process to relieve built-in stresses that arise during the pressing process.

5.5 Biological Properties of Thermally Modified Wood

5.5.1 Microbiological Attack

Heat treatment improves the decay resistance of the modified wood to biological attack. As with other properties, there are differences in performance depending upon the type of treatment and the study undertaken. The mechanism of improved decay resistance has not yet been established, although this is undoubtedly linked to the loss of polysaccharide content, and to the reduced MC of the cell wall. Loss of OH groups from the cell wall polymeric constituents may also affect the ability of enzymes to metabolize the substrate and/or the mode of action of low molecular weight diffusible agents. The possibility that biocidal chemicals are formed in the wood during heating also has to be considered.

Pure Culture Tests

The improved decay resistance of thermally modified wood has been demonstrated in numerous pure culture studies. Stamm *et al.* (1946) exposed heat-treated wood to *Trametes serialis* for 2 months, finding that specimens that had ASE values in excess of 40 % exhibited no weight loss. In a later study, Stamm and Baechler (1960) found that zero weight loss due to decay (according to ASTM D1413–56T) of thermally modified wood by *L. trabea* (*G. trabeum*) was achieved at a weight loss due to thermal treatment of 12–14 %, corresponding to an ASE value of 42 %. It was further noted that the use of $ZnCl_2$ as a catalyst for thermal treatment appeared to result in a slightly improved decay resistance.

Dirol and Guyonnet (1993) studied the decay resistance of retified wood upon exposure to the white rot fungus *C. versicolor* and two brown rotters (*G. trabeum* and *C. puteana*). Samples of poplar, spruce and fir were thermally treated at 250 °C for 10 or 20 minutes. In place of the white rot fungus, a soft rot fungus (*Chaetomium globosum*) was used for poplar. When tested according to the French standard NF-X 41-555) for 6 weeks, the retified wood samples exhibited virtually no loss in weight, indicating high durability. Troya and Navarette (1994) also examined the decay resistance of retified wood that had been heated to a range of temperatures between 220 °C to 260 °C for different times. Samples were placed on wood feeder strips previously infected with *S. lacrymans*. After 5 months' exposure, very low weight losses due to decay were found for wood treated at temperatures of 250 °C and 260 °C, but wood treated at lower temperatures was less resistant to attack. Viitanen *et al.* (1994) reported that heat treatment improved decay resistance in an unsterile soil test and against *C. puteana* in pure culture tests, although few experimental details were given.

Loss of polysaccharides during heat treatment might simplistically be expected to lead to enhanced protection against brown rot fungi compared to white rotters. This has been found in some studies of the effect of thermal treatment upon decay resistance. Tjeerdsma *et al.* (1998a, 2000, 2002) found that Plato heat treatment resulted in significant improvements in decay resistance to *C. puteana*, which were superior to that obtained with the white rot fungus *T. versicolor*. Improved resistance against soft rot was also found. The process conditions of the curing step were found to have a significant influence on decay resistance of the Plato-treated wood to brown and soft rot. The results were also found to depend upon the species studied. Decay resistance of the thermally modified wood against brown and soft rot was attributed to the reduced hygroscopicity of the material.

Welzbacher and Rapp (2002) determined the decay resistance of wood that had been thermally modified by one of the four main European commercial processes (Plato, Thermo Wood, Retified wood or oil heat-treated wood). The thermally modified wood was exposed in pure culture tests (EN113) to the brown rot fungi *C. puteana*, or *Oligoporus* (*Poria*) *placenta*, and the white rot fungus *C. versicolor*. Protection against decay decreased in the order *C. puteana* > *C. versicolor* > *O. placenta*; only with retified wood was protection against the white rot fungus the highest. Although there was an improvement in biological durability, this was not as good as that achieved previously with laboratory heat-treated wood, showing that performance improvements were possible. Evidence that there is enhanced resistance to white rot fungi compared to brown rot is at present inconclusive and would be worthy of a more comprehensive study.

Kim *et al.* (1998) investigated the effect of different treatment times and temperatures upon the decay resistance of wood thermally modified in air against *T. palustris*. The relationship between heating period and weight loss due to decay was modelled using a regression equation. Improvements in decay resistance were found as the heating period and temperature increased. From the numerical analysis employed, it was predicted that heat treatment at 150 °C for about 150 hours would achieve decay resistance comparable to that of a CCA treatment at 1 % retention.

Kamden *et al.* (1999, 2000) suggested that thermal modification could result in the production of toxic by-products, which would inhibit the growth of fungi. This was investigated by analysis of the organic solvent extracts of thermally treated wood using GC–MS and ^{13}C NMR. Results from this study were not conclusive, but the authors did note that further studies should be undertaken to ensure that thermally modified wood was free of toxic by-products that could present a possible environmental risk. Kamden *et al.* (2002) subsequently reported upon the durability of thermally modified samples of beech, spruce, pine and polar that had been heated under a nitrogen atmosphere at temperatures ranging from 200 °C to 260 °C for 1 to 24 hours. Thermally modified samples were tested according to AWPA E10–91, with the brown rot fungi *G. trabeum* and *P. placenta*, and the white rot fungus *Irpex lacteus*. In agar plate tests, it was found that heat treatment of maritime pine resulted in an increased resistance to microbiological attack compared to control samples, and this enhanced decay resistance was still evident in heat-treated samples that had been subject to water extraction before fungal exposure. In soil block tests, moderate protection was afforded to heat-treated samples when exposed to *G. trabeum*, both with samples that were extracted with water or acetone prior to exposure as well as unextracted samples. Heat treatment did not provide good protection to *P. placenta* or *I. lacteus* in soil block tests. It was concluded that the degree of protection against decay afforded by heat treatment was not sufficient to allow for the use of thermally modified wood in ground contact situations. In a study of the thermal treatment of spruce, pine and beech, it was found that thermal treatments at 200 °C in excess of 4 hours resulted in an improvement in decay resistance by two to three durability classes (Hanger *et al.*, 2002).

In a study of the effect of heat treatment upon the durability of spruce and larch heated at 200 °C in a vacuum, both spruce and larch exhibited good resistance to decay by *G. trabeum*, but treated spruce exhibited a decay mass loss of 9 % compared to 1 % with larch against *C. puteana* in an 8-week mini soil block test (Rep *et al.*, 2004).

Unsterile Soil and Field Tests

Tjeerdsma *et al.* (1998a) subjected thermally modified wood to soft rot decay in unsterile soil tests. Although decay resistance was enhanced by thermal treatment, it was not possible to prevent attack entirely. Treatments giving the best performance in terms of decay resistance also exhibited the highest strength losses. Viitanen *et al.* (1994) have also reported on some preliminary experiments on heat-treated wood in unsterile soil.

There appears to be rather limited reporting of results from field trials of thermally modified wood. Welzbacher and Rapp (2004) have given preliminary data on an ongoing field trial of a number of heat-treatment processes with softwood and hardwoods up to 2.5 years' duration. All workers in this area are of the opinion that heat-treated wood is suitable only for out-of-ground applications (Hazard Classes 1, 2 and 3).

5.5.2 Insect Attack

Samples of wood of various species were steam-treated at 150 °C, or oven-heated in air at 150 °C, for various time periods and attack by the termite species *C. formosanus* or *R. speratus* determined (Doi *et al.*, 1997, 1999). Steam heat treatment had the effect of encouraging attack upon the treated wood specimens, whereas dry heat treatment had little effect except for heat-treated Japanese larch with *R. speratus*.

5.6 Compressed Wood

Steaming of wood at temperatures of 180 °C or 200 °C has been used in combination with compression to modify the properties of wood (Inoue *et al.*, 1993a). The modified wood showed no recovery of set during subsequent water-soaking tests, a slight decrease in mechanical properties, increased hardness and a slight darkening. Such treatments have been used to improve the dimensional stability properties of wood-based composites. For example, Kawai *et al.* (1992) used steam-pressing of laminated veneer lumber (LVL) panels to improve dimensional stability of the panels. The mechanism by which steam heat treatment fixes the compressive set of boards is thought to be due to softening of the lignin combined with degradation of the amorphous polysaccharide content. This allows for plastic flow of the cell wall during steam heat treatment, thereby annealing out stresses that arise during the initial pressing process (Sasaki and Kawai, 1994). Navi and Giradet (2000) have reported on a thermohydromechanical process for producing densified wood. This involves the densification of wood under saturated steam at 150 °C and the use of a compressive force of around 130 kg cm^{-2}. Under these conditions, the wood flows plastically, resulting in a permanently fixed deformation of the material. Ito *et al.* (1998a,b) have developed a steam-fixation process for compressively transforming the shape of wood. In order to prevent recovery of set, it was found that at a steaming temperature of 180 °C, 8 minutes' exposure was required, whilst steaming at 200 °C required half that time. Makinaga *et al.* (1997) also found that a temperature of 180 °C was required to ensure that permanent fixation of steam bent wood was achieved. Norimoto *et al.* (1993) have reported on the use of microwave heating of water-saturated wood specimens for bending.

5.7 Oil Heat-treatments

High-temperature treatments of wood in the presence of vegetable oils have been reported in a number of studies. Such treatments are to be distinguished from the Royal, or Royale, process, where impregnation of wood with oil is followed by a low-temperature curing step (60–90 °C), with no significant modification of the wood resulting (Powell, 2003; Treu *et al.*, 2003). The use of oil in thermal modification processes requires far higher temperatures (usually 180–220 °C), which results in chemical changes to the wood constituents. Sailer *et al.* (2000a,b) have reported on a thermal modification process, using refined linseed oil as the heating medium. Wood samples at 6 % MC were placed in a hot oil bath for 4.5 hours. It was found that virtually no oil was absorbed by the wood during the heating process, but this occurred during the subsequent cooling phase where the wood samples remained in the oil bath, resulting in 50–70 % mass increases. Wood surfaces were reported to be dry after treatment. Oil-heat treatment led to a reduction in the volume of samples, as is found with wood heat treated using other processes, showing that the oil is not capable of penetrating the cell wall. The inability of the triglycerides in vegetable oils to penetrate the cell wall has been confirmed in an independent study (Rosenqvist, 2000). Oil heat-treatment results in a slightly higher MOE compared to an identical treatment in the presence of air. Impact resistance is considerably less than that of untreated wood. The decay resistance of wood that had been thermally treated using a hot oil method was found to be superior to that of wood treated under identical conditions in air.

5.8 Conclusions

Although the heating of wood at temperatures only slightly higher than 100 °C for extended periods results in weight losses and property changes, temperatures in excess of 160 °C are generally used for most processes. Thermally modified wood exhibits a decrease in mechanical strength, MOR, impact resistance and abrasion resistance, and is not suitable for load-bearing applications. Heating of wood results in a slight improvement in MOE, but if heating is continued there is subsequently a reduction. Thermally modified wood is much darker than unmodified wood, but the colour change is not permanent when the wood is exposed in exterior situations. Dimensional stability and water vapour sorption properties are considerably improved compared to those of unmodified wood. However, although decay resistance does show some improvement, this is not sufficient for thermally modified wood to be used in ground contact situations. Although limited, the available evidence shows that there is little improvement in resistance to insect attack: indeed, attack may be encouraged in some cases. Property changes are very strongly affected by the treatment method employed, with higher temperatures, longer heating periods, heating in oxygen, or in steam resulting in enhanced degradation. Heating in unvented systems also results in faster rates of degradation. Although the hemicelluloses are most affected by heat treatments (particularly at lower temperatures), the other macromolecular components are nonetheless degraded at temperatures only slightly in excess of 100 °C (albeit to a lesser extent).

6

Surface Modification

6.1 Introduction

Bulk modification can be problematic due to the difficulty of ensuring that the reagent is evenly dispersed throughout the wood material, and the necessity of ensuring that all reagent and by-products are removed at the end of the reaction. However, if the reaction is confined to the surface of the substrate, then accessibility of reagent and subsequent clean-up of the modified material are more easily accomplished. Surface modification of wood has been used to improve the ultraviolet stability of wood, to change the surface energy of wood (to reduce wetting by water, and/or improve compatibility with coatings or matrix materials), and to improve bonding between wood surfaces. Surface modification methods are summarized in Table 6.1.

6.2 Surface Chemical Modification for UV Stability

Wood degrades when exposed to UV light primarily due to the instability of the lignin component. Although clear coatings can be produced that are UV stable, degradation of the underlying substrate results in premature failure of the coating system. Surface chemical modification has been investigated as a means of improving the UV stability of wood. There is a considerable body of evidence showing that chemical modification with certain reagents leads to an improvement in the UV stability of the modified wood, although the mechanism by which this happens is not fully understood. An alternative approach is to graft UV stabilizers on to the wood surface.

Wood Modification: Chemical, Thermal and Other Processes C. Hill

Table 6.1 Summary of surface modification methods

Modification method	Example application(s)
• Conventional chemical modification	• Stability to weathering, compatibilization
• Chemical modification with a bifunctional reagent	• Polymer grafting, self-bonding, stability to weathering
• Surface thermoplasticlzation	• Self-bonding
• Coupling agent	• Compatibilization
• Chemical activation	• Self-bonding
• Enzymatic activation	• Self-bonding
• Plasma or corona discharge	• Compatibilization, stability to weathering

6.2.1 The UV Stability of Chemically Modified Surfaces

Acetylation

Acetylation has been investigated as a potential method for altering the chemical nature of the substrate, so that it is more effectively protected against exposure to solar radiation (Feist *et al.*, 1991a, b; Plackett *et al.*, 1992, 1996; Chang and Chang, 2001). Early studies of the exposure of jute fibres to light indicated that photochemical discolouration of the fibres was inhibited when the fibres had been acetylated (Peill, 1946). Such protection was considered to arise due to the production of acetyl peroxide during exposure, which resulted in bleaching of the fibres (Callow, 1947, 1951). In a study of *Pinus radiata* and paper formed therefrom, it was found that acetylation resulted in significant colour stability when samples were exposed to UV radiation (Leary, 1968). This was attributed to the protection of the lignin phenolic hydroxyl functionalities by esterification, thereby blocking oxidation pathways and leading to the formation of chromophoric quinonoid groups, an explanation also given by several other workers (Kringstad, 1969; Norimoto, 1988; Ota, *et al.*, 1996). It has also been noted that acetylation shifts the absorption maximum of lignin from 280 nm to shorter wavelengths, which might influence UV stability (Hansmann *et al.* 2004). However, any such stabilizing affect would appear to be short-lived. Leary found that prolonged UV exposure resulted in de-acetylation and that UV protection gradually diminished as a consequence. Imamura (1991, 1993) found that when wood was exposed in a natural weathering test, the colour change due to exposure was reduced in proportion to WPG. SEM studies showed that lignin degradation and leaching occurred to an equal extent with both unmodified and acetylated wood, indicating that lignin was not protected as a result of acetylation, although the cell walls of modified wood showed less damage (possibly due to bulking?). Dawson and Torr (1992) also found that lignin degradation occurred at an early stage of UV exposure with acetylation. However, since it has been noted that acetylation results in removal of lignin from the surface of the wood, this may contribute to some enhanced photostability of the modified wood (Dawson *et al.*, 1995). Kalnins (1984) found that acetylation of redwood did not confer additional stability to UV irradiation compared to unmodified wood. However, Ohkoshi (2002) concluded that acetylation did reduce the photochemical degradation of lignin.

It has been noted that the stabilization of wood due to reaction with phenolic OH groups was not considered to be an important mechanism, in view of the observation that the modification of wood with ethylene oxide or butyl isocyanate did not result in an improvement of the UV stability of wood (Rowell *et al.*, 1981). Feist *et al.* (1991a, b) found that acetylation of aspen resulted in a reduction in the rate of degradation (determined by weight loss) upon exposure to UV light. The protection afforded was largely restricted to the hemicellulosic, rather than the lignin, component of the modified wood. It was also found that impregnation of the wood with methyl methacrylate, followed by polymerization, resulted in improved protection, particularly when such a treatment was applied to acetylated wood.

X-ray photoelectron spectroscopy (XPS) studies on acetylated wood exposed to ultraviolet light have shown that although wood modification does result in substantial colour stabilization, de-acetylation undoubtedly occurs (Torr *et al.*, 1996; Kiguchi, 1997). Electron spin resonance (ESR) of acetylated wood exposed to UV irradiation has revealed that active methyl and stable phenoxy radicals are formed (Hon, 1995). However, the view that acetylation prevents discolouration during UV exposure due to free-radical suppression still persists (Sakuragawa, 1996). Evans *et al.* (2000) found that wood samples acetylated to WPGs lower than 10 % were less resistant to weathering than unmodified samples, but samples modified to 20 % WPG exhibited significantly lower mass losses. This protection was confined to the holocellulose portion of the wood. Any photoprotective effect due to acetylation was found to be lost upon extended exposure. It has been shown that Scots pine veneers acetylated to low WPGs (<10 %), exhibit greater mass and strength losses than unmodified controls when exposed to natural weathering (Evans *et al.*, 2000).

Experiments in which clear coatings have been applied to acetylated wood have shown that acetylation does provide additional stability when these samples are exposed to UV light (Plackett and Dunningham, 1991; Plackett *et al.*, 1992; Beckers *et al.*, 1998). In some part, this stability arises due to the higher checking resistance of acetylated wood when exposed outdoors (Dunningham *et al.*, 1992). This is a direct result of the bulking of the cell wall of the modified wood by the acetyl substituents. Acetylation results in a substantial reduction in the polarity of the wood surface, which may affect the adhesion characteristics of surface coatings. Podgorski and Roux (1999) studied the adhesion of an alkyd stain on acetylated wood before weathering and found that there was satisfactory performance. However, after 6 weeks' weathering there was found to be poor adhesion of the coating on acetylated samples, whereas performance with unmodified wood was still satisfactory after this time.

Modification with Other Reagents

Feist and Rowell (1982) modified wood with butylene oxide, or butyl isocyanate, finding that neither modification provided protection when samples were exposed to UV irradiation alternating with water spray. However, protection was afforded when samples were exposed to UV only.

Hill *et al.* (2001) modified wood surfaces with methacrylic anhydride and grafted the activated surface with styrene in order to see if this would improve the UV stability of the modified substrate. There was no evidence to suggest that UV stability was improved either by chemical modification or by modification plus grafting.

Etherification to produce cyanoethyolated, benzylated or allylated wood surfaces does not result in an improvement in UV resistance, but the latter two treatments have been found to improve the performance of clear coatings on the modified substrate (Kiguchi, 1990b). Kalnins (1984) methylated wood by pre-treating with concentrated aqueous

NaOH and then reacting with dimethyl sulphate, and found that such a treatment did not improve the UV stability of the wood.

Pizzi (1990) investigated the potential for covalently bonding alkyd-based coatings to the wood surface to improve durability. Linseed oil and a linseed oil based alkyd resin were reacted with lignin model compounds (β-methyl styrene and guaiacol) under different conditions in order to determine whether reaction was possible with lignin. The use of methyl ethyl ketone in combination with cobalt napthenate peroxide as a free-radical initiator accelerated the rate of reaction with the model compounds. The results were thought to indicate that reaction with wood lignin was possible.

6.2.2 Modification with UV Stabilizers

Williams (1983) reacted the UV stabilizer dihydroxy benzophenone (DHBP) with epicholorohydrin to form the para-substituted product [2-hydroxy-4-(2,3-epoxypropoxy)benzophenone] (HEBP) (Figure 6.1). It was necessary to ensure that the product was not ortho-substituted, since this would have resulted in a product incapable of providing UV protection. Strips of western red cedar (*Thuja plicata*) were modified with the HEBP in 1,4-dioxane or acetone at 120 °C and 150 psi (1 MPa), using TEA as a catalyst. The modified samples were then dried in an oven at 105 °C. Another set of samples were dip-treated in DHBP dissolved in acetone, followed by air-drying, in order to compare the effectiveness of grafting with a nonbonded treatment. Modified samples that were extracted with acetone, methylene chloride and water exhibited identical IR spectra compared with unextracted modified samples, indicating the stability of the modification. Although the IR spectra of the modified wood indicated the presence of HEBP, the weight gain of the samples due to modification was slight. The nonextractibility of the wood samples was taken as evidence that at least a proportion of the HEBP had chemically bonded to the wood but, as noted in Chapter 4, this is unlikely to be the case. The UV stability of the wood (modified, treated and unmodified) was determined in an accelerated weathering test using 20 hours' UV exposure followed by 4 hours' water spraying cycles over periods of time (400 to 1000 hours) using erosion measurements. All HEBP-modified samples exhibited reduced erosion rates compared to control samples: DHBP-treated samples showed some inhibition during the early stages of exposure, but once the stabilizer had leached out, the wood eroded at the same rate as control samples. HEBP modification was also found to improve the performance of clear coatings.

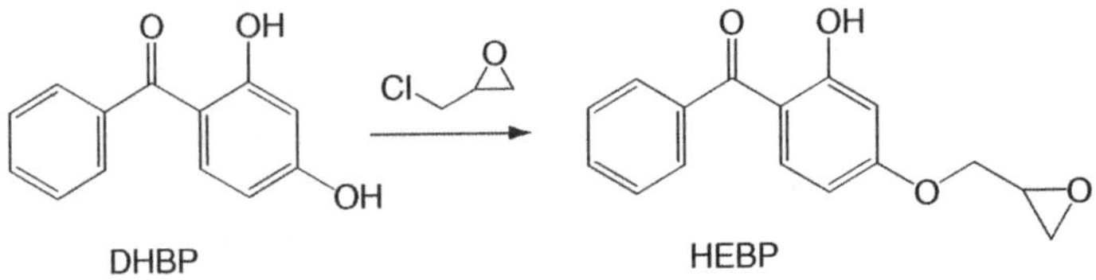

Figure 6.1 Reaction of dihydroxy benzophenone (DHBP) with epichlorohydrin to form [2-hydroxy-4-(2,3-epoxypropoxy)benzophenone] (HEBP).

Kiguchi and Evans (1998) also noted that chemical bonding of the photostabilizer to the wood was required to provide protection in most weathering tests. Kiguchi *et al.* (2001) studied the effect of the modification of wood with HEBP upon the performance of clear-coated wood veneers in outdoor exposure and accelerated weathering tests. The wood was pre-extracted prior to modification and further subjected to solvent extraction after modification, allowing for the determination of the WPG due to modification. Reaction was performed at temperatures ranging from 60 °C to 120 °C for 18 hours in a pressure vessel, using dimethylbenzylamine as a catalyst. The UV stability of samples was determined from mass losses and strength tests. The performance of clear coatings when applied to wood samples modified with HEBP was significantly improved in both outdoor exposure and accelerated tests.

Grelier *et al.* (1997) synthesized a variety of UV stabilizers based upon hydroxyphenylbenzotriazoles incorporating an isocyanate functionality. These were reacted with wood using microwave energy. The modified woods were exposed to UV irradiation and the photoinduced discolouration determined. It was found that the best protection was obtained when the UV stabilizer was used in conjunction with polyethylene glycol, or hindered amine light stabilizers.

6.3 Modification to Render the Wood Surface Hydrophobic

As noted previously, chemical modification changes the surface wettability of wood to water. In most cases, this characteristic is not the primary aim of the modification, but is desirable for wood exposed in exterior situations. Sèbe and Brook (2001) reacted the surface of wood with silicone polymers in order to change the wettability characteristics. Noting that any Si–O–C linkages that may be formed with wood using a conventional reaction were subject to hydrolysis reactions, they utilized an alternative method for ensuring that the bond between the wood and the siloxanes was stable. The wood was first reacted with maleic anhydride and then allyl glycidyl ether, to give oligoesterifed wood. The modified wood was then reacted with either bis(trimethylsiloxy)methylsilane (Figure 6.2a), a hydride-terminated polydimethylsiloxane (Figure 6.2b) or a platinum-divinyltetramethyldisiloxane (Figure 6.2c). The contact angles of water on modified and unmodified wood were measured on the tangential, radial and transverse surfaces. Modification with the silicone reagents was found to render the wood extremely hydrophobic, even though there was no evidence for the reaction using IR spectroscopy, nor any weight gain due to reaction. However, XPS clearly showed the presence of silane on the wood surface, and removal of the top layer of the wood resulted in a significant decrease in the contact angle on the surface. It would seem that the high levels of water repellency achieved required only very low levels of siloxane on the wood surface.

6.4 Surface Chemical Modification for Bonding

There have been a number of studies concerned with the use of chemical modification for the purpose of providing a means of bonding between wood surfaces. Conventionally,

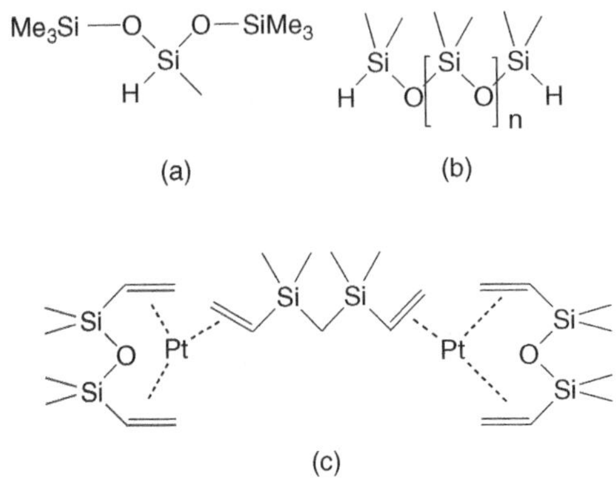

Figure 6.2 Structures of bis(trimethylsiloxy)methylsilane (a), a hydride-terminated polydimethylsiloxane (b), platinum-divinyltetramethyldisiloxane (c).

wood particles, veneers and so on are bonded together using a resin or adhesive. Many of the adhesive systems currently in use are derived from nonrenewable petrochemical resources, and concerns regarding the decline of these resources are driving research efforts based upon the use of renewables, or to find a means of directly bonding wood surfaces without the need for an additional adhesive agent. Several approaches can be adopted:

- Altering the surface energy of wood to improve compatibility with low surface energy materials such as polyolefins.
- Reaction of the wood surface with a functionalized coupling agent to improve compatibility with low surface energy materials.
- Bonding of a reagent to the wood surface in order to thermoplasticize the surface for self-bonding.
- Bonding of a reagent to the wood surface that provides the appropriate functionality for bonding sites.
- Activation of the wood surface by the use of chemical reagents, such as Fenton's systems, to generate surface free radicals.
- Activation of the wood surface using enzymatic means to generate surface free radicals.

6.4.1 Surface Modification for Compatibilization

The use of acetylation to improve the compatibility between wood fibres and thermoplastic matrices has been the subject of a number of studies. In this case, a reduction in the surface energy (polarity) associated with acetylation is advantageous, since this improves the compatibility of the fibre with the matrix material. However, Kishi *et al.* (1988) found that acetylation of refiner ground pulp fibres did not improve the mechanical performance of composites using a thermoplastic polymer matrix. Acetylation of chemithermomechanical pulp (CTMP) fibre was studied as a method to improve interfacial adhesion and reduce water sorption of the fibres in a thermoplastic matrix (Chtourou *et al.*, 1992). Composites were formed by compression moulding or by injection moulding, using weight fractions of 10 %, 20 % and 30 % of unmodified fibre, but a weight fraction of 10 % only for the acetylated fibre. Acetylated fibre was produced with a range of WPGs, meaning that as the WPG increased, a constant weight fraction of modified fibre in the composite would result in a lower fibre content. Acetylation of the fibres improved the mechanical properties of composites as the fibre WPG increased, but at the highest WPGs a decrease in properties was observed. This decrease in properties may be attributable to fibre damage occurring at higher WPGs, plus the lower volume fraction of fibre reinforcement due to the weight taken up by acetyl.

Liu *et al.* (1994) acetylated wood, rayon and cotton fibres, achieving WPGs of 11.3 %, 0.0 % and 0.8 %, respectively. Fibres were also heat treated under identical conditions to acetylation, but with no acetic anhydride present. Fibre surface energies were determined by using dynamic contact angle analysis, which showed that heat treatment reduced the surface energies of fibres in all cases, and for acetylated rayon and cotton. Rather unexpectedly, the surface energy of acetylated wood was found to be higher that that of the unmodified wood. This result was considered, by the authors, to be a direct result of adding polar ester bonds to the wood, but this runs counter to the consensual view (backed up by a wealth of data) that acetylation increases hydrophobicity (reduces surface energy). The interfacial shear strength (ISS) was determined between the fibres and polypropylene in a micro-debond test. This showed that acetylation increased the ISS between wood fibres and PP by a factor of three, and for cotton by a factor of two, a decrease in ISS was recorded for acetylated rayon. The ISS results are in accordance with what would be expected in terms of compatibility between the fibre surface and the PP.

Mahlberg *et al.* (2001) studied the effect of acetylation of wood fibres (20 % WPG) upon the mechanical properties of wood fibre PP fibre composites, with and without the addition of a novalak PF. Acetylation resulted in significant improvements in MOR, MOE and IBS in a composite containing 20 % by weight of PP fibre. With a combination of PF and PP (both 10 % by weight) and acetylated wood fibre, MOR was unchanged, MOE decreased and only IBS showed a significant improvement. Acetylation also contributed to a better dispersion of the wood fibres in the PP matrix.

6.4.2 Reaction of Wood Surfaces with Coupling Agents

In order to improve the compatibility between high surface energy lignocellulosic fibres and low-energy polyolefin matrices, extensive use has been made of coupling agents (Bledzki *et al.*, 1998; Hill, 2000; Lu *et al.*, 2000). The most commonly employed coupling agents are maleated polymers (e.g. maleic anhydride polypropylene, MAPP; Figure 6.3a) or polymethylene polyphenylene isocyanate (PMPPIC, Figure 6.3b). Coupling agents are added to the wood fibre/polymer ingredients and subjected to energetic mixing involving high shear in extruders or high-speed mixers. Evidence has been presented indicating that the attached anhydride moieties are capable of reacting with the surface of the lignocellulosic fibre.

In a study by Felix and Gatenholm (1991), cellulose fibres were heated in a toluene solution of MAPP for 5 minutes at 100 °C. The MAPP was used as received, but also heated at 180 °C for 5 minutes in order to convert any hydrolysed maleic acid groups back to the anhydride form. Treated fibres exhibited contact angles in water of the order

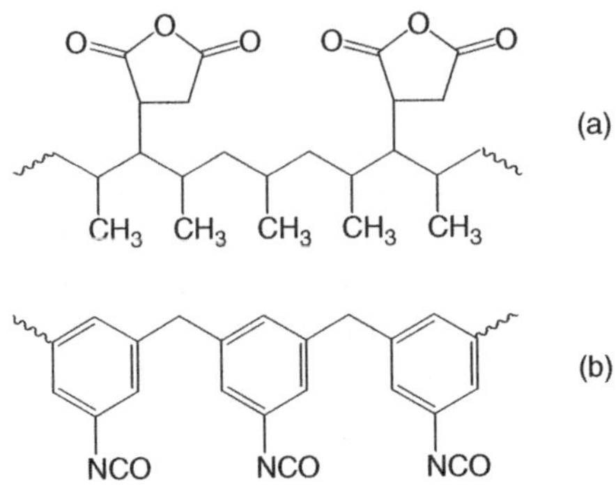

Figure 6.3 Structures of maleic anhydride polypropylene (MAPP) (a) and polymethylene polyphenylene isocyanate (PMPPIC) (b).

of 140°, and this did not decrease when the fibres were Soxhlet extracted with toluene, or with dimethyl formamide (DMF). DMF was used because it is known to be extremely efficient at disrupting hydrogen bond networks. Soxhlet extraction with water decreased the contact angle by 25–30°, and treatment with 0.1 molar aqueous NaOH resulted in a decrease in the contact angle to 67–70°, which was attributed to hydrolysis of ester bond linkages with the surface of the treated fibres.

Fibres reacted with pre-heated MAPP always exhibited larger contact angles compared with fibres reacted with as-received MAPP. FTIR studies showed that there was a higher level of ester linkages when the fibres were reacted with the pre-heated MAPP. Wood polymer composites were made by shear mixing in a twin screw extruder, and the mechanical properties of these composites were found to be superior when pre-heated MAPP was used as the coupling agent.

6.4.3 Surface Chemical Modification for Self-bonding by Thermoplasticization

There have been many studies in which wood has been thermoplasticized using benzylation (e.g. Hon and Ou, 1989). Kiguchi and Yamamoto (1991, 1992) produced hot-melt, self-bonded particleboard from benzylated wood particles. Improvements in dimensional stability, IBS and decay resistance were found compared to conventional PF-bonded boards, in proportion to the WPG due to benzylation. A slight decrease in bending strength was noted, however. Benzylation required the pre-treatment of wood with 40 % aqueous NaOH, resulting in wood degradation that was thought to be responsible for the decrease in bending strength. Hot-pressing produced particleboards with a very smooth glossy surface. Cyanoethylation of wood particles to produce hot-melted self-bonded boards results in products displaying poor water resistance (Kiguchi, 1990a).

6.4.4 Surface Chemical Modification to Provide Bonding Sites

Chemical modification of wood surfaces can be employed in order to provide active sites to allow for self-bonding directly, or to allow for covalent bonding between wood surfaces via an intermediary reagent. One approach is to chemically bond a group that has a double bond present on to the wood surface. It is then possible to graft monomers on to the activated wood surface using free-radical polymerization reactions, or directly bond the activated surfaces together.

Nakagami and Yokota (1981a) bonded methacrylic acid to wood meal using the TFA impelling method, obtaining a WPG of around 28 %. The activated wood was hot-pressed into small discs in the presence of styrene and benzoyl peroxide (BPO). A series of solvent treatments were used to examine the integrity of the formed discs, as well as exposure to aqueous alkali. Although the discs exhibited swelling with some treatments, they did not disintegrate. Analysis of polymers isolated from the grafted wood meal indicated that copolymerization between the styrene and methacrylic acid had occurred (Nakagami and Yokota, 1981b). When methacrylated veneers were bonded using styrene monomer in the presence of BPO, poor mechanical properties were found in tensile shear tests (Nakagami *et al.*, 1983). This was attributed to surface roughness of the wood preventing close contact between methacryl groups, combined with the possibility of inhibited polymerization.

Ohkoshi (1990) treated wood with aqueous NaOH and subsequently reacted the wood with allyl bromide (Figure 6.4a). Allylated wood specimens were bonded together using hot-pressing and the bond strength determined. Apparent good bonding was achieved using this method, as evidenced by the frequent failure of the wood rather than the bond in the tests. Infrared spectroscopy showed that the double bond of the allyl group had not reacted during hot-pressing. It was concluded that bonding occurred as a result of thermoplasticization of the wood. The premature wood failure in tests was probably due to substrate degradation caused by the aqueous alkaline treatment prior to reaction.

The production of MDF from allylated wood fibres has also been reported (Ogawa and Ohkoshi, 1997). The IBS of the 4 mm thick boards was superior to control boards (unmodified fibres bonded with PF resin), provided that the temperature of the allylation reaction and board density was sufficiently high. The MOR was markedly inferior to that of control boards in all cases (*c.* 10 MPa for allylated compared with *c.* 60 MPa for controls, at a board density of 800 kg m^{-3}). Blending of the allylated fibres with acetylated fibres caused a decrease in IBS, but did not affect MOR. It is perhaps significant that no free-radical initiator was used during hot-pressing in either study, which may account for the lack of reactivity of the allylated surfaces.

The possibility of using graft polymerization of styrene on to allylated wood to provide covalent bonding between surfaces was also investigated (Ohkoshi, 1991). In this case, a free-radical initiator (BPO) was present in the styrene monomer that was used to impregnate the allylated wood, prior to hot-pressing. Bond strength was determined by

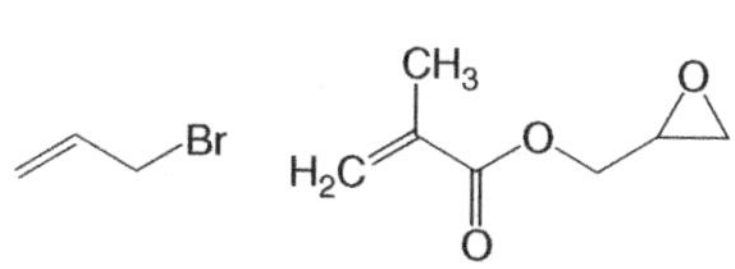

Figure 6.4 *Structures of allyl bromide (a) and glycidyl methacrylate (GMA) (b).*

shear testing under compression in dry conditions and also following water-soaking of the bonded specimens. Bond strength increased with pressing temperature and was reduced by about 50 % following water-soaking. A similar reduction in bonding strength was found when the bonded samples were extracted in benzene for 48 hours. Infrared spectroscopy indicated that grafting of the styrene on to the allylated surface had occurred.

Banks *et al.* (1995) reacted TMP fibres with glycidyl methacrylate (GMA) (Figure 6.4b), maleic anhydride or succinic anhydride and hot-pressed the modified fibres with methyl methacrylate in the presence of BPO. Infrared evidence was presented indicating that copolymerization had occurred between the bonded GMA and methyl methacrylate. However, in a later study, Çetin and Hill (1999) could find no evidence to suggest that a chemical bond had formed between the GMA and wood.

Rozman *et al.* (1997b) reacted wood flour with maleic anhydride (MA) and mixed the modified wood with diallyl phthalate in the presence of BPO. The mixture was then hot-pressed to form well-consolidated boards. Reaction of the wood resulted in significant increases in MOE, MOR and impact toughness of boards, compared to composites made from unmodified wood flour. It was considered that the improved properties arose due to the grafting of the diallyl phthalate monomers on to the double bond of the covalently linked maleic moieties (Figure 6.5).

Hill and Çetin (2000) modified the surfaces of wood veneers with methacrylic anhydride (Figure 6.6a) and formed lap joints using either styrene or methyl methacrylate in the presence of the free radical initiator azoisobutyronitrile. Significant improvements in bond strength were found when compared to unmodified wood,

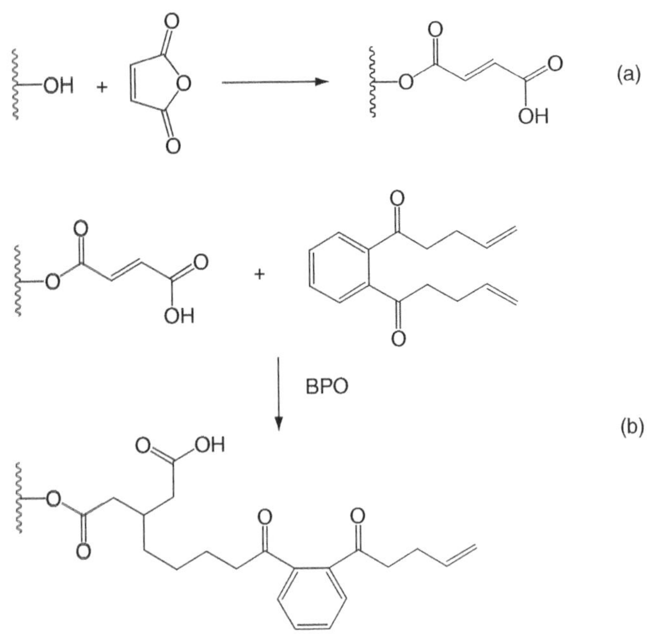

Figure 6.5 *Reaction of wood with maleic anhydride (MA) (a), and subsequent reaction of maleated wood with diallyl phthalate in the presence of benzoyl peroxide (BPO) (b).*

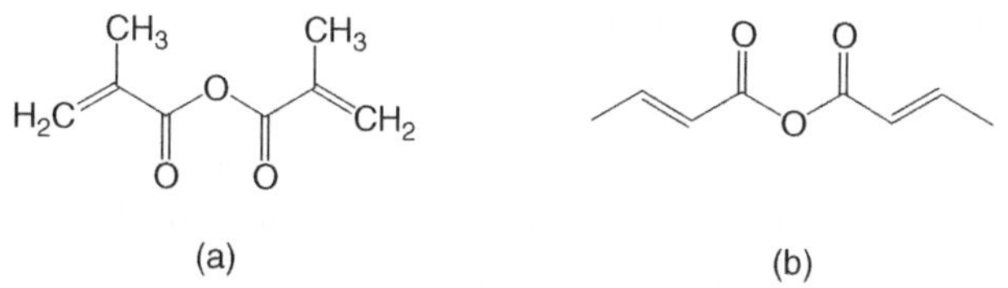

Figure 6.6 *Structures of methacrylic anhydride (a) and crotonic anhydride (b).*

wood modified with propionic anhydride or wood modified with crotonic anhydride (Figure 6.6b). Confirmation of the grafting reaction of styrene monomer on to the methacrylic anhydride modified wood surface was obtained from FTIR and CP-MAS NMR spectroscopy. No reaction was found for the crotonic anhydride modified wood surface, which was thought to be due to steric hindrance of the double bond in this case.

Rather than using free-radical chemistry, other reactions are possible, such as condensation reactions, which are used in certain classes of polymerization. Mallon and Hill (2002) chemically modified wood fibres and veneers with succinic anhydride (SA) and used hexamethylene diamine (HMDA) to provide a flexible covalently bonded bridge between the activated wood surfaces (Figure 6.7). Since it was anticipated that the HMDA would not react with the SA-modified wood surfaces under the relatively mild conditions employed for the reaction, diisopropyl carbodiimide (DIPCI) was employed to promote the coupling reaction (Figure 6.7). SA-modified veneers were placed against HMDA-grafted SA veneers and hot-pressed in the presence of DIPCI to form covalent bonds between the surfaces. The bond strength of the lap joint produced was stated to be comparable to that for a traditional UF resin.

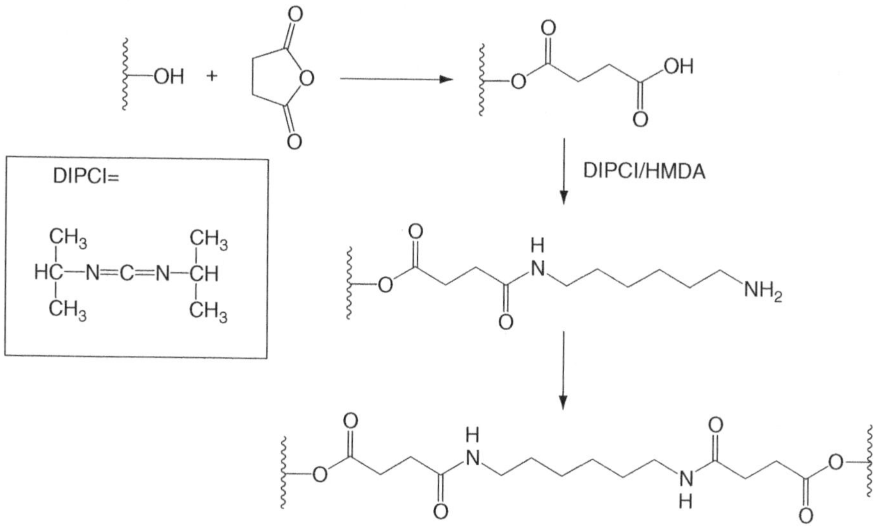

Figure 6.7 *Reaction of wood with succinic anhydride (SA) and subsequent reaction of the activated surface with hexamethylene diamine (HMDA), using diisopropyl carbodiimide (DIPCI) as a reaction promoter.*

6.4.5 Direct Grafting on to Wood Surfaces

Direct grafting on to the surface of acetylated wood fibres has been studied by Narayan and Shay (1986). The method involves the production of an anionic polymer that is then reacted with a suitably modified lignocellulosic material. Initially, wood chips were acetylated using acetic anhydride in the presence of concentrated H_2SO_4. Anionic polymerization techniques were used to prepare polystyryl, or polyacrylonitrile carbanion, which were then reacted with the acetylated wood chips. The highly reactive carbanions of the polymers displace the acetate groups via an SN_2 reaction, to form polymer-grafted wood (Figure 6.8). Analysis of the grafted wood by FTIR and TGA indicated that graft polymerization of the wood had occurred. The use of concentrated sulphuric acid in this study will have undoubtedly resulted in severe degradation of the wood material.

There have been a number of studies investigating the grafting of polymer on to wood fibres that have been treated to produce free radicals on the surface. The methods investigated include treatment with Fenton's reagent, ceric ion, xanthate-mediated grafting, initiation with potassium permanganate or the use of a halide/peroxide system. In many cases, the presence of lignin in the wood has an inhibitory effect upon grafting efficiency. This is mainly due to the high mobility of the formed free radicals, which are able to rapidly migrate from the surface initiation sites to lower energy traps within the cell wall. Thus, although grafting of isolated lignin may be possible in situations in which it is present in a swollen gel-like state, so that the radical trap sites are still accessible, this does not imply that such reactions are possible on the surface of solid lignified wood, since the radicals can readily migrate into the cell wall interior, where they are no longer accessible. Many polymerization techniques require the use of clean conditions to prevent the inhibition of polymerization reactions, and this is particularly so for free-radical, or anionic, polymerization. The introduction of wood fibres into a polymerization reaction presents a number of problems, in that various extractible compounds (e.g. phenolics) can have an inhibitory effect upon polymerization. The formation of homopolymer can also produce erroneous results, particularly if the homopolymer becomes entangled with the cell wall matrix or with the fibrillated surface of fibres, in which case it is very difficult to determine whether grafting has actually occurred.

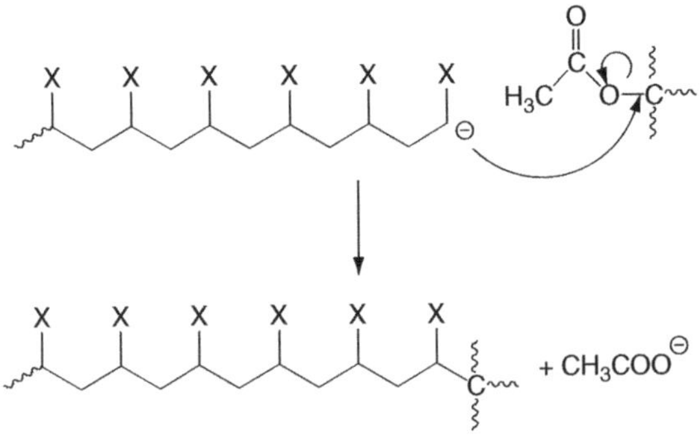

Figure 6.8 Reaction of a carbanion-terminated polymer with an acetylated wood surface to graft a polymer on to the wood surface.

The chemistry of the interaction of Fenton's reagent with wood is complex, but an important reaction is the generation of phenoxy radicals by the abstraction of a proton from the lignin phenolic OH groups by hydroxyl radicals (Figure 6.9) (Gierer *et al.*, 1992; Oniki and Takahama, 1997). Such methods have been used to successfully graft polymer on to clean substrates such as pure cellulose, as well as isolated lignin (Chen and Kokta, 1982), but the presence of lignin in wood inhibits grafting efficiency. There has been some work showing grafting of polymer on to wood that has been activated using Fenton's reagent directly (e.g. Shiraishi *et al.*, 1983), but Fenton's chemistry has been used for the grafting of xanthate-modified bleached hardwood pulp (lignin contents ranging from 4.4 % to 14.8 %) with acrylonitrile (Kokta *et al.*, 1982) (Figure 6.10). Young (1977) also reported on the use of the xanthate grafting method for covalently linking acrylonitrile, styrene or acrylamide to high lignin content wood fibres.

The use of ceric ion initiation to graft polymers on to the surface of wood pulp fibres requires pre-bleaching of the fibres to remove most of the lignin (Mansour *et al.*, 1982). The reaction requires acidic conditions, with a typical system utilizing ceric ammonium

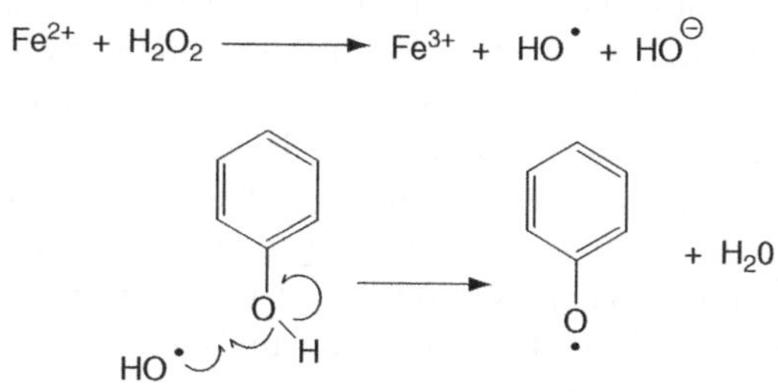

Figure 6.9 *Abstraction of a phenolic hydroxyl proton by a hydroxyl radical generated by Fenton's reaction with hydrogen peroxide.*

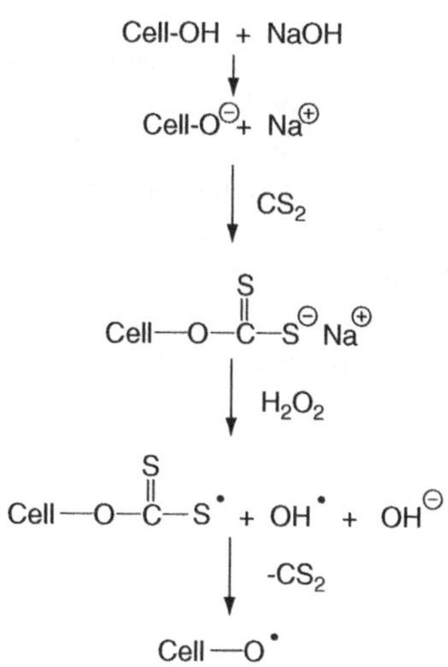

Figure 6.10 *Xanthation of cellulose followed by reaction with hydrogen peroxide to generate a radical.*

nitrate and nitric acid. The use of nitric acid will lead to acid-catalysed degradation of the polysaccharide components of the fibres. The method relies upon the redox properties of cerium to generate radicals in the polysaccharides and requires the elimination of oxygen in order to achieve grafting (Figure 6.11). Ce^{4+} can react with 1,2-diol structures in a disproportionation reaction that results in the formation of free radicals. Other redox metals (e.g. V, Co and Mn) have also been used as grafting initiators with cellulosic substrates.

On the basis of earlier studies of the grafting of methylmethacrylate and acrylontrile on to lignocellulosic fibres, Marchetti *et al.* (1998) used $KMnO_4$ as a free-radical initiator for the grafting of acrylic acid on to *Picea abies* sawdust. Aqueous treatment with $KMnO_4$ requires careful control of the conditions in order to ensure that the lignin is not modified in such a way that it inhibits subsequent grafting of polymer. Grafting was accomplished on to $KMnO_4$-treated fibres dispersed in hexane, since attempts at grafting in an aqueous environment were found not to work. Grafting on to lignin-containing fibres was found to be more efficient than when holocellulose was used, showing the important role of lignin in such reactions.

A halide/peroxide initiating system has been developed for the grafting of polymers on to lignocellulosic surfaces (Chen *et al.*, 1995). Although the mechanism for the reaction does not appear to be fully understood, successful grafting requires the presence of chloride ions, hydrogen peroxide and a lignocellulosic substrate. Reaction of chloride ions with hydrogen peroxide is thought to produce chlorine radicals that then abstract protons from the cell wall polymers, generating radical sites where polymerization occurs. There is very little homopolymerization if the lignocellulosic material is not present. Reactions are usually performed in dimethylsulphoxide, in which the wood pulp forms a gel. Successful grafting on to isolated lignin has been demonstrated using this system.

6.4.6 Surface Activation for Self-bonding

The principle of surface activation is to generate free radicals on the surface of the wood: this then provides bonding sites on the wood surfaces for resin-free board production. The use of biological systems to generate surface radicals will be considered later, this section being concerned with the use of chemical systems for surface activation.

Fenton's reagent chemistry has been used for the direct self-bonding of wood particles. Indirect methods, involving the use of surface-activated fibres in combination with other bonding agents such as furfural and lignosulphonates, have also been studied (Philippou, 1982; Philippou *et al.*, 1982). Since the use of Fenton's reagent requires the presence of

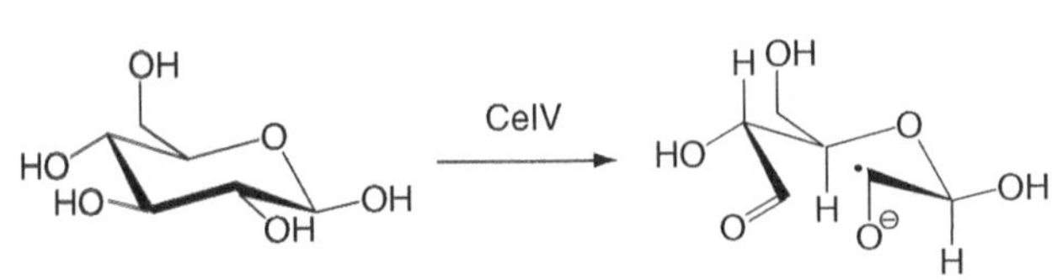

Figure 6.11 Ring opening of glucose with a cerium IV salt to generate a radical.

hydrogen peroxide, this leads to a great reluctance on the part of composite manufacturers to use such systems, because of safety concerns.

Widsten *et al.* (2003) used Fenton's chemistry to bond wood fibres for the production of 12 mm thick medium density fibreboard (MDF). An aqueous solution of hydrogen peroxide and ferrous sulphate was sprayed on to fibres at the beginning of the refiner blowline. Fibreboards were then produced using a range of pressing temperatures and times, and the MOE, MOR and IB of the boards determined, as well as thickness swell (TS) in a 24-hour cold water-soak test. The bulk of the work was performed using spruce fibres, but there was a lesser study on the production of MDF from beech fibres. It was found that there was an increase in mechanical properties as a result of increasing the defibration temperature during fibre production. This was associated with an increase in radical production in the fibres (determined using ESR), although the relationship between free-radical concentration and mechanical properties was not straightforward. The mechanical properties of boards produced from fibres produced at a defibration temperature of 171 °C were below those required by the standards, whereas most boards pressed from fibres defibrated at 202 °C met the European standards. None of the boards produced in this study met the European standards for exterior boards in the TS tests.

Yelle *et al.* (2004) used a biomimetic system comprising a synthetic chelator (2,3-dihydroxy benzoic acid) in combination with iron salts and hydrogen peroxide to generate free radicals on the lignin-rich surface of TMP fibres for self-bonding. Boards of 4 mm thickness were produced from fibres activated in a wet process, involving pressing of fibres having 300–400 % MC at 200 °C for 4 minutes; or from fibres activated in a dry process, where the activation chemicals were sprayed on to the fibres, followed by pressing at 170–180 °C for about 6 minutes. The highest IBS achieved using boards produced from wet-process activated fibres was 0.69 MPa, and 0.58 MPa with dry-processed fibres (0.45 MPa for UF-bonded fibres). The best MOR achieved with boards produced from dry-processed fibres was only 50 % of that recorded from UF-bonded boards (20.6 MPa compared with 43.4 MPa). Several dry-process treatments gave comparable MOE values to those of the UF-bonded boards. Although bonding was achieved using surface activation, it was thought that even with the 'dry' process, the MC of the fibres was sufficiently high to allow for diffusion of the reagents into the cell wall.

6.5 Enzymatic Modification

The use of enzymatic systems for surface activation of wood particles to form binderless composites has been the subject of study for over 20 years. Enzyme-catalysed bonding of wood can be achieved either by activation of lignin, which is mixed with wood particles, or by surface activation of the wood particles directly (Grönqvist *et al.*, 2003).

The phenol oxidases, peroxidase and laccase, are capable of catalysing one-electron oxidations of phenolic OH groups, whilst simultaneously reducing H_2O_2 or O_2, to produce phenoxy radicals and water. Oxygen radicals, such as superoxide and hydroxyl radicals, have also been detected during fungal degradation of lignin (Faison and Kirk, 1983). Kharazipour *et al.* (1997) found that laccase treatment produced a highly oxidized lignin surface with a high content of carboxylic groups. Although free radicals in lignin model compounds have half-lives of the order of 30 minutes (e.g. Caldwell and Steelink, 1969), free radicals generated in lignin in the cell wall are long-lived entities. For example, Felby *et al.* (1997a) reported a free-radical half-life of more than 2 weeks for beech fibres oxidized with laccase. Due to the inability of enzymes to penetrate the cell wall, activation is confined to fibre surfaces only (Srebotnik and Messner 1991). However, it is certain that migration of the free radicals through the lignin network into the cell wall will occur.

Felby *et al.* (1997a) noted that in previous studies of the use of enzymatic activation for auto-adhesion of lignocellulosics, interference from added protein and carbohydrate may have affected the results obtained. They activated a 5 % consistency aqueous suspension of wood fibres by treatment with laccase. Boards of 3 mm thickness were made either in a wet process, where the activated fibre suspension was transferred to a former directly, or a dry process, where the activated fibres were air-dried at 40 °C before hot-pressing. A moderate increase in MOR from 38.3 MPa (untreated) to 44.6 MPa was found for wet-process boards and a rather larger increase (from 27.4 Mpa to 41.7 MPa) was found for dry-process boards. Felby *et al.* (2002) further developed the technique and undertook a pilot-scale study of the production of fibreboards from laccase-oxidized beech wood fibres. The strength properties of the boards manufactured from enzyme-treated fibres were found to be comparable to boards bonded using UF resin, but the dimensional stability properties were inferior. The use of wax was found to interfere with the bonding process for the enzyme-treated fibres.

Barsberg and Hassingboe (2003) noted that the use of enzymes for surface activation of fibres for board production can produce highly variable results and that the reasons for this are not understood. In order to further understand the process, they treated TMP fibres with a laccase from *Trametes villosa* for 1 hour and dried the fibres. A control set of fibres was subjected to an identical protocol, but with no enzyme present. Air-laid fibre mats were produced from the fibres, which were then hot-pressed to form 3 mm thickness boards. Varying amounts of wax were sprayed on to the fibres prior to board production. Both the MOE and the MOR of the composites increased with board density. Boards produced from enzyme-treated or control fibres exhibited no difference in MOE, but the MOR of boards formed from enzyme-treated fibres was higher above a density of 800 kg m^{-3}. Wax addition resulted in a decrease in mechanical properties. At a board density of *c.* 930 kg m^{-3}, the MOR was of the order of 23 MPa and the MOE 11 GPa.

Felby *et al.* (1997b) undertook a study to identify the free radicals formed in beech wood during laccase-catalysed oxidation. ESR spectroscopy indicated the presence of at least two radical species in the suspension liquid, with the spectrum evolving with time. The radical concentration in the suspension liquid remained constant until the fibres were removed, when it subsequently decayed rapidly. The ESR spectrum of the suspension liquid showed hyperfine structure, but this was not observed for the treated fibres, which exhibited a typical broadened solid-state spectrum. The radicals in the suspension liquid were identified as phenoxy radicals situated on solubilized lignans and colloidal lignin. It was proposed that the wood fibres were modified by direct laccase oxidation of the surface, as well as a mediation process involving colloidal lignin fragments in a cyclic phenol/phenoxy reaction pathway (Figure 6.12).

Phenoxy radical species were believed to be present on the fibre surface, and superoxide was also identified in a cytochrome *c* assay. However, the presence of superoxide

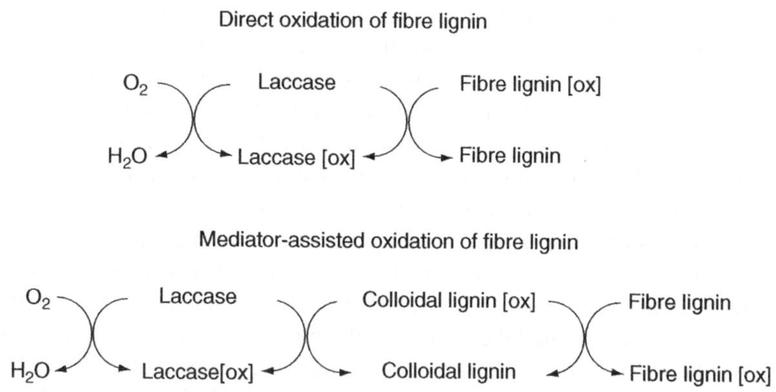

Figure 6.12 *A mechanism for the generation of lignin phenoxy radicals due to direct oxidation by laccase and a mediation process involving colloidal lignin fragments, according to Felby* et al. *(1997b).*

was not confirmed by the use of an ESR spin-trapping technique. This anomaly was explained as being due to the reduction of cytochrome *c* by the presence of semiquinone anions in the lignin. Barsberg (2002) and Barsberg and Thygesen (1999) used a number of spectroscopic techniques (ESR, IR, UV–Visible and fluorescence) to identify the changes occurring to lignin as a result of a laccase-catalysed treatment. The results were interpreted as indicating that primary oxidation products (cation radical/hole states) were produced in the lignin matrix, or extractives. These evolved to produce final oxidation products (assumed to be quinonoid structures) produced by the reaction of oxygen with the primary oxidation species. However, a proportion of the primary products were considered to migrate from the fibre surface through the lignin network to low-energy sites, where they were stabilized due to the limited diffusion of oxygen into the cell wall.

Kharazipour *et al.* (1998) used a peroxidase enzyme in combination with H_2O_2 to activate the surface of TMP fibres for self-bonding. Fibres were activated in a wet system, then dewatered and 'fluffed out' before pressing at 190 °C for 5 minutes. The best IBS recorded for boards of 5 mm thickness made from the activated fibres was 0.55 MPa. The IBS was found to be dependent upon the pH of the treatment solution, the time of treatment and the board density. The authors noted that phenoloxidase gave comparable results to laccase, which was unexpected, since it was thought that the phenoloxidase would lead only to depolymerization of the lignin.

6.6 Corona or Plasma Discharge

Atmospheric corona discharge has been used to increase the surface energy of wood surfaces by oxidative activation (Back and Danielsson, 1987). Improved bonding of a water-based acrylate lacquer was found following corona treatment of the wood surface, with no improvement found when a solvent-based alkyd system was used.

Mahlberg *et al.* (1998) modified the surface of birch veneer with a hexamethyldisiloxane (HMDSO) plasma in order to improve the compatibility between the wood surface and polypropylene. Plasma treatment resulted in an increase of hydrophobicity of the wood surface, but there was no improvement in adhesion to polypropylene. Denes and Young (1999) coated wood with polydimethylsiloxane (PDMSO) containing various UV stabilizers, absorbers or reflectors, and polymerized the coating using an oxygen plasma. Samples were exposed to an accelerated weathering protocol consisting of UV exposure and UV exposure combined with water-spray cycles. The use of such coatings significantly improved the weathering resistance of the coated wood. Denes *et al.* (1999) used a HMDSO plasma to modify the surface of pre-extracted pine wafers. The wood surface became highly hydrophobic as a result of this treatment, with water contact angles increasing from about 5° to about 120° following plasma treatment.

Podgorski *et al.* (2000) studied the influence of a number of variables upon the surface wettability of plasma-treated wood. The contact angle was measured using a two liquid phase method, in which the samples were dipped in octane before a water droplet was applied to the surface. Plasma treatment of unmodified and heat-treated wood with N_2, O_2, N_2/O_2, CO_2 and Ar/O_2 resulted in significant reductions in the water contact angle. Treatment with an Ar plasma resulted in a moderate reduction in contact angle, whilst treatment with an ammonia gas plasma resulted in an increase in contact angle. For nonheat-treated samples, increases in contact angle were found when the plasma-treated samples had been exposed to air for a few weeks. Corona treatment was found to result in a lesser decrease in contact angle compared with plasma treatment. The adhesion of coatings to wood was not improved by plasma treatment. Large decreases in water contact angle were found when wood was treated with a CF_4/C_3F_6 plasma. Podgorski and Roux (1999) also studied the use of an oxygen plasma to increase the surface polarity of heat-treated wood. This did not lead to improved performance of coatings after weathering.

A disadvantage with conventional plasma treatment techniques is the requirement for treatments to take place in a vacuum, adding to the equipment costs. However, if a dielectric material is placed between the electrodes of the plasma equipment, then treatment can be performed at atmospheric pressure. This method is known as a dielctric barrier discharge treatment and has been the subject of some recently reported studies.

Rehn and Vlöl (2003) investigated the use of a cold plasma discharge in air, helium, nitrogen or argon at atmospheric pressure to modify the surface properties of wood. The use of a gas mixture of methane or ethane in argon imparted hydrophobic surface properties and resulted in significantly reduced water uptake, whereas hydrophilic properties were imparted with the use of an oxygen plasma. Bente *et al.* (2004) further reported on the use of a dielectric barrier discharge apparatus, or a hand-held ionized jet generator, to modify the wood surface under ambient conditions. Plasma treatment with ethane, methane or silane/N_2 was used to modify the wood surface. The surface of the wood was made hydrophobic by all of the above treatments, with the silane treatment exhibiting the greatest improvement. The surface tension of the wood surface, as determined by the contact angle with water, was reduced from 37.5 mN m^{-1} to 4.5 mN m^{-1} after silane plasma treatment. Atomic force microscopy showed that the wood surface was considerably smoother following plasma treatment, which was thought to contribute to the improved hydrophobicity of the surface.

6.7 Conclusions

The most promising areas for further research into wood surface modification remain self-bonding and weathering stability. For any application, it is essential that the cost of treatment can be fully recovered in terms of added value to the final product. Most of the processes discussed in this chapter require expensive equipment or reagents, which suggests that they are unlikely to be used on a large scale in the foreseeable future. Although, conceptually, the use of enzymatic surface activation for self-bonding is attractive, it is unlikely that there will be commercial use of this technology in the short term at least. The adoption of this technology is only likely if the costs of conventional resins increase dramatically, or the cost of the enzymatic treatment is reduced significantly. However, this area does appear to offer considerable potential, with several pilot-scale operations now running.

Little advantage appears to result from directly covalently bonding coatings to the surface of wood, in terms of long-term improved weathering performance. However, directly covalently linking UV stabilizers does appear to lead to improved performance. The costs and difficulties involved in performing such reactions on a large scale appear to limit these modifications to the laboratory at present. However, should a viable UV stabilizer for wood surfaces ever become available, there would a ready market for such a product. Only acetylation appears to offer the prospect of medium-term weathering stability at this time, particularly if a clear coating is applied to the acetylated wood.

6.7 Conclusions

7

Impregnation Modification

7.1 Introduction

The principle behind impregnation modification, as defined in this text, is to impregnate the cell wall of wood with a chemical, or a combination of chemicals, that then react so as to form a material that is 'locked' into the cell wall. For this to occur, it is necessary that during the impregnation phase that the cell wall is in a swollen state, so as to ensure accessibility to the impregnant. It is self-evident that the molecular components of the impregnant should be small enough that they can gain access to the cell wall interior. The fixation of the impregnant can occur using two main mechanisms:

- Monomer (or oligomer) impregnation, with subsequent polymerization within the cell wall.
- Diffusion of a soluble material into the cell wall, with a subsequent treatment so as to render the material insoluble (immobile).

To be effective, it is essential that the impregnant is nonleachable in service conditions. However, it is not a primary requirement that the impregnant be chemically bonded to the cell wall polymeric constituents, although this may occur in some circumstances. It is essential that the fixed impregnant is nontoxic whilst in the cell wall and under any circumstances in which it is released from the cell wall, such as disposal by incineration or composting, or due to any recycling process.

As noted in Chapter 2, solute exclusion studies indicate that the maximum diameter of the cell wall micropores is of the order of 2–4 nm. This would then be the expected upper limit for the diameter of molecules that are able to penetrate the swollen cell wall. Studies of the penetration of homologous series of liquids into the wood cell wall have shown that liquids possessing a molar volume greater than 100 $cm^3 mol^{-1}$ (roughly corresponding to a molecular diameter of 0.68 nm) have a limited ability to penetrate the cell wall (Mantanis *et al.*, 1994). However, this property depends very strongly upon the hydrogen-bonding ability of the molecules of the penetrating liquid. Molecules with a greater tendency to form/break hydrogen bonds are able to swell the cell wall and show greater penetrating

Wood Modification: Chemical, Thermal and Other Processes C. Hill

ability. Additionally, molecules can be introduced in carrier liquids, which swell the cell wall to a greater extent than the penetrant molecule is able to do in isolation.

A very important factor in ensuring that full cell wall penetration has occurred is to allow sufficient time for the impregnant molecules to diffuse into the intracellular spaces. Many workers allow several days (weeks in some cases) for this to occur. It is important to emphasize that pressure treatment will aid penetration of larger wood samples, but will not in any way result in cell wall penetration, which is a purely diffusion-controlled process.

Treatment of wood with multi-component systems is likely to result in separation of the components when large wood samples are treated. This has been likened to the action of a chromatography column (Schneider, 1995). This is a significant problem that is often only encountered during scale-up of laboratory-based studies, where satisfactory results were previously obtained on small wood samples. Similarly, treatment of large wood samples can often lead to considerable variability in results due to inhomogeneous distribution, which again may not be evident with small samples treated under laboratory conditions.

The presence of nonleachable material in the wood cell wall can affect the physical and biological properties of wood through a number of mechanisms:

- The material swells the cell wall, resulting in dimensional stability being imparted to the treated wood.
- The material occupies space within the cell wall that would otherwise be available to water molecules, which leads to a reduction in the hygroscopicity of the wood.
- The blocking of cell wall micropores reduces diffusion of water and other molecules into the cell wall. If effective blocking of the cell wall micropores occurs close to the lumen, then complete filling of the cell wall may not be required in order to provide decay protection, or reduce hygroscopicity. However, this barrier may be breached in time.
- Masking of some of the cell wall hydroxyl content occurs, which can be a contributory factor in reducing water vapour sorption. The efficiency of OH masking depends upon the efficiency of diffusion of the cell wall bulking agent.
- Reaction of the cell wall bulking agent may occur with the cell wall polymeric constituents, resulting in some cross-linking.
- If the molecules are too large to penetrate the cell wall, this will result in a barrier forming on the lumen surface preventing ingress of degradative agents. Again, this barrier may be breached after some time.

7.2 Resin Treatments

7.2.1 PF, MF, MMF and UF Resins

As with so many research areas in wood modification, much of the early work in resin impregnation was performed by Alfred Stamm and co-workers at the Forest Products Laboratory in Madison (Stamm and Seborg, 1941). On the basis of this work, two veneer-based products were developed (Impreg and Compreg), which will be described in

Chapter 8. Stamm and Seborg (1936) treated relatively small wood blocks of white pine by immersion in the treating solutions (water or methanol) at room temperature and applying a vacuum. Once the blocks were fully saturated, they were left for 24 hours in the treating solution to allow the resin molecules to diffuse into the cell wall. Various resin formulations were studied, including phenol formaldehyde, urea- and thiourea-based resins and furfural. Resin curing was accomplished by heating for 3 days at 70 °C or 1 day at 105 °C. Resin contents from 36 % to 121 % by weight were obtained, although this was not solely due to cell wall penetration, particularly with more viscous resins. ASE values up to 70 % were obtained, which increased with the resin concentration in solution up to about 20–30 % by weight in methanol. Higher ASE values (approaching 100 %) were obtained by repeating the resin treatment process a number of times. Prepolymerization of the resins prior to treatment resulted in drastic reductions in ASE, due to the limited ability of the larger molecules to penetrate the cell wall. A series of experiments was also performed on larger wood specimens, where it was found that only permeable species could be treated in this way to give satisfactory penetration. Resin treatment resulted in appreciable increases in both hardness and compression strength at the proportional limit perpendicular to the grain. Good results were obtained when the resin-treated wood was bonded in a hot-press using phenolic resin adhesives.

Stamm and Seborg (1939) listed three essential criteria for effective resin treatments of wood:

- The molecular size of the resin molecules must be sufficiently small to allow for penetration of the cell wall, requiring that the resin should be unpolymerized or only slightly polymerized.
- The molecules of the resin should be soluble in polar solvents, so that the cell wall is in a swollen state to allow for diffusion into the interior of the cell wall.
- The resin molecules should exhibit sufficient polarity, so that they exhibit a high affinity with the cell wall macromolecular components.

On the basis of earlier studies, Stamm and Seborg (1939) selected PF-based resins as being the most promising for further study into using resin-treated wood for plywood manufacture. Although previous studies had shown that both water and methanol treatment solutions gave satisfactory results, only water-delivered resins were used in this later study. Douglas fir veneers 1/16 in thick and 14 in square were vacuum impregnated with resin for 15 minutes and then stacked in a high-humidity environment for 3 days to allow for redistribution of the resin. The veneer stack was then slowly dried at 50 °C and 75 % RH for 3 days and then the temperature was raised to 75 °C to complete the drying process. Following drying, the temperature was raised to 95–100 °C for a day to polymerize the resin. Water-soaked veneer was found to take up a much higher amount of resin, compared to veneer that had been dried to 6 % MC. ASE was found to increase up to about 30–40 % resin content in the wood (58 % maximum ASE), above which no further increases were noted. A variety of adhesives were used to bond the plies, but a PF resin was found to give superior performance, especially in wet-strength tests. Good decay resistance was found in pure culture tests.

Goldstein *et al.* (1959) studied the effect of solution aging on the effectiveness of PF resins to improve wood properties. An analysis of a laboratory prepared PF solution was made over a total of 135 days, which showed that there was initially an increasing mono-methylol phenol content up to about 10 days' aging, but that over longer time intervals this decreased in favour of the formation of di-methylol and tri-methylol substituted products. With continued aging, the total methylol substituted phenol content decreased to produce unreactive products. This change in composition was reflected in changes of the properties of wood treated with PF resins aged for different time intervals. The ASE of treated wood initially increased as the PF resin was aged up to about 20 days and then began to decrease again. This effect was accelerated when the PF solution was aged at a higher temperature. The improvements in dimensional stability were strongly correlated with the mono-methylol phenol content of the PF resin.

Ryu *et al.* (1991) found that little change in ASE was observed above resin loadings of 40 % with a PF resin treatment. Furthermore, as the molecular weight of the treatment resin was increased, there was a significant reduction in the effectiveness of the resin in improving dimensional stability (Ryu *et al.*, 1993). Ohmae *et al.* (2002) treated wood with a low molecular weight PF resin in water and obtained ASE values as high as 74 % at around 30 % WPG. This increase in dimensional stability appears to be caused both by bulking of the cell wall and a cross-linking of resin components within the cell wall. Cross-linking was more important at low WPGs.

Inoue *et al.* (1993b) impregnated small specimens of sugi with a solution of methylolated melamine and formaldehyde (MMF) (Figure 7.1), followed by air-drying and gradual heating up to 105 °C. As the resin loading was increased, the wood volume increased to a maximum of 5 % (about half of the theoretical maximum), showing that any further increase in resin loading resulted in the resin residing in the lumen only. ASE of specimens increased asymptotically up to around 40 %, with little change in ASE occurring above about 15 % resin loading. Resin-impregnated wood samples were compressed at temperatures ranging from 120 °C to 180 °C for 1 hour. The recovery of set of the specimens was determined by soaking in water at 20 °C and 98 °C. No recovery of set was found at the lower temperature above a resin loading of 5 %, but heating in water at 98 °C required a minimum resin loading of 20 % before the set recovery was zero. The use of PF resin to stabilize compressed wood has also been studied (Inoue and Norimoto, 1991; Inoue *et al.*, 1992).

Galperin *et al.* (1995) has reviewed the work undertaken in the Byelorussian Institute of Technology, including work with PF, UF and melamine resins. Apart from resin curing using conventional heating, microwave curing was also studied.

Vasishth (1983) performed a study of cell wall impregnation using a variety of systems and identified a water-based alkyd system that could be used to dimensionally stabilize wood. Militz and Peek (1993) also studied the impregnation of wood with alkyd resins, and Rapp and Peek (1995) reported on a study of the impregnation modification of wood with over 30 different water-based resin systems. Most emulsion-based systems showed rather poor penetration of the wood samples, with the solutions showing the best performance. Of the systems studied, only solution resin systems imparted significant improvements to the ASE of wood samples. Further studies indicated that treatment of wood with MMF resins gave the best performance in terms of dimensional stability (Sailer and Rapp, 1997). In order to reduce formaldehyde emissions from MMF-treated wood samples, Lukowsky *et al.* (1998) and Lukowsky (2002) investigated the effect of formaldehyde content upon the properties of MMF resin treated wood. The highest ASE values were obtained when using resins with either a high or a low formaldehyde content. Resins

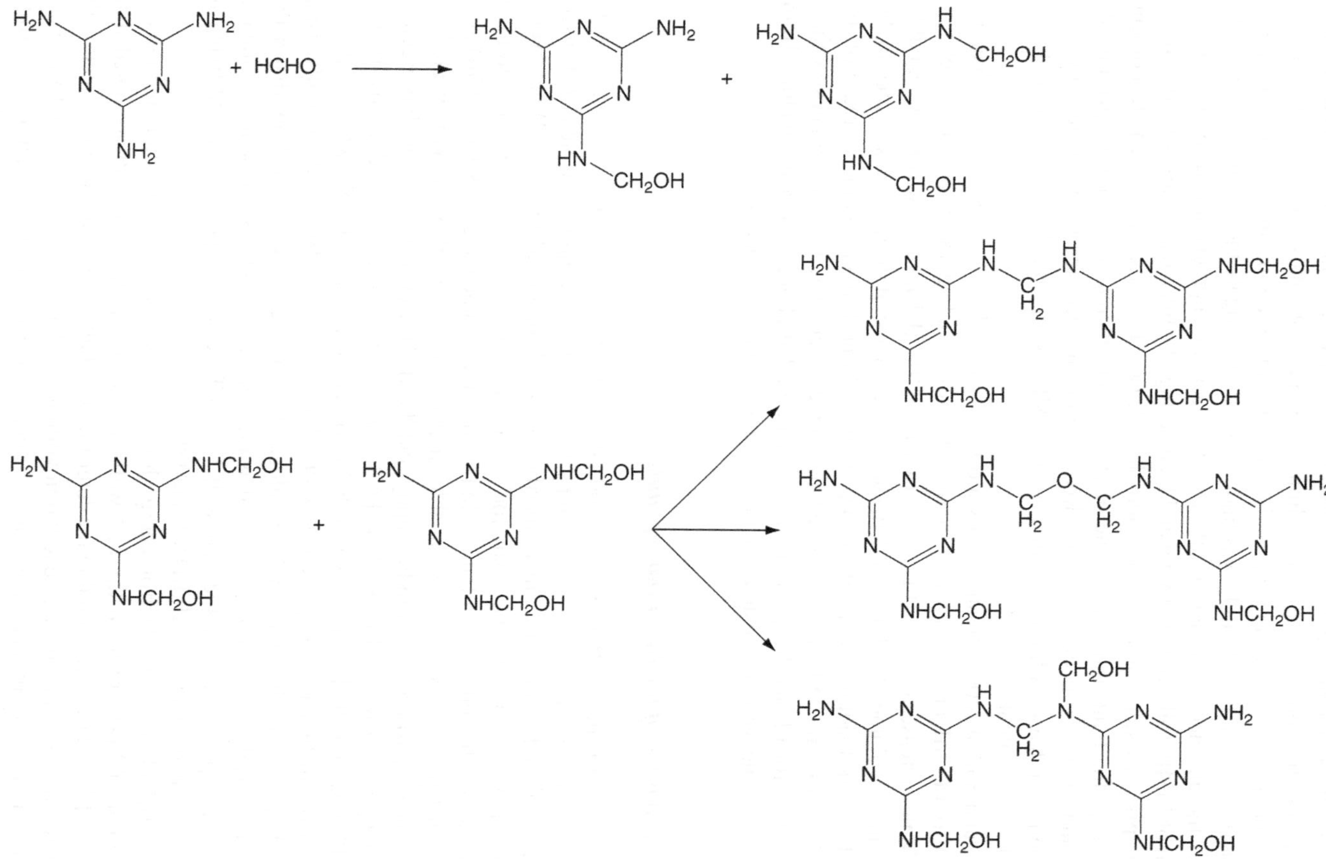

Figure 7.1 Reaction of melamine with formaldehyde and subsequent dimerization of the methylolated melamine monomers.

of intermediate formaldehyde content were of higher molecular weight and were less able to penetrate the cell walls. The curing of a low formaldehyde content resin was the subject of a more detailed study. The authors noted that a commercial treatment process would require the reuse of the resin solution a number of times, and that it would not therefore be possible to use an acid catalyst in the resin solution, since this would drastically shorten the pot life. A study was undertaken to investigate the curing of the resin in wood in the absence of an acid catalyst. After vacuum impregnation with an aqueous solution of MMF, followed by drying and then curing, the treated samples were stored for 1 week prior to determining formaldehyde release. The free formaldehyde content decreased as the temperature and time of curing were increased, with a temperature of 140 °C and time of 6 hours being required in order to meet the German standards for formaldehyde release. Addition of urea to the MMF resin resulted in a reduction of free formaldehyde content.

Deka and Saikia (2000) treated softwood (*Anthocephalus cadamba*) with methanolic solutions of PF, MF or UF resins. At around 34 % resin loading, the wood samples exhibited volume increases of 14 % (PF), 12 % (MF) and 9 % (UF), which was nearly equal to the calculated volume of polymer added, showing that most of the resin was located in the cell wall. Resin loadings higher than 34 % resulted in no further volume increase of the wood. Increases in ASE also exhibited the same behaviour, with the highest values (around 70 % for PF and MF, and 50 % for UF) found at about 34 % resin loading. Both the MOE and the MOR of treated samples increased with resin loading.

Microdistribution of the Resin Impregnants

The evidence for penetration of the wood cell wall by resin systems was discussed in some detail in a review of wood adhesion by Wellons (1977). Smith and Côté (1971) used SEM–EDXA to determine the microdistribution of a brominated PF resin. When wood was pressure treated with the resin for 12 minutes, this resulted in a concentration gradient across the cell wall, with no resin found in the middle lamella. Nearn (1974) also showed that phenolic resins were capable of penetrating the cell wall, as indicated by fluorescence microscopy, autoradiography and TEM. However, the use of fluorescence microscopy was criticized by Smith and Côté (1971), who stated that sodium hydroxide in the PF resin resulted in fluorescence being observed in the cell wall, which could be erroneously attributed to the presence of PF. Smith and Côté (1972) subsequently used SEM–EDXA of metabromophenol formaldehyde resin treated wood to show that a PF resin was capable of penetrating the cell wall. Bolton *et al.* (1988) presented evidence showing that penetration of the cell wall by a UF resin was possible, by utilizing a thiourea-labelled UF and SEM–EDXA analysis.

Rapp *et al.* (1999) used electron energy loss spectroscopy (EELS) to study the microdistribution of MMF resin in the cell wall in Norway spruce. The authors noted that although SEM–EDXA was suitable for determining whether penetration into the cell wall had occurred, the relatively large excitation volume precluded accurate quantitative determination of the distribution of material in the cell wall. They therefore examined the suitability of TEM–EELS as a method for determining the cell wall distribution of resin. The use of EELS in combination with TEM allows for determination of light elements, such as nitrogen, and potentially has a resolution of the order of 1 nm. Resin-treated wood samples were embedded in epoxy resin and sectioned on a microtome to give cross-sections 80–100 nm thick. The results clearly showed that resin penetration into the cell wall had occurred, and that the S_3 and ML layers contained rather higher levels of resin compared to the S_2 layer. There was also some variation noted in the localized distribution of the resin within the cell wall, as determined by this technique, which had a spatial resolution of 50 nm in these experiments. A more recent study has also shown the usefulness of UV microscopy and confocal Raman microscopy to determine the cell wall penetration of MF resin (Gierlinger *et al.*, 2005).

The Mechanism of Bonding of the Resin in the Cell Wall

Penetration of the swollen cell wall by the low molecular weight resins occurs by diffusion and it is a requirement that during the subsequent drying process the resin constituents are not transported to the lumen surface. Ensuring that a sufficient amount of resin is located within the cell wall is greatly facilitated by allowing sufficient time for the components to diffuse into the cell wall. Preventing the resin from transferring to the exterior of the cell wall requires that some polymerization takes place within the cell wall even before the onset of a curing regime. Two mechanisms are responsible for preventing the loss of the cell wall impregnant once curing has taken place. Polymerization within the complex cell wall micropore geometry effectively 'locks' the resin in place due to entanglement of the resin network structure with the cell wall polymeric constituents. The formation of chemical bonds between the resin components and the cell wall polymers is also a distinct possibility: indeed, Wellons (1977) considers this to be a certainty. Irrespective of the mechanisms, it is essential that the bonds formed (either intramolecular within the resin network or intermolecular between the resin and the cell wall polymers) are stable to hydrolysis; otherwise, the resin will leach and the cell wall bulking effect will be lost.

The Resistance of Resin-treated Wood to Biological Degradation

Stamm and Baechler (1960) studied the microbial resistance of wood modified by five different techniques, including PF resin impregnation. The study was performed in order to determine whether decay protection due to modification was caused by a reduction in the moisture content of the cell wall. Accordingly, the weight loss due to decay of PF-treated wood was plotted against ASE, rather than WPG due to treatment. The plot showed that weight loss due to decay became negligible when the ASE was 70 % or higher. The authors concluded that the decay resistance of PF-impregnated wood arose due to a physical blocking of cell wall OH groups.

As noted above, the physical properties of the resin-impregnated wood also depend upon the molecular weight of the resins used, with higher molecular weight fragments being unable to penetrate the cell wall. If cell wall penetration is an important factor in the decay protection mechanism, it would be expected that similar effects should be seen with decay resistance. The effect of the molecular weight of resin upon the decay resistance of PF-treated wood samples has been the subject of a number of investigations. Takahashi and Imamura (1990) and Ryu *et al.* (1991) reported on a study in which small test blocks of Japanese cedar, western hemlock and Japanese beech were vacuum impregnated with a water-soluble, or ethanol-soluble, PF resin and then heated in an oven to ensure curing of the resin. Treated blocks were exposed to *T. palustris*, or *C. versicolor*, according to JWPA 3–1979 and tested for termite resistance according to JWPA 11–1981. Protection of the treated softwoods in pure culture tests was observed for samples having a water-soluble resin loading of about 12 %, but higher levels of resin loading were required for the ethanol-soluble resin. It was thought that this was due to the lower molecular weight of the water-soluble resin (*c*. 170) compared to the ethanol soluble resin (*c*. 300), resulting in better cell wall penetration of the former. Protection of the hardwood in general required much higher resin loadings than for the softwoods, which was also attributed to differences in cell wall penetration by the resin. Resin treatment was found to impart a significant degree of resistance to attack by termites. Deka and Saikia (2000) also reported that resin-treated wood was resistant to attack by termites. Ryu *et al.* (1993) undertook an extended study of the influence of the molecular weight of the PF resin components upon the biological resistance of treated wood. In this case, seven resins were studied, having molecular weights ranging from 369 to 1143. Higher molecular weight resins imparted a lower degree of decay resistance at lower resin loadings, although differences in efficacy became less apparent as resin loadings were increased. Similar results were also found with respect to resistance to termite attack.

Noting that methylol-substituted melamine-based resins had proved to be the most suitable for wood treatment in an earlier study, Rapp and Peek (1996) went on to investigate the resistance to fungal attack of wood treated with those resins. Small samples were treated with aqueous resin solutions and, following curing, were exposed to *C. puteana* and *T. versicolor*. Some of the samples were also leached according to EN84 prior to fungal exposure. Good protection against fungal attack was achieved with the resin treatment, even after leaching, although treatment with the lowest resin concentration did result in some slight mass loss following decay. The importance of allowing sufficient time for the resin system to diffuse into the wood cell wall was noted by Lukowsky and Peek (1998), who observed an improvement in the decay resistance of MMF resin-treated wood when resin immersion periods were extended to 24 hours and above. Additions of urea to an aqueous MMF solution resulted in improved decay resistance in unsterile soil tests and to *C. puteana* in pure culture tests (Lukowsky *et al.*, 1998). This may be attributable to enhanced cross-link formation in the urea-containing resins. Although pure urea was also found to exhibit an inhibitory effect. Sailer *et al.* (1998) treated Scots pine and beech with aqueous MMF resin, in the presence and absence of NH_4NO_3 as a catalyst, with little effect upon the decay resistance. Van Acker *et al.* (1999) determined the decay resistance of MMF-treated Scots pine and beech in EN113 and ENV807 tests. Resin retentions in excess of 50 kg m^{-3} were found to be required in order to provide adequate protection to the wood. Westin *et al.* (2004b) reported on unsterile soil tests according to an extended ENV807 test on MMF-treated wood samples, with weight losses of 4 % and lower being found. When MMF resin treated wood was exposed in outdoor field trials in Sweden for 8 years, it was found that resin loadings as high as 30 % by weight were required to provide protection, and some decay was observed even at this level at one site. Ritschkoff *et al.* (1999) studied the fungal degradation of a number of resin treatments, including a water-soluble MF resin. Both polymerized and unpolymerized MF resin did not reduce production of cellulase or hemicellulase by *P. placenta*, but did reduce the rate of hydrolysis of wood cellular components by isolated cellulase (Econase CE) enzymes. Production of xylanase was, however, inhibited by resin-treated wood substrates. Lukowsky *et al.* (1999) noted that the presence of free formaldehyde in MMF-treated wood could have an inhibitory effect upon the activities of enzymes.

7.2.2 Dimethyloldihydroxyethyleneurea (DMDHEU)

Unlike the resins discussed in the previous section, DMDHEU (Figure 7.2) is a resin that was developed for the textile, rather than the wood processing, industry. There has been interest in using this resin for wood modification for some years. Weaver *et al.* (1960) appear to have been the first researchers to have studied the modification of wood with this compound, finding rather poor dimensional stabilization. Nicholas and Williams (1987) studied the use of DMDHEU to improve the dimensional stability of wood. They investigated the effect of different curing agents and temperatures upon the dimensional and strength properties of wood specimens. Of the four metal salt catalysts studied, $AlCl_3$ in conjunction with tartaric acid was selected for further investigation, as this metal salt had given the highest ASE in the preliminary investigation and, furthermore, results from textile work had shown that curing could be accomplished at relatively low temperatures and low catalyst loadings. ASE values were determined after leaching, but over one cycle only, so it was not possible to determine the stability of this treatment to hydrolysis. In most cases, although treatments imparted varying degrees of dimensional stability, there was a large reduction in MOR following curing, due to acid degradation of the polysaccharide components of the cell wall from HCl released by the aluminium chloride due to hydrolysis.

Videlov (1989) treated wood samples with an aqueous solution of DMDHEU, using $MgCl_2$ as the curing agent. The modified wood was tested for biological resistance against *C. puteana*, where it was found that good protection was achieved at about 10 % WPG, and that this protection remained even after the modified wood had been subjected to a number of water-leaching cycles.

Ashaari *et al.* (1990a,b) treated sweetgum and southern pine wood samples with aqueous solutions of DMDHEU containing methanesulphonic acid as a catalyst. Following air-drying, the samples were cured at 55 °C, or 80 °C. For treatment at the lower temperature, with sweetgum there was no change in MOE, but there was a 12 % reduction in MOR. There was no significant change in mechanical properties for treated southern pine. Curing at 80 °C resulted in significant reductions in MOR for sweetgum,

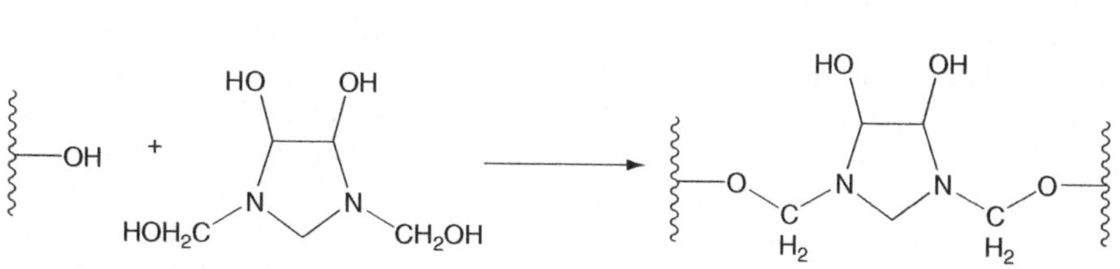

Figure 7.2 Cross-linking of cell wall polymers by reaction with dimethyloldihydroxyethyleneurea (DMDHEU).

and in both MOE and MOR for southern pine. These changes were attributed to acid degradation of the polysaccharide cell wall components. ASE values in the region of 50–60 % were obtained due to DMDHEU treatment, which increased slightly with curing temperature and were higher for the softwood. There was no study of the stability of the treatment to hydrolysis.

Ahmed Kabir *et al.* (1992) treated wood with DMDHEU as well as DMDHEU combined with a vinyl polymer, and determined the dimensional stability of the wood. Methane sulphonic acid was used as a catalyst in both cases. DMDHEU treatment resulted in a 50 % reduction in radial swelling following immersion in water for 100 minutes, with the combined treatment being superior. However, the ASE (one cycle) of DMDHEU treated wood (30 %) was superior to that found for the combined treatment (17 %). DMDHEU appeared to be stable to hydrolysis over a number of wetting cycles. When DMDHEU-treated samples were exposed in outdoor weathering trials, they exhibited considerable variation in moisture content and developed severe surface checks, whereas the combined treatment showed superior performance.

Isocyanate-bonded particleboards were made from DMDHEU-treated albizzia and tested for dimensional stability and biological durability (Yusuf *et al.*, 1995b,d). A WPG of 6.9 % was sufficient to give an ASE of nearly 77 % and resulted in a large improvement in decay resistance. Yusuf *et al.* (1994, 1995b) modified Japanese cedar and beech with DMDHEU and exposed the treated samples to *T. palustris* or *C. versicolor*, according to JWPA 3–1992. Weight loss due to decay was reduced to 5 % at 10 % WPG, except with treated Japanese beech exposed to *T. palustris*. DMDHEU treatment also improved resistance to attack by termites. Yasuda *et al.* (1994) studied the moisture-sorption properties of wood modified with DMDHEU, finding a reduction in EMC as a result of treatment. The creep deformation was also investigated, with the results obtained suggesting that cross-links had formed as a result of treatment. Evidence for cross-linking has also been found in measurements of the dynamic mechanical properties (Minato and Yasuda, 1992; Yasuda and Minato, 1994).

Militz (1993) treated European beech with DMDHEU and evaluated the effectiveness of a variety of commercial catalysts. It was found that temperatures of 100 °C were necessary for effective curing of the resin. The ASE was calculated from the dimensions at 100 % RH in comparison to the sample dimensions at 30 % RH, rather than in a water-soak test, so that any possible effect of leaching was not determined. The highest ASE measured in this way was 75 %. The EMC of the treated samples was also determined and found to be higher than unmodified samples in all cases.

Resistance to fungal deterioration was improved by treatment with the resin, although full decay protection was not observed, contrasting with the studies of Videlov (1989).

Simonsen (1998) investigated the hydrolytic stability of wood treated with DMDHEU and wood impregnated with polyethyleneglycol (PEG) (MW=400), or glycerol cross-linked with DMDHEU. Treatment with DMDHEU alone resulted in a maximum ASE value of 50 %, irrespective of retention. About 25 % by mass of the DMDHEU was lost in the first leach cycle, but the treatment appeared to be stable to leaching subsequently. Leaching of PEG was also reduced by cross-linking using DMDHEU. However, ASE values of PEG- or glycerol-impregnated wood cross-linked with PEG were lower than those obtained with DMDHEU treatment. In one example of a PEG–DMDHEU treatment, the ASE assumed large negative values after the first leach cycle. This was explained on the basis of the DMDHEU–polyglycol cross-linked polymer swelling during the first leach cycle, resulting in internal (i.e. cell wall) damage to the wood.

Zee *et al.* (1998a) studied three different catalyst systems (based on $MgCl_2$, $AlCl_3$ and citric acid) and temperature (100–150 °C) upon the dimensional stability imparted to wood by DMDHEU treatment. The use of catalysts markedly improved dimensional stabilization, with ASEs of the order of 40 % being recorded at a resin loading of 20 % by weight. Higher curing temperatures improved dimensional stability above a resin loading of 30 %, although resin loadings above 80 % resulted in a decrease in ASE. An increase in catalyst concentration also resulted in an increase in dimensional stability. The results indicated that cross-linking was not the mechanism responsible for imparting dimensional stability. The effects of possible leaching of DMDHEU were not discussed.

Van Acker *et al.* (1999) reported on the decay resistance of DMDHEU-treated pine and beech wood determined in EN113 and ENV807 tests. Significant reductions in weight loss due to decay were found in some tests, which was strongly dependent upon the treatment parameters, but not upon resin retention. Krause *et al.* (2004) found that DMDHEU treatment provided good decay protection against *C. puteana*, although there was a mass loss of 2 % when samples were exposed in sterile control leaching studies for 6 weeks.

7.2.3 Impregnation with Furfuryl Alcohol

There has been recent renewed interest in the modification of wood with furfuryl alcohol (FA) (Figure 7.3). One major benefit of using this chemical in wood modification is that it is derived from a renewable resource (primarily corn cobs). In the systems used for wood modification, there are two competing reactions (Figure 7.3, schemes A and B) (Westin *et al.*, 2003). Reaction A usually dominates, with higher reaction temperatures and FA concentrations resulting in a reduction in the formation of ether bridges according to reaction scheme B. The ether bridge is unstable and it is believed that this undergoes further reaction, as shown in scheme C of Figure 7.3. Further cross-linking reactions are

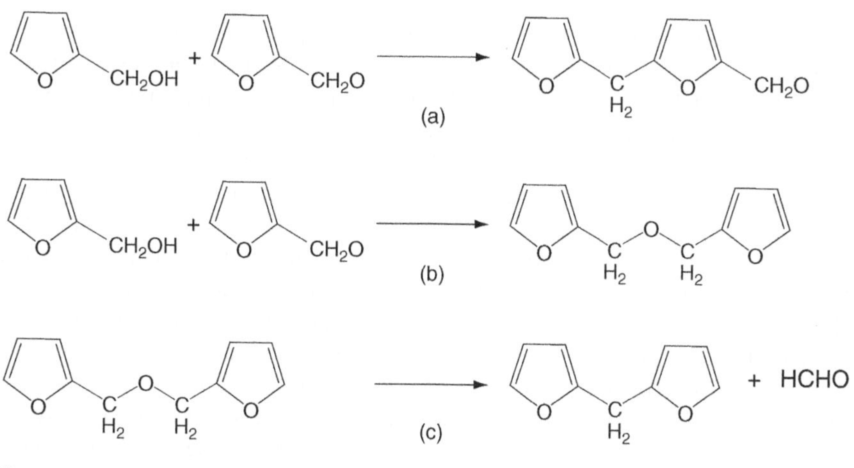

Figure 7.3 *Proposed reactions of furfuryl alcohol, according to Westin* et al. *(2003).*

also possible, resulting in the formation of a three-dimensional polymeric network within the cell wall. The possibility of the furfuryl polymer reacting with the lignin has been suggested, although it has not been proven.

The kinetics of FA polymerization in the presence of wood bound nitric acid has been studied by Balaba and Subramanian (1985).

Early work with FA treatment was reported by Goldstein (1955) and Goldstein and Dreher (1960). Wood was pressure-treated using a solution of monomeric FA with a catalyst present (zinc chloride, or an organic acid). Heating at around 100 °C resulted in polymerization within the wood, to produce acid and alkali-resistant treated wood. Resin contents up to 120 % by weight were obtained, at which level of treatment a 70 % ASE was reported. Treatment of wood with FA resin directly did not result in satisfactory dimensional stability properties, because the resin molecules were too large to penetrate the cell wall. The colour changes of the wood varied from dark brown at lower resin contents to black at higher resin contents. Stamm (1977) reported on the dimensional stabilization, decay resistance and mechanical properties of wood vacuum-treated with aqueous solutions of FA, using zinc chloride, citric acid or formic acid as polymerization catalysts. Curing was accomplished by heating the treated wood specimens at 120 °C. An increase in ASE was noted with increasing resin loadings up to a WPG of 50 %, where a maximum ASE of 70 % was found. Further increases in resin content did not improve the ASE further. Resin treatment resulted in a decrease in MOR, abrasion resistance and toughness and an increase in MOE, which was attributed to acidic degradation of the wood by the catalysts used. FA treatment also resulted in significant improvements in decay resistance, as determined in pure culture tests. A study by Ryu *et al.* (1992) showed that pine and cedar were more effectively protected from microbiological attack by FA treatment than was beech. The treatment was also found to be more effective against white rot (*C. versicolor*) than brown rot (*T. palustris*). Some protection against termite attack (*Coptotermes formosanus*) was also afforded by FA treatment. A pre-impregnation of the wood with boric acid was found to render the subsequent FA treatment more effective in providing decay protection. Furfuryl alcohol modification has recently been used in combination with borate treatment to reduce leaching of the preservative (Baysal *et al.*, 2004).

Apart from the study by Ryu *et al.* (1992), little interest appears to have been shown in furfurylation by the wood research community until the mid-1990s, when research groups in Canada and Sweden both began to assess the possibility of developing a wood modification process based upon furfurylation with a view to commercial exploitation. Schneider (1995) noted that although the use of $ZnCl_2$ as a curing agent in FA treatments was satisfactory in many respects, there remained the problem of a change in hardness and colour from the outside to the inside of the treated wood. Two FA systems (composition not reported) were developed to minimize the formation of such gradients within the wood. Westin *et al.* (2004b) also developed new FA treatments (using citric acid or cyclic anhydrides as catalysts) and determined the biological resistance of FA-treated wood in pure culture tests, unsterile soil, in-ground field tests, subterranean termite tests and marine tests. A WPG of resin of around 100 % provided excellent performance in all cases. Westin *et al.* (2004a) subsequently reported on the mechanical properties of furfurylated wood, finding an increase in hardness of the wood as the level of resin treatment increased, but a decrease in impact toughness. The authors also reported on the commercialization of this process in this paper. Lande *et al.* (2004) investigated the environmental aspects of using furfurylated wood. The growth of fungi was not inhibited by water leachates from the treated wood, nor on growth medium containing furfuryl alcohol. Only low levels of VOCs were detected in emissions to air from furfurylated wood, comprising furfural during early stages of the experiment and FA subsequently. VOC emissions in fire tests were found to be lower for FA-treated wood compared with untreated control samples. Furfurylated wood also performed better in ecological toxicity testing compared to untreated pine controls. It was concluded that FA-treated wood did not have a significant environmental impact in service.

7.2.4 Maleic Acid with Glycerol

Fujimoto *et al.* (1987, 1988, 1991) treated wood flakes with aqueous solutions of maleic acid – glycerol (MG). Particleboards were then made from treated flakes by hot-pressing at 210 °C. Water sorption and thickness swelling decreased with increased loading of MG. This treatment has also been applied to solid wood (Fujimoto and Yamagishi, 1991; Fujimoto, 1992) (Figure 7.4). Volume increases due to MG treatment were higher when water was used as a delivery solvent compared with methanol, due to the greater capacity of water for cell wall swelling. Good dimensional stability was obtained, even with low levels of MG treatment (a 10 % WPG resulted in an ASE of 80 %), and the treatment was stable to leaching by water. MG-treated wood exhibited good performance in outdoor weathering trials. However, the use of the high curing temperature of 160 °C did lead to the wood becoming brittle.

7.2.5 Maleic Anhydride with Polyglycerol

Roussel *et al.* (2001) reacted polyglycerol with maleic anhydride and impregnated wood with aqueous solutions thereof. Methyl ethyl ketone peroxide and cobalt napthenate were also present to promote polymerization. After heating of the treated wood to form bound polymer, there was a significant weight loss observed due to loss of water. Treated blocks exhibited good dimensional stabilization (around 48 % ASE). Impregnation of the wood followed by polymerization resulted in an increase in MOE. The impregnation modification improved the decay resistance of the wood, as determined in pure culture tests against *C. versicolor* and *P. placenta*.

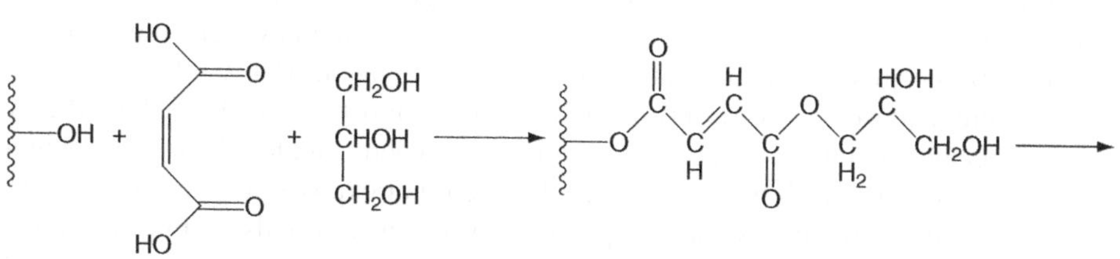

Figure 7.4 Reaction of maleic acid and glycerol with cell wall polymers.

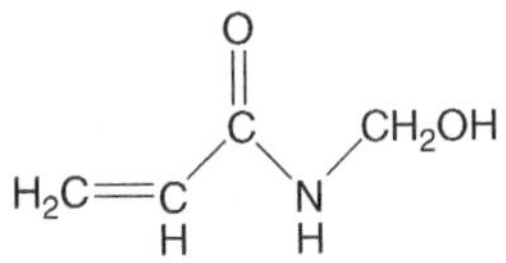

Figure 7.5 The structure of N-methylolacrylamide.

7.2.6 *N*-methylolacrylamide

Rijckaert *et al.* (2000) reported on the potential of *N*-methylolacrylamide (NMA) (Figure 7.5) as a wood modification agent. Since NMA is a bifunctional molecule, this allows for the possibility of the hydroxy methyl group reacting with the cell wall polymers, with subsequent cross-linking via the vinyl functionality. However, a significant problem with this reagent is that it is a neurotoxin and is carcinogenic, although it was noted that the polymer formed therefrom is nonhazardous. Wood samples which had been treated with an aqueous solution of NMA and then cured were subjected to ecological toxicity testing, in order to determine the potential environmental hazards that such modified wood presented in service. Leachate from the treated wood was found to exhibit ecotoxicity, and NMA was found in the leachate, although this was strongly influenced by the curing regime employed. In view of the handling difficulties associated with using this reagent in large quantities, it is unlikely that this modification system will be used in a commercial process.

7.3 Impregnations using Silicon-containing Compounds

7.3.1 Inorganic Silane Impregnations

Chlorosilanes

Owens *et al.* (1980a, 1980b) suggested that chlorosilanes might be suitable for use as wood preservatives. These authors noted the high reactivity of $SiCl_4$ towards hydroxyl groups in water and alcohols, resulting in the formation of silicic acid. This could then be decomposed upon heating to give a cross-linked cell wall bound silica gel with Si–O–C linkages to the cell wall polymers. The production of HCl as a by-product was noted, but the potential for acid-catalysed degradation of the polysaccharide content of the wood was not discussed. Oven-dry wood samples were immersed in $SiCl_4$ for 60 seconds and then heated to constant weight at 104 °C. Treated wood samples were exposed to a variety of fungi and in most cases showed markedly improved decay resistance compared with untreated wood samples. However, weight losses were also noted with treated samples that were not exposed to fungi, which was thought to be due to loss of residual trapped HCl over longer periods of time. The presence of this trapped HCl would also have a significant inhibitory effect upon wood decay fungi.

Stevens (1981, 1985) also reported on the potential for the use of chlorosilanes as possible wood preservatives. Wood samples were treated with $SiCl_4$, methyltrichlorosilane (CH_3SiCl_3), dimethyldichlorosilane ($(CH_3)_2SiCl_2$), trimethylchlorosilane ($(CH_3)_3SiCl$) or methyldichlorohydrogensilane (CH_3SiHCl_2). Due to the potential for degradation of the polysaccharide components of the cell wall by evolved HCl during the reactions, the wood was in most cases pre-treated with basic amine acid scavengers for vapour-phase silicon reagent treatments, or the silicon reagent was dissolved in amine solutions. After treatment and curing, the treated wood samples were leached in water for 3 days. With many of the treatments, only small weight gains were found: indeed, with some $SiCl_4$ treatments, weight losses were recorded, which were attributed to acid-catalysed degradation of the wood. In other cases, variable weight gains were recorded due to incomplete removal of solvents. The treatments were not found to result in significant improvements in decay resistance. In view of the poor performance coupled with the need for dry wood and the difficulty of handling the reagents, the authors concluded that such treatments were not suitable candidates for use in wood preservation.

Zollfrank (2001) reacted beech wood with trimethylsilyl chloride under anhydrous conditions in tetrahydrofuran in the presence of pyridine. Pyridine was present to scavenge the HCl produced during the reaction, thus limiting polysaccharide degradation. After reaction at room temperature for 12 hours, the wood was extracted with methanol to remove the pyridinium chloride by-product. SEM of the modified wood indicated that there was no evidence of cell wall degradation.

Alkoxysilanes

Treatment of wood with tetraalkoxysilanes followed by hydrolysis and curing will result in wood impregnated with SiO_2 sol-gel networks, without the production of destructive acidic by-products (Figure 7.6) (Mai and Militz, 2004a). In conventional sol-gel manufacture, very high temperatures are used (800 °C), which is clearly impractical for a wood modification method. Saka *et al.* (1992) studied the impregnation of wood with a solution of tetramethoxysilane (TMS) in methanol, tetraethoxysilane (TES) in ethanol or tetrapropoxysilane (TPS) in *n*-propanol, containing catalytic amounts of acetic acid. The wood was in contact with the impregnation solutions for up to 2 weeks to allow sufficient time for the alkoxysilane to diffuse into the cell wall. After diffusion treatment, the samples were dried and subsequently cured in an oven at 105 °C. SEM–EDXA of the

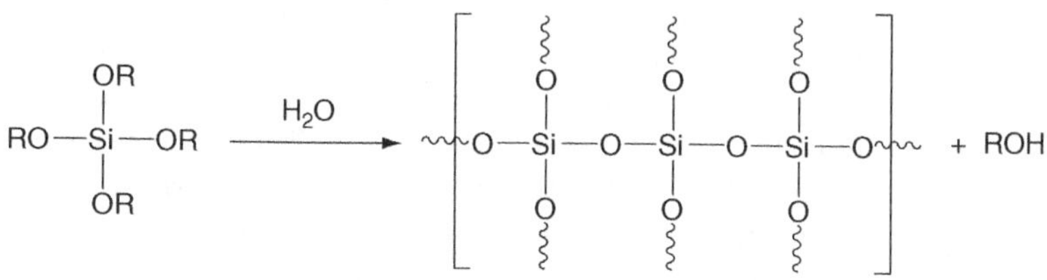

Figure 7.6 Reaction of a tetraalkoxy silane with water and subsequent polymerization to form an SiO_2 sol-gel network.

treated samples showed that when the wood was water saturated prior to impregnation with the alkoxysilane, the SiO_2 gel was mainly found in the cell lumen, whereas if the wood specimens were moisture conditioned before treatment, the SiO_2 gel was located in the cell wall. The WPG increased in proportion with the MC of the wood up to the FSP, but decreased in the order TMS > TES > TPS. The decrease was explained as being due to the reduced swelling capacity of the alcohols as the molecular weight decreased and the increase in molecular size of the alkoxysilanes as molecular weight increased. The WPG due to treatment also increased with longer immersion times in the alkoxysilane solution. The ASE of the treated samples increased in proportion with WPG, reaching 40 % ASE at 10 % WPG. The presence of the silica gel in the cell wall, or in the lumen, reduced the flammability of the modified wood.

In a subsequent paper, the effect of an ultrasonic irradiation during impregnation upon the distribution of silicon in the wood sample was studied (Ogiso and Saka, 1993). The use of ultrasonic irradiation during treatment appeared to have little effect upon the distribution of Si in samples, but did result in a slight increase in WPG with moisture-conditioned wood samples. With moisture-conditioned wood samples, significant improvements in ASE were found at WPGs less than about 15 %, but with water-saturated wood samples, negative ASEs were found up to about 70 % WPG. Increases in WPG resulted in improved resistance to flammability, particularly if the SiO_2 gel was located in the cell wall. The treated wood was also found to exhibit enhanced resistance to termite attack (*R. speratus*).

Tanno *et al.* (1997, 1998) treated moisture-conditioned wood samples with TES. The impregnated wood specimens were subsequently cured in an oven to achieve a WPG of 6.5 %. Although the rate of water absorption was reduced by this treatment, the ultimate extent of absorption was the same as for untreated wood. However, the use of co-additions of other silanes containing hydrophobic groups did reduce the total amount of water absorbed. In fungal decay tests, the presence of SiO_2 gels in the wood reduced hyphal penetration into the cell walls. The treatment did not, however, offer any protection to the wood, because the rate of weight loss due to fungal attack in unsterile soil was only slightly less than that for untreated wood. In pure culture tests, weight loss due to fungal attack by the white rot fungus *C. versicolor* was actually greater than in the unmodified wood, but was lower for brown rot attack (*T. palustris*). Some researchers have used SiO_2 gel formation in combination with other reagents, such as P_2O_5/B_2O_3 (Miyafuji and Saka, 1996; Saka and Tanno, 1996). Although these systems exhibit good resistance to flammability, they are not stable to leaching (Miyafuji *et al.*, 1998; Miyafuji and Saka, 2001).

Saka and Ueno (1997) considered that TES was not a suitable reagent to use in operational environments and suggested the use of methyltrimethoxysilane (MTMS) instead. Wood specimens were prepared at MCs of 20–25 % and impregnated with MTMS in methanol and acetic acid for 3 days. The impregnated wood samples were then cured in an oven for 2 days. A WPG of around 8 % was obtained, resulting in an increase in the volume of the treated wood of nearly 2 %. This result was comparable with a TES treatment, but with a much lower concentration of MTMS in the treatment solution. However, treatments with MTMS or TES were found to be susceptible to leaching by water. Accordingly, a variety of co-additives were investigated, of which 2-heptadecylfluorooctylethyltrimethoxysilane was found to be the most effective at limiting leaching.

Miyafuji and Saka (2001) treated moisture-conditioned wood with a methanolic solution of MTMS in combination with sodium ethoxide, or sodium acetate, to form Na_2O–SiO_2 gels. The gels were found to be mainly located in the cell wall and improved the fire resistance of the wood. The use of benzophenone in combination with a Si-gel treatment has been found to improve the UV stability of wood (Miyafuji *et al.* 2004).

Donath *et al.* (2004), treated moisture-conditioned wood with ethanolic solutions of the alkoxysilanes or aqueous sols of TES, methyltriethoxysilane (MTES) or propyltriethoxysilane (PTES). Bulking of the wood due to silane treatment was determined and was found to be greater for the ethanol-delivered alkoxysilanes, compared to the equivalent sols. This trend was mirrored in the ASE determinations, in that the samples exhibiting high bulking also showed high ASEs. The presence of an organic moiety attached to the silicon atom improved bulking and ASE. Both bulking and ASE decreased as a result of curing of the silane-treated wood. The durability of the treated wood was only slightly improved by these silane treatments in unsterile soil test, although in pure culture tests (*T. versicolor*), silane treatments performed rather better than sol treatments.

7.3.2 Silicate Impregnations

Furuno *et al.* (1991, 1992, 1993) impregnated wood with aqueous solutions of sodium silicate, followed by treatment with aqueous solutions of aluminium sulphate, barium chloride or calcium chloride to form a nonleachable precipitate. Although most of the precipitated material was found in the lumens of the wood samples, there was evidence of cell wall penetration from SEM–EDXA studies. Treatments increased the hygroscopicity of wood, due to the presence of the inorganic material, imparted limited decay resistance and improved resistance to combustion, but reduced the bending strength of the wood.

Noting that previously reported studies into inorganic impregnations of wood either required a two-step process or the use of expensive reagents or solvents, Yamaguchi (1994a, b) proposed the used of silicic acid, or cation-exchanged sodium silicate, in a one-step process. Preliminary studies indicated that a colloidal silicic acid preparation containing phosphoric acid, or a cation-exchanged sodium silicate solution, might have potential for wood treatment. Sugi wood specimens were impregnated with either of these systems for 7 hours. Treatments improved the dimensional stability of the wood, particularly with a low concentration cation-exchanged sodium silicate treatment, where it was thought that a degree of cell wall penetration had occurred. Yamaguchi (2002) also investigated the decay resistance of wood treated with the two treatment systems and the systems in combination with boric acid.

7.3.3 Impregnation with Silicones

Silicones represent a class of compounds consisting of chains of Si – O – Si units, upon which various organic substituents (R) are attached to the Si atoms, with reactive groups situated at the chain ends (Figure 7.7). The generally high molecular weight of these systems precludes the possibility of cell wall penetration.

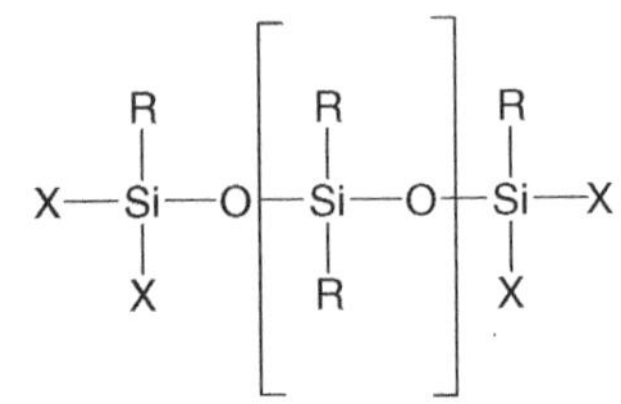

Figure 7.7 The generic structure of a silicone.

Hager (1995), noting that water-based silicone systems had been used on masonry and concrete to provide water repellency, investigated their potential as wood protection agents. He mentioned the work of E.G. Rochow, who built a house in 1958 using cedar that was treated with a silicone solution as a water repellent. After 28 years, the silicone protective layer was still sound. Hager treated wood with water-based silicone micro-emulsions and found that good water repellency could be obtained.

Lukowsky *et al.* (1997) studied five water-based silicone systems as possible candidates for the improvement of wood properties. Four of the systems were micro-emulsions and one was a silicone solution. EN113-sized samples with end grains sealed were vacuum impregnated with the silicone emulsions, or treated by dipping in the silicone solution. After impregnation, they were stored at 20 °C and 65 % RH for 4 weeks. The swelling of the treated wood samples in water was measured. Treatment using the silicone solution was only partially effective, in that extended water-soaking resulted in loss of water repellency. This was because this silicone formed a film on the wood that became detached during the tests. One commercial system showed promise in terms of water repellency.

7.3.4 Organo-silane Impregnations

Organo-silane impregnations require the presence of an organic group attached to the silicon atom that may be inert, or may be capable of participating in a chemical reaction either with the cell wall polymers, with other silane molecules or with a co-additive (Mai and Militz, 2004b). It is often assumed that hydrolysis of methoxy, or ethoxy, groups of the organo-silane produce silanols, which then react with the cell wall (Figure 7.8). However, it is important to note that, even if such a reaction takes place, the Si—O—C bond between the silane and the cell wall polymeric groups can be hydrolysed very

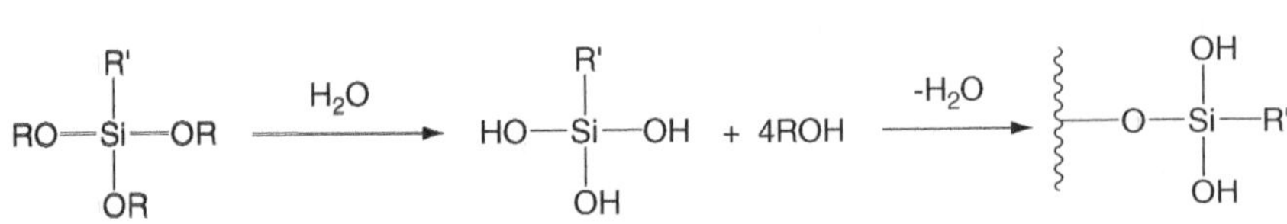

Figure 7.8 The proposed reaction of a hydrolysed mono-organo trialkoxysilane with cell wall polymers.

easily, resulting in the regeneration of silanol. The resistance of the organo-silane to leaching by water then relies upon either bond formation with the cell wall polymers mediated by the organo-functional group, or due to entanglement of the silane polymer within the cell wall matrix.

Schneider and Brebner (1985) impregnated wood samples with the organo-silane γ–methacryloxypropyltrimethoxysilane (TMPS) (Figure 7.9a) and determined the dimensional stability as a result of this treatment. Wood samples were conditioned to 6–8 % MC before being impregnated with a methanolic solution of the silane. The methanol was allowed to evaporate from the treated samples and they were then submerged in water to

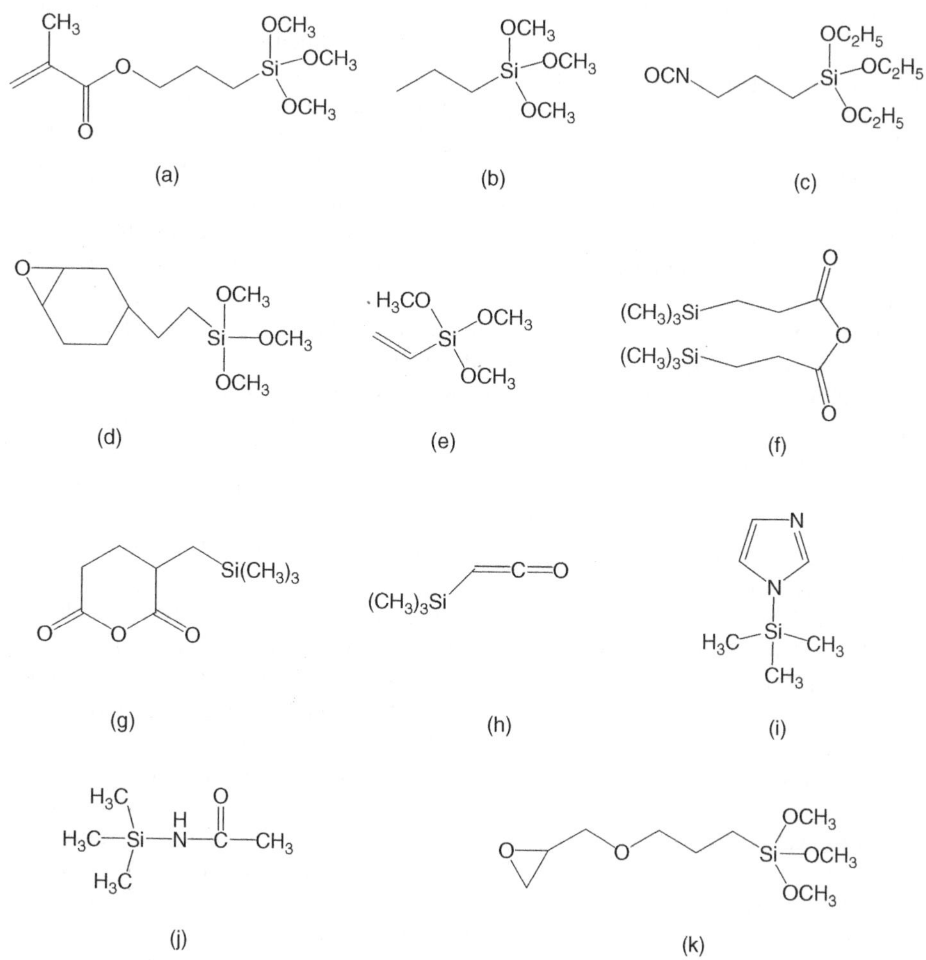

Figure 7.9 *Structures of γ–methacryloxypropyltrimethoxysilane (TMPS) (a), n-propyltrimethoxysilane (PTMS) (b), 3-isocyanatopropyltriethoxysilane (IPTES) (c), β-(3,4, epoxycyclohexyl)trimethoxysilane (EETMS) (d), vinyltrimethoxysilane (VTMS) (e), 3-trimethylsilylpropanoic anhydride (f), 2-trimethylsilylmethylglutaric anhydride (g), trimethylsilylethenone (h) 1-(trimethylsilyl)imidazole (i), N-(trimethylsilyl)acetamide (j) and 3-glycidoxypropyltrimethoxysilane (k).*

hydrolyse the methoxy groups. The samples were then cured by heating at 110 °C. ASE increased with WPG due to treatment, with values as high as 70 % being found for a WPG of 25 %. Calculated volume increases were higher than observed for the highest WPGs, showing that some of the chemical remained in the lumen. However, it is important to note that relying upon the measurement of external dimensions to calculate changes in cell wall volume is not a dependable method (Hill and Ormondroyd, 2004). When TMPS-treated wood was subjected to repeated WS/OD cycles there was some reduction in ASE, but the treatment showed good stability to hydrolysis (Brebner and Schneider, 1985).

Goethals and Stevens (1994) impregnated small beech and pine wood samples with an aqueous emulsion of *n*-propyltrimethoxysilane (PTMS) (Figure 7.9b) for 30 minutes at room temperature, followed by curing at 100 °C for 48 hours. No solvent extraction was employed after treatment. The treated samples were subjected to five water-soak/oven-dry cycles, with only slight loss of the silane. ASE improvements due to treatment were moderate, ranging from 21 % (at 12 % WPG) to 27 % (at 24 % WPG) for treated pine, although treated beech showed rather higher ASE values at lower WPGs (ASE 31 % at WPG 8 %; ASE 35 % at WPG 11 %), indicating that cell wall penetration of the beech by the silane was more efficient. Only slight reductions in EMC were found, compared to unmodified samples at the same RH. This poor performance in terms of improved hygroscopicity was considered to be due to a low level of substitution of the cell wall polymeric OH groups. The general ineffectiveness of the PTMS treatment was also reflected in the high weight losses recorded in decay tests.

Saka and Yakake (1993) chemically modified wood samples with 3-isocyanatopropyltriethoxysilane (IPTES) (Figure 7.9c). Wood was impregnated with acetone solutions of IPTES of different concentrations. Where the wood had been pre-swollen by treating with pyridine vapour, a maximum WPG of 27 % was obtained, whereas without pre-swelling of the cell wall a WPG of only 7 % was obtained. Pyridine also serves as a catalyst for the reaction of isocyanate with wood. SEM–EDXA showed the presence of silicon in the cell walls of the treated wood. Because of the lower WPG obtained without cell wall swelling, ASEs of the order of 20 % were obtained, compared to about 60 % when swelling of the cell wall was used prior to reaction.

Ogiso and Saka (1994) impregnated wood with IPTES, β-(3,4, epoxycyclohexyl) trimethoxysilane (EETMS) (Figure 7.9d), vinyltrimethoxysilane (VTMS) (Figure 7.9e) or TMPS. BPO was used as a curing catalyst for VTMS and TMPS, and triethylamine with EETMS. Oven-dry samples were used for reaction and the silanes were dissolved in either acetone (IPTES, EETMS), or methanol (VTMS, TMPS) for the impregnation phase (7 days). Curing was accomplished by heating of the samples for various time intervals, followed by solvent extraction to remove unreacted silane. With the exception of TMPS (WPG=43 %), only low WPGs (below 10 %) were found after curing and solvent extraction (using either methanol or acetone). Subsequent extraction with the nonswelling solvent toluene did not remove any of the silanes from the wood. However, water extraction did significantly reduce the WPG of VTMS-treated wood, although weight reductions with the other silanes were slight. Although low WPGs were recorded for three of the silane treatments, nonetheless ASE values as high as 30 % were reported (no data was given for TMPS), indicating that the silane was located in the cell wall. Further increases in ASE were found when the silane-modified woods were subsequently impregnated with tetraethoxysilane to form cell wall bound SiO_2 gels.

Rozman *et al.* (1997a) treated rubberwood by impregnation with a methanolic solution of TMPS for 24 hours, followed by curing at 110 °C for 5 hours. No solvent extraction was employed after treatment. Volume increases due to treatment were considerably lower than predicted theoretically, showing that little cell wall penetration had occurred. ASE measurements were made over two cycles, and it was found that ASE increased in the second cycle. This odd behaviour was not explained and the experimental details for ASE determination were not given.

Sèbe and De Jéso (2000) reacted wood with 3-trimethylsilylpropanoic anhydride (Figure 7.9f), 2-trimethylsilylmethylglutaric anhydride (Figure 7.9g) or trimethylsilylethenone (Figure 7.9h). Dimensional stability due to reaction was attributed to bulking of the cell wall, rather than the number of OH groups reacted. The wood reacted with trimethylsilylethenone was not stable to hydrolysis.

Zollfrank and Wegener (2002) modified triethylamine impregnated beech wood with 1-(trimethylsilyl) imidazole (Figure 7.9i) in tetrahydrofuran under scrupulously dry conditions. White fir samples were reacted with *N*-(trimethylsilyl) acetamide (Figure 7.9j), using trimethylchlorosilane as a catalyst. FTIR was used for confirmation of reaction, and the distribution of silicon in the wood was determined using SEM–EDXA, and in the cell wall using TEM–EDXA. TEM showed that the silicon was concentrated on the lumen side of the cell wall and that the reaction front progressed into the cell wall from the lumen.

Hill *et al.* (2004a) modified wood with TMPS or VTMS and studied its physical and biological properties. On the basis of a preliminary study, different treatment protocols were used for the two silanes. Reaction of TMPS was found to occur via free-radical initiated polymerization of the double bond of the methacrylic moiety, requiring a free-radical initiator and a low wood moisture content. The highest WPGs achieved with VTMS required a high wood moisture content, but did not require a free-radical initiator. VTMS did not polymerize via the double bond of the vinyl group, but by condensation of silanol groups formed by hydrolysis of the methoxy groups. The wood increased in volume due to treatment up to about 20 % WPG, with little change thereafter. Maximum swelling was approximately half of the water-saturated volume of the wood, showing that full bulking of the cell wall was not achieved. Maximum ASE values obtained were of the order of 40 %, again a reflection of the incomplete filling of the cell wall. Weight losses during the ASE tests were relatively low, indicating that the treatments were stable to hydrolysis (samples were Soxhlet extracted after modification to remove unreacted silane and the TMPS-treated wood was pre-hydrolysed to remove OCH_3 groups prior to ASE tests). The resistance of treated wood was determined in pure culture tests using EN113 protocols (*C. puteana*, *T. versicolor* and *P. chrysosporium*) and in unsterile soil (ENV807), where soft rot was the predominant form of attack. Neither treatment afforded protection against *C. puteana*, but protection against the two white rot fungi was achieved, albeit at high WPGs (40–50 %). Protection in unsterile soil was realized at about 15 % WPG with VTMS treatment, but full protection was not achieved with TMPS treatment, even at 70 % WPG.

Sèbe *et al.* (2004) reacted wood with IPTES, 3-glycidoxypropyltrimethoxysilane (Figure 7.9k) or PTMS. Slight swelling of the wood samples was noted after modification, with moderate improvements in ASE (no greater than 25 %). Some loss of material was found during a series of water-soak tests. The authors concluded that there had been reaction between the reactive organic portion of the silanes and the wood. The reactions appear to have been carried out in essentially dry conditions, although the use of DMF as a solvent would presumably have swollen the cell walls sufficiently to allow access of reagent. The rather poor ASE values may indicate that there had been a low amount of cell wall penetration by the reagent molecules.

7.4 Other Inorganic Cell Wall Precipitation Treatments

Ohmae *et al.* (2002) treated wood using a double diffusion treatment, by immersion in a solution of $BaCl_2$ and boric acid, followed by a second treatment using $(NH_4)_2HPO_4$ and boric acid. WPGs of the order of 30 % were obtained, but ASE was not improved significantly and became negative in some cases.

Apart from alkoxysilanes, several other metal alkoxides (e.g. Ti, B and Zr) have been used to form metaloxane gels in wood (Miyafuji *et al.*, 1998). Treatment of wood with tetraisopropoxytitanium in order to form cell wall bound TiO_2 gels was not found to be successful, due to rapid hydrolysis leading to gels forming in the lumen only (Saka and Yakake, 1993). Miyafuji and Saka (1997) therefore undertook a study of a range of titanium compounds that might exhibit a lower rate of hydrolysis, in order to see whether it was possible to form TiO_2 networks in the cell wall. The titanate reagents were dissolved in isopropanol or butanol, and moisture-conditioned wood samples impregnated with these solutions for 3 days. The impregnated samples were then cured in an oven for 2 days. By using SEM–EDXA to map the titanium distribution, it was concluded that formation of titanate gels in the cell wall was facilitated when the rate of hydrolysis of the titanium compound was lower. The fire resistance of the titanate-treated wood was superior to that of unmodified wood.

7.5 Cell Wall Impregnation with Monomers

There has been extensive research into the impregnation of wood with monomers followed by a polymerization reaction to form wood polymer composites. In many cases, this results in lumen filling of the wood only. This section will be concerned with those studies in which impregnation into the cell wall has occurred. Since cell wall penetration is accompanied by swelling, an increase in the external dimensions of the wood has been taken as evidence that at least some of the polymer is located within the cell wall. Where bulking of the cell wall occurs, this is accompanied by a reduction in the swelling of the wood when immersed in water, or exposed in a high-humidity environment. Lumen filling would not be expected to lead to swelling of the wood, or any increase in dimensional stability. This was found in a study by Schneider *et al.* (1991), who investigated the swelling properties of a lumen-filled and a cell wall filled wood polymer composite. Polymerization within the cell wall resulted in a lower degree of swelling when the modified wood was immersed in water, whereas the lumen filled wood exhibited the same ultimate swelling as unmodified wood. FSP was determined from the wood MC at maximum swelling and was significantly lower for the cell wall filled material.

Shiraishi *et al.* (1972) studied the polymerization of methylmethacrylate (MMA) in the cell wall of wood. The presence of small amounts of CCl_4 in the MMA–water system increased the degree of polymerization, as well as the weight increase in the wood, with a decrease in both values with higher concentrations of CCl_4. SEM studies showed that the MMA-treated wood had a high degree of cell wall swelling. Although the hygroscopicity of the wood was reduced by over 50 % at high WPGs (90 %), the reduction in moisture content of the wood substance (taking account of the weight gain due to polymerization) was minimal. The dimensional stability of the treated wood was not reported.

Nakagami and Yokota (1983) impregnated wood with a solution containing methacrylic acid, trifluoracetic acid and sulphuric acid, to form a covalent bond with the cell wall polymers. The methacrylic-reacted wood was then impregnated with styrene, or methylmethacrylate, to form cross-links with the reacted cell wall polymers. Improved dimensional stability was obtained, although degradation of the wood was also observed.

A solution of styrene in methanol to impregnate wood samples, followed by polymerization, was used by Furuno and Goto (1979). Penetration of the monomer into the cell wall was determined by solvent extraction of samples after polymerization. This removed lumen located polymer, whilst leaving the cell wall bound polymer in place. This showed that the concentration of cell wall bound polymer increased in proportion to the monomer content in methanol, up to a maximum of 80 % of the monomer in the solvent. No cell wall penetration was observed for treatment with neat monomer. This was also found for bulking of the wood, as determined from external dimensions of the samples. Improvements in ASE were obtained as a result of the presence of cell wall bound polymer. To achieve similar ASE values with lumen located polymer required very high polymer loadings.

Timmons *et al.* (1971) used SEM–EDXA (with dichlorostyrene monomer) and autoradiography (with tritium-labelled methylmethacrylate) to investigate the microdistribution of vinyl polymer in the cell wall of the monomer-treated wood. One set of samples was solvent exchange dried using ethanol, another set dried was from anhydrous pentane and a third set was oven-dried. Solvent exchange allowed for the most efficient penetration of the cell wall by the vinyl monomer, drying from pentane resulted in a slower rate of penetration and oven-drying resulted in very little monomer penetration. Furuno and Goto (1979) used SEM–EDXA to examine cell wall penetration of bromostyrene. Impregnation with a 100 % concentration of monomer resulted in essentially no cell wall penetration, whereas cell wall penetration was observed when an 80 % monomer solution in methanol was used for treatment.

7.6 Cell Wall Impregnation with Polymers

Fujimura *et al.* (1990) synthesized acrylic copolymers from 2-hydroxyethyl methylmethacrylate (HEMA) and methylmethacrylate (MMA) monomers (Figure 7.10) having an average molecular weight of 20 000 (range of molecular weight from 3000 to 60 000). Wood specimens were treated with a solution of copolymer dissolved in acetone and then mixed with a blocked isocyanate, which acted as a cross-linking agent. The ASE values of the treated wood were dependent upon the composition of the copolymers, reaching a

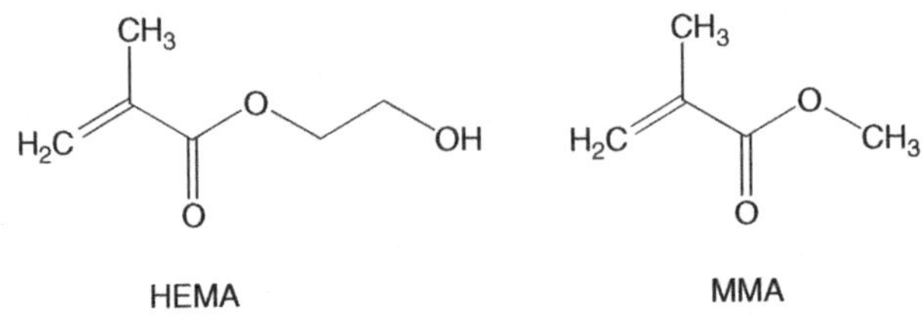

Figure 7.10 *Structures of 2-hydroxyethyl methylmethacrylate (HEMA) and methylmethacrylate (MMA) monomers.*

maximum (up to 60 %) at around 10 % HEMA in the copolymer. It was thought that a major contributor to the dimensional stabilization was penetration of the acetone-swollen cell wall by lower molecular weight components of the copolymer. Affinity with the wood cell wall was improved by increasing the OH content of the copolymer and dimensional stability was improved by increasing the isocyanate content. There is some indication that leaching of samples may have occurred during the ASE determination. Changes in ASE were linearly correlated with bulking and were dependent upon the copolymer composition (Fujimura and Inoue, 1992). The decay resistance of copolymer-treated wood blocks was also found to be dependent upon the composition of the copolymer (Fujimura *et al.*, 1993a). The weight average molecular weights of the copolymers were of the order of 2×10^4, and the radius of gyration of the copolymers (from intrinsic viscosities) was found to range from 2.9 nm to 3.6 nm (Fujimura *et al.*, 1993b). This would appear to be close to the maximum dimension (as determined by solute exclusion) to allow for cell wall penetration, although conformational changes of the polymer could lead to improved cell wall accessibility. Additionally, lower molecular weight fractions of the copolymer would be able to penetrate the cell wall with greater efficiency. The reduced efficiency of acetone as a cell wall swelling solvent compared to water would have also had a limiting effect upon the ability of the copolymers to enter the cell wall. Fujimura and Inoue (1991) noted that the presence of unreacted OH groups on the HEMA-based copolymers would probably limit the durability of treatment. They accordingly studied a copolymer system of GMA and MMA that, it was thought, would not be subject to this disadvantage. The copolymers (average molecular weight 13 000) were prepared with different GMA contents, and acetone solutions of these were used to impregnate wood at various moisture contents. The weight gain due to polymer treatment increased as the wood MC was increased, with ASEs in the region of 50 % being found for the highest copolymer loadings.

Soulounganga *et al.* (2004) impregnated wood with polyglycerolmethacrylate (PGMA) (Figure 7.11) in aqueous conditions, in the presence of methyl ethyl ketone peroxide as a free-radical initiator, followed by curing at 80 °C for 3 days. By using different solution concentrations of PGMA, a range of WPGs from 10 % to 70 % were obtained. ASE increased with WPG (indicating that the polymer was located in the cell wall) and the modified wood was found to be relatively resistant to leaching by water. Decay resistance, as determined in pure culture tests, was much improved by PGMA treatment.

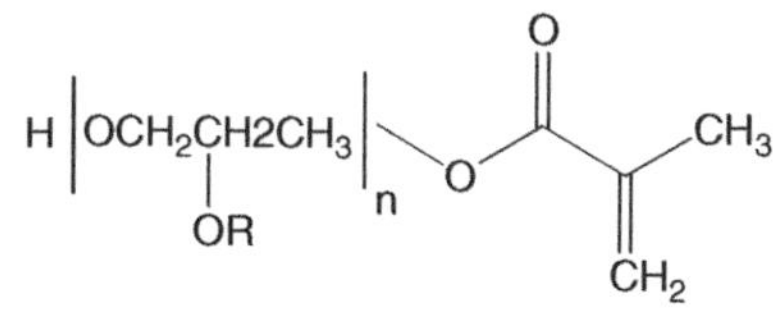

R = H, or $CH_2CHOHCH_2OH$

Figure 7.11 *The structure of polyglycerolmethacrylate.*

7.7 Conclusions

The only treatments that are likely to be viable commercially are aqueous delivery systems. Of the systems described in this chapter, furfurylation is the most advanced commercially and appears to show great promise. There has also been the recent introduction of the DMDHEU-based modified wood ‘Belmadur’ on the market by BASF. At the present time, no other systems appear to offer any immediate prospects for commercial exploitation. The use of silicone treatments has apparently received little attention, which is very surprising due to the ready commercial availability of these systems for masonry treatment. Whether this apparent lack of activity is due to an oversight, represents a lack of real potential or perhaps is due to commercial sensitivity will become clearer in the future. However, silicone treatments are confined to the wood surface only and are not capable of penetrating the cell wall, and would therefore provide little improvement in dimensional stability. Similarly, no significant improvement in biological durability would be expected, since the relatively thin envelope of the treatment would be breached easily. However, the use of silicones in combination with other treatments that may be leachable in service (e.g. borates) would be an area well worth exploring.

Impregnation modification is an area of research that is relatively unexplored compared to other wood modification methods and there are undoubtedly many other systems that remain to be studied in the future.

8

Commercialization of Wood Modification

8.1 Introduction

A long established wood modification industry has been in place since the 1930s manufacturing densified wood niche products. The past decade has seen considerable commercial activity in the commodity product wood modification sector, largely focused in Europe. This reflects the change in the legislative climate affecting the wood protection industry. In Europe, Finland has seen the greatest advances, with commercialization of thermal modification, France, Austria, and The Netherlands have also developed commercial processes. There have been developments at the pilot-scale level with acetylation (The Netherlands and Sweden) and furfurylation (Norway).

8.2 Thermal Modification

Stamm and co-workers developed a heat-treated wood material known as Staybwood, on the basis of research undertaken at the Forest Products Research Laboratory in Madison, Wiseconsin, USA (Stamm *et al.*, 1946). However, this process was not subject to any commercial development. In contrast, the development of processes in which there is a combination of heat with compression has resulted in industrial production. Stamm and co-workers also developed a process in which wood was heated under pressure (Seborg *et al.*, 1945). An essential part of this process was the need to press wood at a sufficiently high process temperature and moisture content to ensure that the lignin flowed during processing, to relieve internal stresses during compression. The compressed wood was then held under pressure whilst the wood cooled. Aircraft propeller blanks were prepared by this method, although there does not appear to have been any commercial development. Compression of solid wood and laminated compressed wood has been performed in Germany since the 1930s under the trade name of Lignostone. Lignostone is still being

Wood Modification: Chemical, Thermal and Other Processes C. Hill

manufactured in Germany by Röchling. Similar products, known as Jicwood and Jablo, have been manufactured in the UK.

8.2.1 ThermoWood®

ThermoWood is the registered trademark name owned by the Finnish ThermoWood Association, applied to heat-treated wood produced according to a methodology developed by VTT in Finland. With the exception of the long-established Compreg, heat treatment is the most advanced wood modification process commercially, with the first heat-treatment plant being built at Mänttä in Finland in the early 1990s. By the year 2000, there was production capacity of 50 000 m^3 and this had increased to 60 000 m^3 by 2001: it currently stands at around 75 000–80 000 m^3. Sales production, according to the Finnish ThermoWood Association, was about 18 800 m^3 in 2001, 24 600 m^3 in 2002 and 21 600 m^3 in 2003, and exceeded 31 000 m^3 in 2004. Raw material consumption was 25 800 m^3 in 2003 and 35 000 m^3 in 2004. Sales of ThermoWood products have been strong in Finland and there have been increasing quantities sold in Europe, in which the main consuming countries are France, the Benelux group and Germany (Figure 8.1). The ThermoWood process is licensed to members of the Finnish ThermoWood Association, the members of which are listed in Table 8.1.

The Finnish ThermoWood Association was formed in December 2000. The main roles of the association are quality control of the material, product classification and R&D activities. Although ThermoWood production is primarily located in Finland, there is limited production in Austria and Estonia; a small-scale production facility was also due to start producing in Canada at the end of 2004.

ThermoWood is produced by a heat-treatment process in the presence of steam, and is thus a hygrothermal treatment. The steam acts as a blanket to limit the oxidative degradation of wood, and there are also additional reactions occurring as a result of the presence

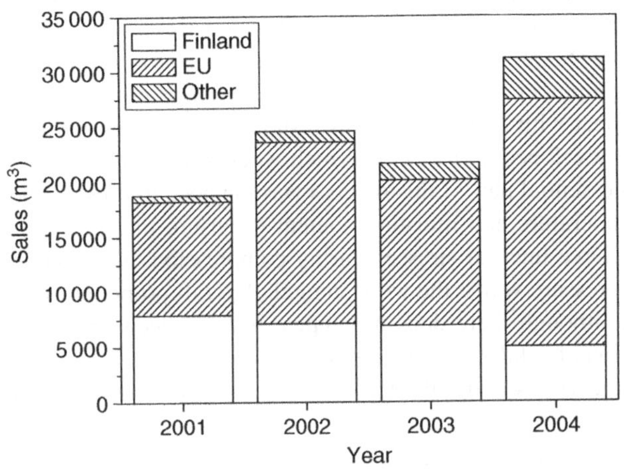

Figure 8.1 Sales of ThermoWood products for the years 2001–2004.

Table 8.1 *Members of the Finnish ThermoWood Association (2005)*

Production plants	Kiln manufacturers	Offices
Ekosampo Oy	Stellac Oy	Finnforest Oyj, head office
Finnforest Oyj, Kaskinen	Tekmaheat Oy	Stora Enso Timber Oy Ltd
Finnforest Oyj, Hammaslahti	Valutec Oy	UPM-Kymmene Wood Products
Heinolan Ruskopuu Oy		Finnish ThermoWood Association
Laukaan Lämpöpuu Oy		
Oy Brown Wood Ltd		
Oy Lunawood Ltd		
HJT-Holz Oy		
Stellac Wood Mikkeli Oy		
Stora Enso Timber Oy Ltd		
Suomen Lämpöpuu Oy		

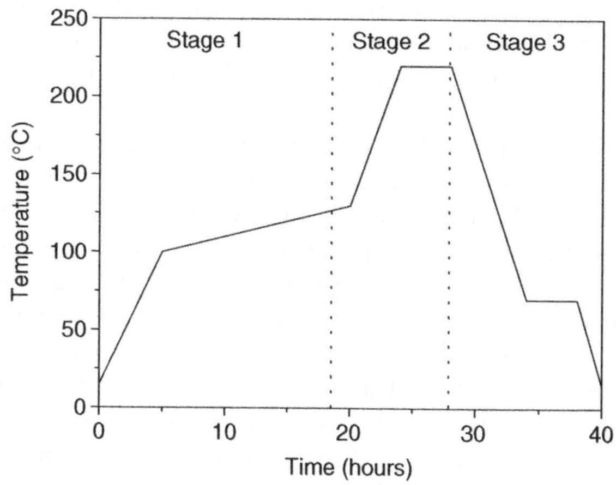

Figure 8.2 *A diagram of the ThermoWood thermal modification process stages.*

of moisture (see Chapter 5). Because of the presence of steam, the air content in the kiln is limited to 3–5 % during the heat-treatment process. The ThermoWood production process is divided into three phases (Figure 8.2):

- **Temperature increase and wood drying**: A combination of heat and steam is used to raise the temperature of the wood to 100 °C. The temperature is then ramped to 130 °C for the high-temperature drying phase, which reduces the wood moisture content to approximately 0 %.
- **Heat treatment**: When the high-temperature drying is complete, the temperature of the kiln is then increased to 185–230 °C, and maintained at the target temperature for 2–3 hours. The temperature and time used for this step depends upon the heat-treatment class required.

- **Cooling and moisture conditioning**: In this stage of the process, the temperature is gradually lowered using water-spray systems, until the temperature is in the range 80–90 °C. The wood is then conditioned to a moisture content ranging from 4 % to 7 %.

Although, originally, three treatment classes were defined for ThermoWood products, the industry has now decided upon two standard treatment classes, called Thermo-S and Thermo-D, although the material can be produced using higher or lower temperatures for special uses (Table 8.2).

Thermo-S, which is produced by a relatively mild heat treatment, is a light brown colour and is suitable for interior applications, and for other products where exposure to water is not prolonged (equivalent to Hazard Classes 1 and 2, EN335–1), where improved dimensional stability would be of benefit. Thermo-D, a product of a more intensive heat treatment, is more dimensionally stable, darker and exhibits a degree of resistance to biological attack, and is suitable for uses in Hazard Classes 1–3. Heat-treated timber is not suitable for use in Hazard Class 4 or 5. Softwood species treated are pine and spruce, and hardwoods are birch and aspen. Most of the production is in thermally treated softwood species, for end-uses in decking, garden accessories, and exterior and interior claddings. Exterior claddings are expected to have a 30-year service life. Hardwood production (about 15 % in 2004) is used for sauna benches and panels and for flooring (birch).

Pine is the preferred species for thermal treatment, although resin exudation can cause problems. There can be some problems with knots becoming loose, and with big knots there may be some twisting of the wood with small-dimension timber. As with pine, heat-treated spruce is also used for exterior applications, although this species is considered to be less satisfactory for heat treatment, mainly due to knots cracking and loosening. However, larger amounts of spruce are, in fact, heat treated compared to pine. Heat-treated birch is used for interior applications, where decay resistance is not an issue. The main desired property is the colour change due to heat treatment. Twisting of the wood can be a problem with birch. Heat-treated aspen is also used for interior applications, but there can be uneven colour changes and the wood is more likely to split as a result of heat treatment, especially when it contains both heartwood and sapwood. Treatment costs are of the order of 100–150 Euros per m^3, depending upon the extent of treatment. The cost of ThermoWood will fall as the industry matures.

A comprehensive series of tests have been performed on ThermoWood, which are detailed in the ThermoWood handbook, available from the Finnish ThermoWood Association (Mayes and Oksanen, 2002). A synopsis of the material properties is given below.

Due to the more intensive treatment conditions, ThermoWood-D exhibits higher dimensional stability and biological durability, compared with ThermoWood-S. Thermo-D

Table 8.2 *ThermoWood standard class treatment temperatures*

Class	Softwoods	Hardwoods
Thermo-S	190 ± 3 °C	185 ± 3 °C
Thermo-D	212 ± 3 °C	200 ± 3 °C

is produced in greater quantities than Thermo-S, mainly for cladding applications. ThermoWood is of a lower density and has a reduced bending strength compared with unmodified wood. The MOE is not changed by the thermal treatment. ThermoWood is not recommended for load-bearing applications. Toughness and abrasion resistance are reduced, and the wood has a greater tendency to split. The EMC of ThermoWood is approximately 40–50 % lower at a given RH compared with unmodified wood. The wood also exhibits reduced permeability to moisture and greater dimensional stability. Thermal conductivity is reduced by about 20–25 % compared with normal wood. Although ThermoWood performs less well in fire tests compared with unmodified wood, the difference is stated as being 'not significant'. ThermoWood exhibits improved durability when tested in pure culture, but performs less well in ground contact tests. The modified wood is claimed to be less susceptible to insect attack.

ThermoWood is not resistant to exterior weathering and the colour will gradually change to the grey that is characteristic of outdoor exposed wood. In addition, exposure in exterior conditions results in the formation of small cracks on the surface of uncoated wood. Unpigmented or low-build stain coatings do not protect the surface of the wood, but solvent-borne alkyds and water-borne acrylic paints have been found to exhibit better performance than on unmodified wood. VOC emissions from the heat-treated wood are lower compared to unmodified wood and the compositions of the emissions differ. The level of emissions is lower when the wood is treated at a higher temperature. Emissions of terpenes are reduced to very low levels, and the VOC content is dominated by furfural, hexanal and acetic acid (treated at 180 °C), and by acetic acid (treated at 230 °C). ThermoWood passes ecotoxicity tests.

There are three manufacturers of heat-treatment kilns in Finland:

- **Stellac Oy** produce tailor-made kilns that range in annual capacity from 1000 to 50 000 m^3.
- **Tekmaheat Oy** manufacture three kiln systems: a compact chamber of 15–25 m^3 per batch, having an annual production capability of 1600–3000 m^3; a drive-through chamber having a volume of 40–60 m^3, with an annual capacity of 5000–10 000 m^3; and a multi-stage production line, in which the timber is pre-heated, dried, heat-treated and conditioned in a continuous throughput process, which has a capacity of 30 000–50 000 m^3 per annum.
- **Valutec Oy** produce two kilns, a small 20 m^3 capacity kiln with an annual production of 2000–3000 m^3 and a larger kiln of 100 m^3 with an annual capacity of 10 000–15 000 m^3.

8.2.2 Plato® Wood

The Plato (Proving Lasting Advanced Timber Option) process is derived from a technology that was originally developed by Royal Dutch Shell, and was a spin-off from research into the hydrothermal conversion of biomass into liquid fuels. The Plato thermal modification process involves four stages (Figure 8.3):

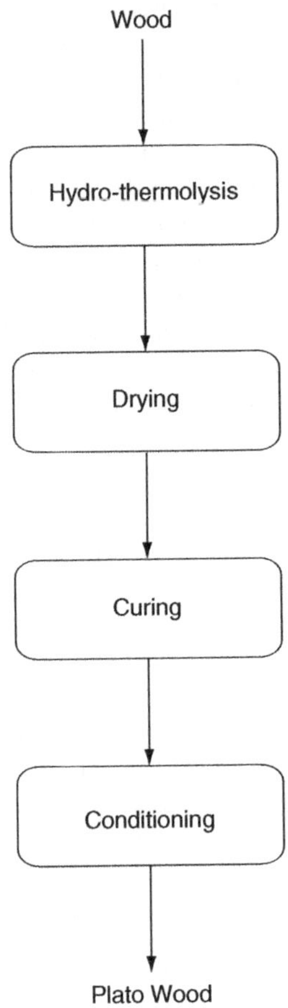

Figure 8.3 *A schematic of the Plato Wood production process.*

- **Hydrothermolysis**: The wood is heated to a temperature of 150–180 °C in steam under a pressure of 6–8 bar (0.6–0.8 MPa), for 4–5 hours. The moisture content of the wood entering this stage is 14–20 % and is only slightly reduced by the treatment. This stage involves degradation of the hemicelluloses to form aldehyde and organic acid containing compounds, and activation of the lignin. The cellulose component is not altered.
- **Drying**: The activated wood is dried in a conventional kiln to an appropriate moisture content (8–10 %) for the curing stage. This step takes from 5 days to 3 weeks.
- **Curing**: The wood is thermally treated at a temperature ranging from 150 °C to 190 °C under dry conditions, for 12–16 hours. During this stage, reactions occur between the hemicellulose breakdown products and the activated lignin to form cross-links. The wood leaves this part of the process at less than 1 % moisture content.
- **Conditioning**: A kiln is used to increase the moisture content of the wood to a desired level (about 4–6 %). This stage requires about 3 days.

It is thought that the hydrothermolysis stage is responsible for activating the lignin in addition to causing degradation of the hemicelluloses (cleavage of acetyl furfural formation etc.). In the second heat-treatment stage, it is believed that autocondensation of the lignin occurs via formation of methylene linkages between the aromatic rings, as well as further hemicellulose degradation.

At the time of writing, there is one Plato plant operating, located at Arnhem, in the Netherlands. This plant has a capacity for handling 35 000 m^3 of wood per year and is currently producing 12 000 m^3 of Plato Wood per year. The plant consists of two stainless steel vessels for the hydrothermolysis stage, 20 m in length, capable of treating 20 m^3 of material per batch. There are eight conventional drying kilns of 160 m^3 capacity, and one dry curing oven of 80 m^3 capacity. The main wood species treated is Norway spruce, although Scots pine, Douglas fir, poplar and birch are also treated. Refractory species, such as spruce, are more difficult to treat satisfactorily.

The process produces some organic waste derived from the thermal breakdown products of the wood, which is dealt with in a biological treatment facility. Volatile by-products are removed in a scrubber associated with a steam-condensing unit. Energy consumption is 2.8 MJ to produce 1 kg of modified wood and the cost of production is about 150 Euros per m^3, with operational costs of 20 Euros per m^3. A plant capable of an annual production capacity of 75 000 m^3 would require an initial investment of some 10–15 million Euros.

Plato Wood has two divisions: Plato Wood products, responsible for production, and Plato International Technology, responsible for research and development and licensing of the technology. The main end-uses for Plato Wood are garden furniture, fencing, cladding, window frames and doors. The company is currently developing methods for the use of Platonized wood chips in the production of particleboard, and for the production of Plato veneer and plywood.

Plato wood is a dark brown colour, but will discolour if exposed to exterior weathering conditions. The change in wood properties depends upon the species treated and the exact conditions employed in the process. The MOE of Plato Wood is 0–10 % higher than for untreated wood, but the MOR is decreased between 5 % and 20 %. The lower strength losses reported for Plato-treated wood are a consequence of the milder conditions employed compared with other thermal treatments. The resistance of wood to fungal attack is improved by Plato treatment. The EMC of the treated wood is reduced by 30–40 %, depending upon the species treated and the exact conditions employed for the treatment. The ASE of the treated wood is approximately 50 % of that found for unmodified wood. Paintability of Plato Wood with alkyd paints is claimed to be superior compared with untreated wood, with longer periods required between repainting. There is also reported to be improved adhesion between Plato Wood and common wood adhesives (PVAc and PU). The treated wood is suitable for laminating and finger-jointing, and can be sawn and planed in the same way as untreated wood. LCA studies commissioned by Plato Wood have shown that the product has superior environmental performance compared with materials such as concrete and PVC, as well as preservative-treated wood products (creosote and CCA).

8.2.3 Retification®

Retification is a process developed in France by the École des Mines de St. Etienne. It involves the heating of wood (previously dried to about 12 % MC) in a nitrogen atmosphere (containing less than 2 % oxygen) at temperatures between 180 °C and 250 °C. Industrialization of this process by a company called New Option Wood (NOW) SA (also known as Retitech) started in 1995, with the first prototype 2 m^3 kiln, capable of treating boards up to 3 m long, being commissioned in 1997. An industrial kiln was developed by the company Four et Brûleurs REY, based near St. Etienne. Three kilns were subsequently ordered by NOW, each of 8 m^3 capacity, with a production capacity of 3500 m^3 per year per kiln. The planned production capacity was to be of the order of 10 000 m^3 per year by 2001. The cost of the retification process was of the order of 150–160 Euros per m^3 in 2001. New Option Wood were subsequently acquired by EPMBH, who are now using the name Retitech for the wood treatment company, with a kiln installed at La Rochelle.

8.2.4 Le Bois Perdure

The Perdure process has been commercialized by PCI Industries Inc., based in Quebec. The Perdure process involves the drying and then subsequent heating of wood at temperatures ranging from 200 °C to 230 °C in a steam atmosphere. The company produces two kilns for thermal modification, the PC5 (of 8.75 m^3 capacity) and the PC6 (of 10.5 m^3 capacity). Process costs are of the order of 100 Euros per m^3.

8.3 Oil Heat Modification/Treatments

These fall into two categories:

- A true wood modification, in which the oil treatment is carried out at a sufficiently high temperature so that the molecular components of the wood are modified.
- A wood treatment in which the temperature is sufficient to cure the oil to some extent, but leaves the wood substrate unaltered.

8.3.1 Menz Holz

A pilot-scale oil heat-treatment plant, with a capacity of 2 900 m^3 per annum, was installed by Menz Holz in 2000, located in Reulbach, Germany. The process consists of a process vessel that is charged with wood (spruce or pine) and then hot oil is introduced into the vessel. The vegetable oil used (rapeseed, sunflower or linseed) serves as a heat-transfer medium and also excludes oxygen during the process. Wood up to 4 m in length can be treated at a temperature ranging from 180 °C to 220 °C, with a treatment time of 18 hours. The treatment involves the wood taking up oil at levels in the region of 20–60 kg m^{-3}, although this is of secondary importance to determining wood properties compared with the actual heat treatment. The treatment costs have been estimated as being of the order of 60–90 Euros per m^3.

8.3.2 The Royal/Royale Process

The Royal (or Royale) process was originally developed as a method for drying timber, in which the wood is heated in oil under vacuum. The temperatures used are low (60–90 °C) compared to other thermal oil treatments, and although sufficient to lead to some curing of the oil itself, there is no direct modification of the wood as a result of this process. The oil does not penetrate the cell wall. In this process, wet timber is placed in a treatment vessel and oil is then introduced, which is heated to the desired temperature, whilst a vacuum is applied. Water is removed from the timber and the vapour is transported away by the vacuum system. When the wood has reached the desired MC, the oil is removed from the treatment vessel. After this, a vacuum is applied to removed excess oil from the wood. Some dimensional stability is imparted to the timber due to the water repellency of the oil. This treatment is marketed by Osmose as the Royale process.

8.3.3 Ecotan

This process uses a thermal treatment of wood impregnated with a chemically modified vegetable oil. It is claimed that wood treated using this process does not exhibit the usual reduction in mechanical properties associated with a thermal treatment. This process is at a pre-commercial pilot stage at the time of writing.

8.4 Acetylation

There have been several attempts at commercialization of the wood acetylation process, but these were not successful because the economic climate was not favourable. Compared with most other wood treatment and modification processes, acetylation is expensive and requires significant capital investment. In the 1960s, the Koppers company in the USA attempted to produce wood on a commercial scale, and there were later attempts in the former USSR, but both of these were discontinued (Rowell, 1983a). Evans (2003) and Kiguchi (1995) reported on a commercial solid wood acetylation process developed by Daiken Wood Industry of Japan. The commercial product was called α Wood. Other processes for wood etherification by alkyl halides (EX Wood) (Tokiwa Industry Ltd), oligo-esterification (Okura Industry Ltd) and for the preparation of wood inorganic composites (Mu-tech Wood) (Matsushita Denko Ltd) were also developed to a commercial scale. However, because of deteriorating economic conditions, commercial interest in wood modification in Japan declined in the 1990s.

8.4.1 BP Chemicals

BP Chemicals developed a pilot-scale plant for the acetylation of wood fibres in collaboration with A-Cell acetyl cellulosics AB, Depac Engineering and the BioComposites

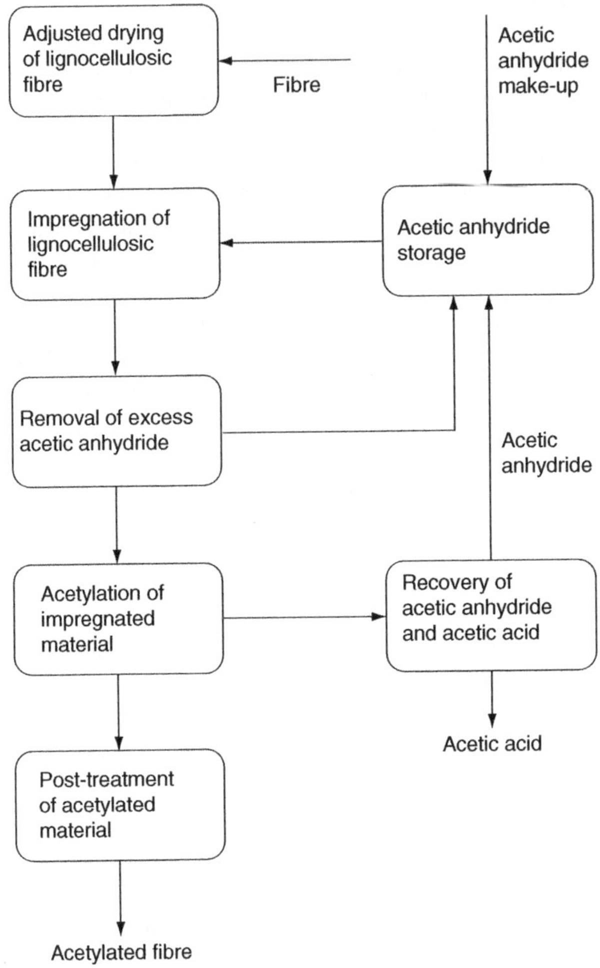

Figure 8.4 *A schematic of the fibre acetylation process developed by BP Chemicals (Sheen, 1992).*

Centre (Sheen, 1992) (Figure 8.4). In this trial, 11 tonnes of acetylated fibre was produced for evaluation, and this led to the development of a continuous fibre acetylation process. The process used a solvent and catalyst-free acetylation system, because the use of either would complicate a commercial process and introduce unnecessary costs. A range of reaction conditions were studied and the following conclusions were drawn:

- It was preferable to use fibre having a low moisture content (up to 5 %), since the presence of high levels of moisture moisture would lead to the production of large quantities of acetic acid by-product. However, if oven-dry fibre was used, this led to clumping of the fibre.
- The presence of low levels of acetic acid accelerated the initial rate of reaction, but levels above 30 % by volume inhibited the reaction.

- Fibre damage was noted at acetylation temperatures above 130 °C.
- Removal of excess reagent and by-product was best accomplished by using an air flow at 50 °C.

BP Chemicals developed a batch process capable of producing 100 kg of acetylated fibre per day, and a larger batch process, capable of producing 1 tonne per day, was to be constructed by Depac Engineering Ltd. Sheen (1992) also reported on an analysis of the costs for batch and continuous plants producing 10 000 tonnes of acetylated fibre per year at a WPG of 20 %. An essential part of the process to ensure commercial viability was the recovery of by-product acetic acid. The calculated costs in pounds sterling, based on fully dried fibre and in 1992 prices, are shown in Table 8.3.

The capital cost of the batch acetylation plant was calculated at £12 million and that of the continuous process plant at £5.5 million. It was assumed that the acetylation facility would be integrated with a traditional fibreboard plant, so that fibre production and drying costs were not included in the analysis. The analysis showed that wood fibre could be acetylated to a WPG of 20 % for a cost of £562 per tonne in a batch process and £384 per tonne in a continuous process. BP subsequently lost interest in the wood acetylation process and sold the pilot plant.

Table 8.3 *Process costings for a batch and continuous fibre acetylation process*

Raw materials	Cost (£ per tonne)
Batch process	
Acetic anhydride	223
Wood fibre	64
Acetic acid credit	−60
Total raw materials	227
Total utilities	30
Total variable cost	257
Operating costs	155
Cash cost	412
Depreciation	150
Cash cost + depreciation	**562**
Continuous process	
Acetic anhydride	223
Wood fibre	64
Acetic acid credit	−60
Total raw materials	227
Total utilities	3
Total variable cost	230
Operating costs	99
Cash cost	329
Depreciation	55
Cash cost + depreciation	**384**

8.4.2 A-Cell

On the basis of the development work undertaken by BP Chemicals in Hull and research performed at Chalmers University in Gotenborg, Sweden, and the USDA Forest Products Laboratory in Madison, Wisconsin, USA, a fibre acetylation pilot plant was commissioned in 2000 at Kvarntorp in Sweden. The plant has a capacity of 4 000 tonnes of acetylated fibre per year (Simonson and Rowell, 2000). The process and plant are jointly owned by A-Cell Acetyl Cellulosics AB and GEA Evaporation Technology AB. A schematic of this process is shown in Figure 8.5.

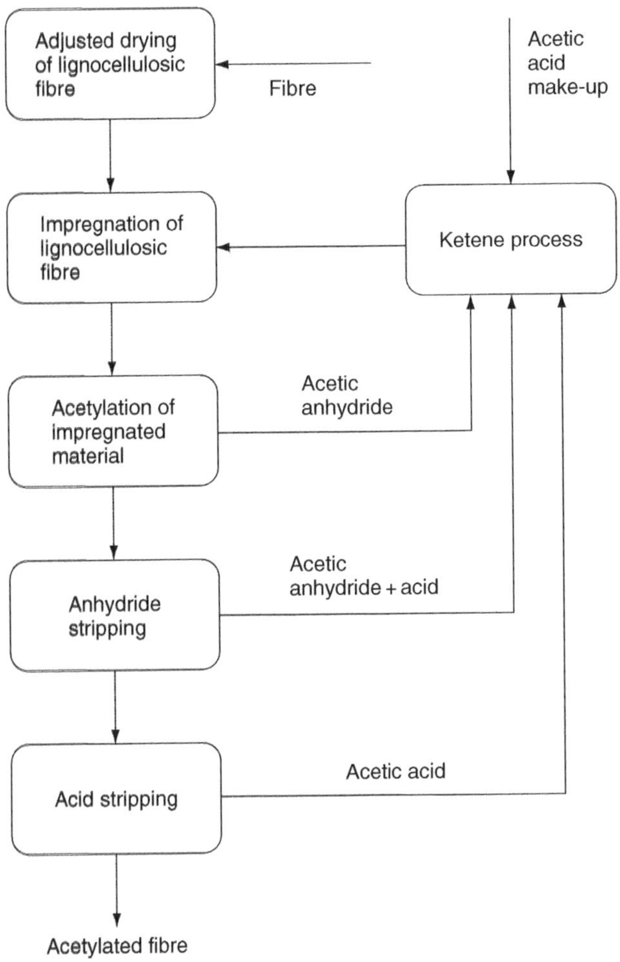

Figure 8.5 A schematic of the A-Cell wood fibre acetylation process.

The first stage of the process involves the drying of fibre to an optimum moisture content to minimize the production of acetic acid, but to allow for sufficient swelling of the fibres to accelerate the rate of reaction. After drying, the fibre is introduced into the reactor using a screw feed and the acetic anhydride added. The reaction temperature is set within the operating range of 110–140 °C, and the reaction time is between 6 and 30 minutes. The acetylated fibre is then fed into an anhydride stripper, which is essentially a long tube heated in the temperature range of 185–195 °C. The stripping stage takes about 1 minute. Although the primary function of this stage is to remove excess reagent and by-product acetic acid, further reaction of the fibre also takes place. After this stage, the reacted fibre and superheated vapour are separated in a cyclone. The hot vapour is split, with a proportion returning to the inlet of the stripper and the rest being fed into the chemical recovery system. Although most of the process chemical and by-product is removed in the first stripping stage, there remains some residual chemicals, which are removed in a second stripper. The secondary stripper involves a long tube in which the fibre is transported and treated with superheated steam. This removes residual chemicals and hydrolyses the unreacted acetic anhydride present to form acetic acid, which is transferred to the chemical recovery stage. Acetic acid recovered in this process is converted to acetic anhydride using the ketene process. The make-up chemical is acetic acid, which is also fed into this unit to produce acetic anhydride. The economics of the process were presented in the paper by Simonson and Rowell (2000), and this data is reproduced in Table 8.4.

8.4.3 Titan Wood

Titan Wood limited was formed in April 2003, by Accys Chemicals PLC, a UK-based company. The company is in the process of building a full-scale production facility for the acetylation of solid wood located at Arnhem in The Netherlands, which will become operational at the end of 2005. The company acquired equipment and intellectual property owned by the former AKBV, based in Arnhem. AKBV developed a pilot plant acetylation reactor of 2 300 litre capacity, 85 cm in diameter and 4 m long, capable of acetylating 0.9 m^3 of solid wood per batch.

Table 8.4 *Costs of acetylated fibre production in Swedish crowns (SEK) (Simonson and Rowell, 2000)*

	Annual production (tonnes per year)		
	8 000	20 000	100 000
Total investment (MSEK)	78	142	365
Fixed costs (SEK per tonne)	1 500	1 100	550
Operating costs (SEK per tonne)	3 000	2 900	2 450
Personnel costs (SEK per tonne)	200	100	50
Total cost (SEK/kg)	4.70	4.10	3.05

8.5 Impregnation Modification

There have been a number of examples of the production of wood polymer composite materials where the wood is impregnated with vinyl monomers (such as methyl methacrylate), which are then polymerized by the use of gamma irradiation (Schneider and Witt, 2004). Wood modified using this technology has been used in applications such as flooring. Products based upon the impregnation of wood with resin have been available since the Second World War, and these are still manufactured. More recently, there has been a renewed commercial interest in furfurylation, as well as the development of a wood-hardening process, using impregnation with polysaccharides.

8.5.1 Impregnation with Phenolic Resin

Compreg and Impreg are products manufactured with a process that involves the impregnation of wood veneers, normally using a water-soluble phenolic resin. This is followed by a drying stage at a temperature low enough to prevent resin pre-cure. The impregnated veneers are placed in stacks and resin curing then takes place at a higher temperature, with sufficient pressure to bond the veneers but not compress the wood to manufacture Impreg, or with a combination of heat and veneer compression to manufacture Compreg (Stamm and Seborg, 1941). Similar products, known as Kunstharzschichtholz, or Lignofol, have been manufactured in Germany. Lignofol was used for the manufacture of aircraft propellers, such as that for the radical Horton H Va 'flying wing', built in 1936–7. Phenolic resin impregnated densified wood was used extensively during the Second World War, particularly for the production of propellers for aircraft. Production decreased dramatically in the 1950s with the advent of synthetic polymers and man-made fibre-reinforced composites.

Manufacture of Compreg is currently taking place at a number of sites worldwide (Fibron and C-K composites in the USA, Permali in the UK, Dymonwood in Pakistan and Western India Plywoods in India). Compreg and related products have very good dimensional stability in the presence of moisture, and high compression strengths, hardness and abrasion resistance, although the bending strength is only slightly better than timber of a comparable density. Products are specialized and low volume compared with the markets at which other wood modification technologies are aimed.

8.5.2 Indurite

The Indurite process was developed by Engineered Wood Solutions Ltd of New Zealand, and the rights to this process have been obtained by Osmose. The process involves the impregnation of wood with a water-soluble polysaccharide solution (soy and corn starch), followed by a curing stage that renders the polysaccharide insoluble. The impregnated wood is held in a covered area for 24 hours before the curing step. The curing step takes place in a temperature-controlled kiln, and is facilitated by including catalysts in the impregnation solution. A major advantage with this process (described in the sales literature as a wood 'enhancement') is the ability to use conventional wood treatment and drying facilities, thereby avoiding significant investment in new plant. Although primarily intended for internal applications, where dimensional stability is required (such as flooring), Indurite-treated wood can be used in exterior applications where water is shed rapidly, such as certain forms of cladding. Its main application is in the improvement of the hardness of softwood for flooring applications. A product called Maro flooring, produced using this process, is available on the market.

8.5.3 Furfurylation

The development of furfurylation began with the research by Alfred J. Stamm at the Forest Products Laboratory in Madison, Wisconsin, in the 1950s. This led on to an industrialization of the process, with production in the USA of furfurylated wood in the mid-1960s. Products included laboratory bench tops, pulp mixer rotorblades and knife handles. However, commercial production had ceased by the early 1970s.

In the late 1980s, Professor Marc Schneider, of the University of New Brunswick, and associates took up development work on the furfurylation process (Schneider and Witt, 2004). They found that uneven levels of treatment occurred with the initiators used previously (e.g. $ZnCl_2$) and they developed a two-stage process in which the wood was impregnated by the initiator prior to treatment with furfuryl alcohol. This process was then commercialized, along with a diallyl phthalate impregnation process, by the formation of a company called Wood Polymer Composite Processes Ltd. The products were trademarked as 'WISTIwood'. The processes were licensed to a company called Wood Polymer Products Inc., which was to manufacture the modified wood, but by 1990 this attempt had failed. One problem with the licensed process was the need for a double treatment schedule for furfurylation, which was not satisfactory for full-scale commercial production. A superior process was developed by Marc Schneider, which led to an improved technology to be licensed by a Canadian company (Woodtech Inc.), formed in 1993. A collaboration was then developed with Norwegian researchers at Norsk Institutt for Skogforskning, leading to the formation of Wood Polymer Technologies (WPT) ASA, based in Norway, in 1997. WPT acquired the rights to the technology from Woodtech Inc. Following this, a trial pilot plant was established at Høylandet in Norway. In 2000, marketing of the products began in Europe and the USA (by Chemblock Industries AS). Work on the furfurylation process had also taken place at Trätek and the University of Chalmers in Sweden, leading to the development of similar processes, and the research efforts of the two groups subsequently merged. Research and development of the furfurylation process is undertaken by WPT, whilst marketing of furfurylated wood products is the responsibility of Kebony Products DA.

The process involves an impregnation step (using either a Kebony or VisorWood treatment solution), followed by intermediate drying prior to curing, and final kiln drying (Figure 8.6).

The furfuryl alcohol used for the treatment is derived from corn cobs or sugar cane residues.

Three products are available from this process, VisorWood (20–40 % WPG), Kebony 30 (10–50 % WPG) and Kebony 100 (70–100 % WPG). Uses of VisorWood include decking, marine applications, cladding, window joinery, poles, roofs, garden furniture and building materials. Both Kebony 30 and VisorWood have a light golden brown

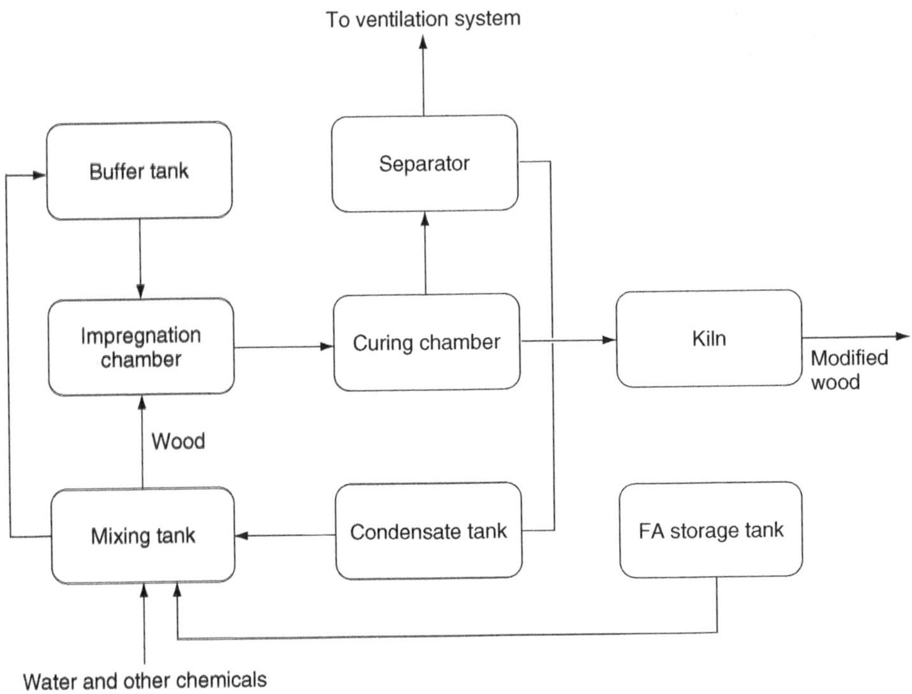

Figure 8.6 A schematic of the WPT impregnation modification process.

colouration, whereas Kebony 100 has a very dark brown, almost black colouration. Kebony is used for flooring, garden furniture, boat decking, decking, cladding and decorative joinery. Kebony 100 is classed as highly durable, and has a very low swelling coefficient (S) of 1.9 %. Kebony 30 is classed as durable, and has a modest S value of about 10 %.

A license to produce Kebony has been acquired by the parquet floor manufacturer Boen Bruk AS, who have production facilities in Lithuania. Kebony Products DA have a production facility of 5 000 m^3 per annum capacity, which has been established by WPT, Boen Bruk and Ole Schrøder & Co. AS at Herøya Industripark, Porsgrunn, Norway. A further plant is planned with a production capacity of 15 000 m^3 per annum.

8.6 Conclusions

The past decade has seen the successful commercialization of thermal wood modification in Finland. Developments in France and The Netherlands have been more hesitant, although significant progress is now being made. An industry based upon wood furfurylation is being established in Norway. The next few years will hopefully see the successful commercialization of acetylation by Titan Wood, the Indurite impregnation process by Osmose, the Plato process, and thermal modification in France and Quebec. This is a very exciting time for all who are involved in wood modification.

9

Wood Modification: Environmental Considerations and Future Developments

9.1 Introduction

Wood modification has recently become a topic of great interest, largely as a response to increasing legislative pressure driven by environmental considerations. As environmental concerns have featured more prominently, there has been a response from industry to both examine and demonstrate the environmental worthiness of products. Such claims are used as powerful marketing tools. The term 'environmentally friendly' is used far too frequently and inappropriately. *All* industrial processes have an associated environmental impact and it is not possible to describe any as environmentally friendly. What we should seek to do is minimize the environmental impact associated with any particular activity, or choose an activity that has the lowest environmental impact out of a range of choices. Although tools are available to quantify environmental impacts, it is important to note that our choice should be dictated following all of the principles of sustainability. Our choice should be dictated by determining the process that exhibits the greatest relative environmental, social and economic benefit.

Towards the end of the 20th century, a number of techniques were developed in order to determine the environmental impact of industrial processes. These analyses are inevitably complex and sometimes obscure in their methodology. Two often used methods for reporting environmental impacts quantitatively are:

- embodied energy;
- life cycle assessment.

Wood Modification: Chemical, Thermal and Other Processes C. Hill

9.2 Principles of the Determination of Environmental Impact

Although the methodology of assessment methods can be criticized, it is essential that we have accurate tools at our disposal in order to make the correct decisions regarding the appropriate materials/processes to use. An important aspect of making any determination of the environmental impact of a process, or as a consequence of the use of a specific material, is the determination of where the boundaries of the analysis should lie. A simple diagram of the principle elements of an environmental impact analysis is shown in Figure 9.1.

For a full life cycle assessment, the basic principle is that each material and energy input into the system should be traced back to natural resources obtained from the environment, or to releases into the environment. These are termed 'elementary flows', and they represent inputs into or outputs from the system being analysed. In an analysis of this type, it may be relatively straightforward to assign a material value to a flow of (for example) water effluent into the environment, but what may be less certain is the environmental impact of such a flow in a quantitative sense.

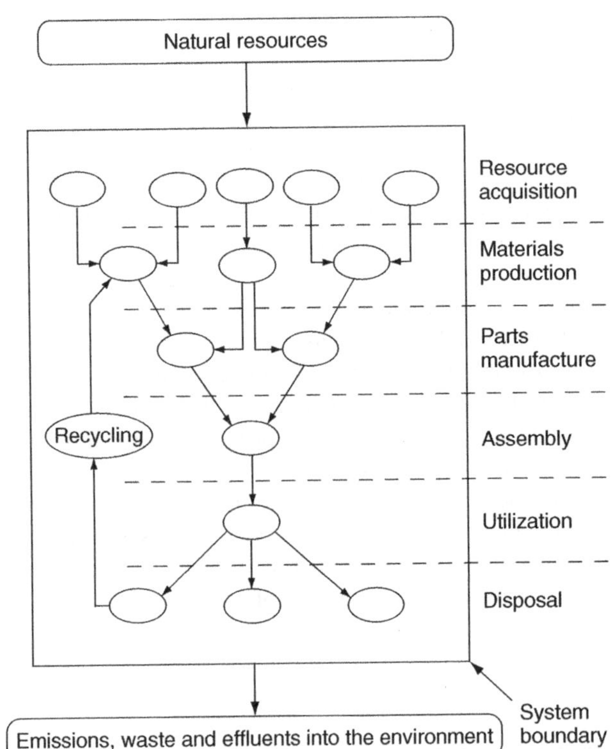

Figure 9.1 A schematic showing the various components considered in a life cycle analysis.

9.3 Methods of Determining Environmental Impacts

9.3.1 Embodied Energy

Embodied energy is divided into two forms:

- **Initial embodied energy** represents the quantity of *nonrenewable* energy that is used in the winning, transport and processing to produce a material to be used for a specific application.
- **Recurring embodied energy** is the amount of energy used to maintain a material in service, or the amount of energy required to replace that material if it fails in service.

Embodied energy is expressed in units of energy (Joules) per unit mass (kg, tonne), or volume (m^3). It gives an approximate estimate of the global warming potential of a particular material, but it does not measure all environmental impacts. For example, although the energy required to produce cement would be accounted for in an embodied energy calculation, the production of carbon dioxide as a result of the decomposition of limestone would not be included. As with all such analyses, the system boundary is all-important. For example, if we examine the impacts of timber production, we might include in the analysis the fuel use in planting, silvicultural management, harvesting, processing and transporting of the material. But resources are required to produce the machines required, and the impacts of this should also be included. The analysis rapidly becomes very complicated and, for this reason, it is easy to find discrepancies in the literature.

9.3.2 Life Cycle Assessment (LCA)

Life cycle assessment (analysis) involves the determination of the total environmental impact of the use of the material from when the raw materials are extracted, through the processing and utilization of the material, possible recycling and ultimate disposal.

A full LCA is expected to involve the following stages:

- goal definitions and scope;
- preparation of an inventory of resources used and pollutants emitted for each stage of the life cycle;
- a quantitative assessment of the environmental impacts;
- An interpretation of the results and an improvement analysis.

In 1992, the International Organisation for Standardisation established a technical committee on Environmental Management (TC207) to develop a series of standards on environmental management. This was subsequently published as the ISO14000 series of standards. Although this is a field that is rapidly maturing, LCA has been criticized on a number of points.

Hillier *et al.* (1996) note that there is an observed tendency for LCAs to support the case of those funding the study, leading to suspicion of the veracity of the analysis.

LCAs tend to be complex and often difficult to analyse independently, even though the conclusions or output data may be presented in relatively simple terms. Because of the often contentious nature of the findings and the potential economic impact of the analyses, it is absolutely essential that a worthy LCA satisfies the conditions of transparency and completeness (Hillier and Murphy, 2000).

LCAs also focus on the environmental impacts of processes and do not address the social or economic impacts. Although there is still debate about the usefulness of such analyses, LCAs do provide a means of examining the environmental impact of a process in a holistic sense.

9.4 The Environmental Impact of Wood Modification

9.4.1 LCA Studies of Wood Modification Processes

There is little published literature in the public domain regarding LCAs of wood modification processes. All wood modification processes (as with all industrial processes) will undeniably have an environmental impact associated with them. Apart from the environmental impacts associated with forestry operations, common to all wood products, additional impacts associated with wood modification are as follows:

- Process energy, which may be obtained from renewable or nonrenewable sources.
- Process chemicals, if used, which may be obtained from renewable or nonrenewable sources.
- Environmental impacts associated with process plant construction and maintenance.
- Impacts associated with processing, transport and maintenance of modified wood.
- Impacts associated with ultimate disposal of modified wood.

Of these, the latter would be comparable with, or lower than, those associated with untreated timber. Furthermore, modified wood would be expected to have lower maintenance requirements than conventional timber products. Impacts associated with the production, use and disposal of wood preservatives are obviously no longer present.

Most wood modification methods use temperatures higher than ambient (significantly so in some cases) and a significant impact will be recorded where the process energy is derived from nonrenewable sources. An obvious solution to this would be to use timber residues as a source of energy. With modification processes using chemicals, there will be an associated negative environmental impact, particularly if these chemicals are derived from fossil fuel resources. With furfurylation and oil-thermal treatments, the process chemicals are derived from renewable plant-based resources. Acetic anhydride for acetylation is currently derived from fossil-based resources, as are the components for resin treatments and silane treatment. The magnitude of the associated impact will depend upon the quantity of chemicals, their origin and the processing required to render them suitable to be used for the modification reactions.

9.5 Industrial Ecology and Wood Modification

The concept of industrial ecology first rose to prominence in an article in *Scientific American* in 1989 (Frosch and Gallopoulos, 1989). The idea behind this concept is to mimic the materials flows that are found in natural metabolic processes in industrial technologies. This involves the use of closed-loop materials utilization strategies, in which the waste from one industrial process is used as a feedstock for another. One often quoted example of this is the location of a plasterboard production facility next to a flue gas desulphurization unit at a power plant. Removal of sulphur from the flue gases using limestone produces gypsum (calcium sulphate, $CaSO_4$), which is used in the production of plaster. Proponents of this concept envisage mutually dependent enterprises operating in an 'eco-park' industrial facility, where all wastes are used within the complex for production of other materials. This is industrial ecology viewed in a narrow sense, but in a broader sense industrial ecology means *fully* integrating industrial activities within natural materials flows, so that the industrial processes essentially become a part of the biological ecological processes. There are many texts dealing with this subject (e.g. Soclow *et al.*, 1994; Ayres and Ayres, 1996), and the purpose of this section of the chapter is to briefly examine how it might be possible to integrate wood modification processes into an industrial ecology, using acetylation as an example. Since wood modification is essentially a new industrial process, it would seem sensible to plan the infrastructure of the industry now, rather than have to make changes later.

9.5.1 Wood Modification and the Flow of Energy and Materials

Let us examine a simplified flow diagram for the production of acetylated wood (Figure 9.2).

Inputs into this process include energy (derived from a mix of renewable and nonrenewable sources) and acetic acid for the production of anhydride, which is at present derived entirely from nonrenewable sources. Thus, if all the process energy can be derived from biomass, then the chemical feedstock is now the origin of the greatest environmental impact. However, it is possible to obtain acetic acid from wood using a pyrolysis process, derived from the acetyl groups associated with the hemicelluloses. Furthermore, acetylated wood contains a significant acetyl content, from which far higher yields of acetic acid can be obtained. The flow diagram in Figure 9.2 can thus be modified, so that acetic acid is now derived from recovered acetylated wood and the inevitable losses of acetyl can be replaced by pyrolysis of other wood-processing residues (Figure 9.3). The pyrolysis residues can then be incinerated to produce process energy, with the carbon dioxide returning to the atmosphere, where it is then available to be sequestered by forest growth. It may even be possible to incorporate a wood thermal modification plant in such a facility, and recover acetic acid generated during thermolysis of the wood during treatment.

It would not be possible to have such an infrastructure in place in the start-up phase of a wood acetylation industry. Such a process would rely upon there being sufficient quantities of acetylated wood available as a feedstock, and this would require many years of production. Furthermore, the costs involved in setting up such a plant would in all probability make the acetylation process uneconomic at present. Only when the industry matures would it be possible to bear the costs of building the necessary extra processing facilities.

The point is that if the longer-term sustainability of a process can be demonstrated, then a reliance upon fossil resources in the initial stages of the industrial cycle should not be used as an argument against the technology.

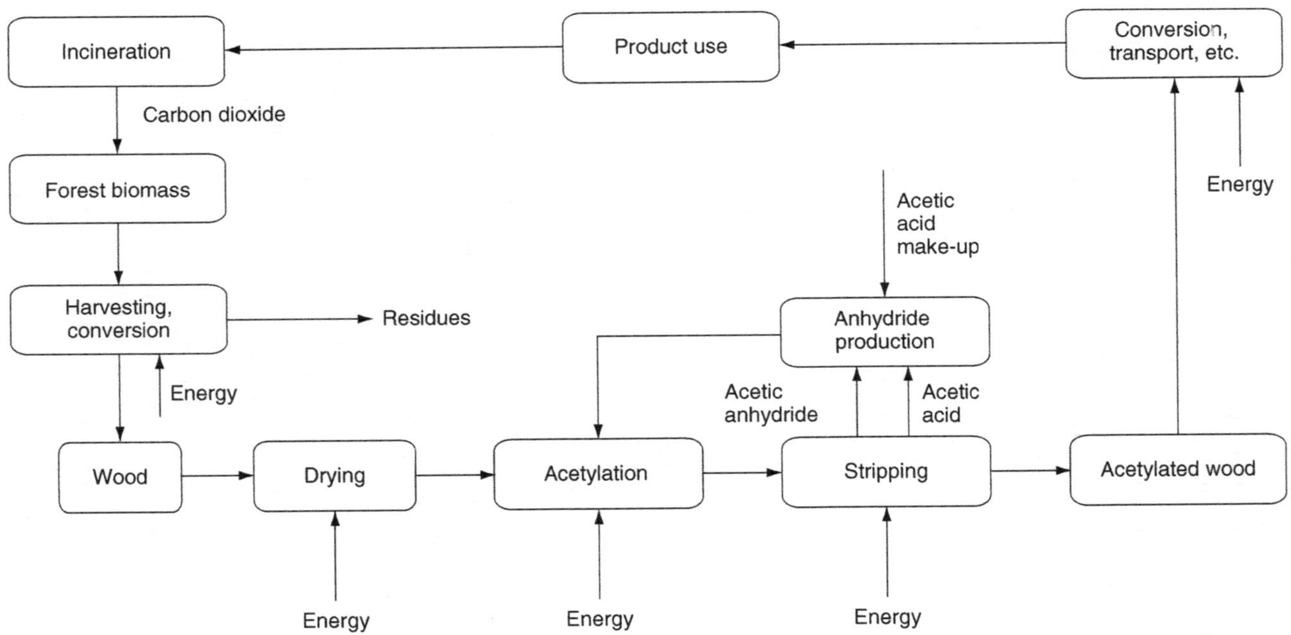

Figure 9.2 A schematic showing materials and energy flows through a conventional acetylation process.

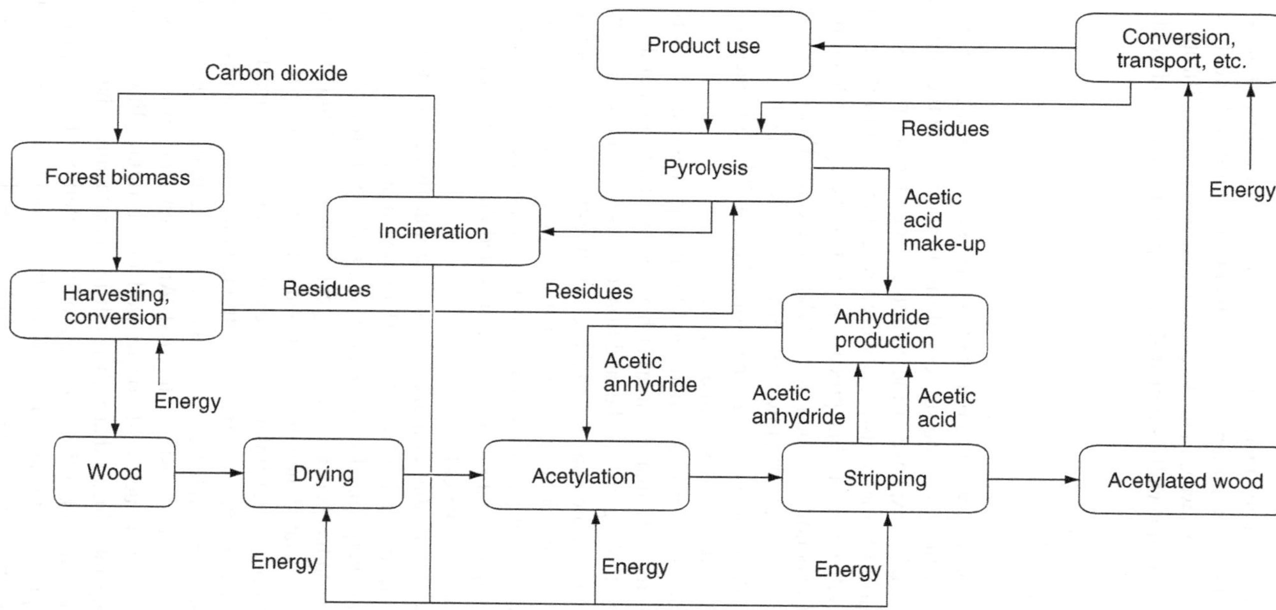

Figure 9.3 A schematic showing a possible future acetylation process, with acetic anhydride derived from wood and energy supplied by biomass.

9.6 The Future of Wood Modification

9.6.1 Economics

There has been a dramatic increase in interest in wood modification over the past decade. The economic and environmental background is now quite different from the situation in the past when attempts were made to commercialize wood modification. The wood modification industry is not well established at the present time, and is characterized by the following:

- lack of market penetration;
- low public profile;
- high expense.

The best-established wood modification process is the Finnish ThermoWood process. One of the major factors in the success of this technology in the commercial arena is the relatively low capital expense of the equipment required for processing. The capital cost of plant for furfurylation is also relatively low compared to full chemical modification, and commercialization of this process is well developed. Other processes involving higher capital expenditure (such as acetylation and the Plato process) have seen rather more hesitant development. However, once these processes have been established for a few years and the capital costs are paid off, there will be a reduction in the price of the products, leading to an increase in demand.

The introduction of a new wood modification process involves risks and it is only comparatively recently that the market has become more favourable for such ventures. There are inevitably going to be additional costs associated with wood modification, when compared to more conventional preservative-treated wood. It is obviously essential that these costs are not so excessive that any product made from the modified wood cannot be sold. However, it may be possible to sell products made from modified wood at a premium compared with competitors, provided that the benefits associated with using modified wood are apparent.

At the present time, the economic cost of a product is determined by the manufacturing costs. These are determined by the cost of raw materials, capital costs of processing equipment and energy used in manufacturing and transportation, and so on. As environmental legislation has become more stringent, then any part of the manufacturing process that has an associated negative environmental impact has *tended* to become more expensive. However, the linkage between environmental impact and economic cost is sometimes weak, or even nonexistent, although the whole area of environmental taxation is slowly developing. What has not, until now, featured in the cost of a product in the market-place is the full costs of disposal. At present, the costs of disposal fall upon the consumer, and often such costs occur well in the future. Environmental legislation is currently being introduced that will result in manufacturers having to bear the costs of disposal of the material at the end of a product's lifetime. This approach to materials use is known as product stewardship. Ultimately, it could be suggested that manufacturers will only lease products to consumers. Since one of the primary aims of wood modification is to reduce the environmental impact associated with disposal, it follows that this element of the material's cost will be low compared to competitor materials. This will result in modified wood becoming more economically competitive, even when the manufacturing costs are higher.

Forestry provides other benefits besides the production of timber. Forests, if managed correctly, also provide social benefits (tourism, leisure and recreation); they can be used to improve the quality of water resources, and to reduce flooding by reducing run-off and preventing soil erosion. By planting a suitable mix of species and utilizing appropriate management techniques, forests can be used to enhance biodiversity. Finally, of course, forests can be used to sequester atmospheric carbon. All of these functions can be considered as economic benefits (so-called externalities), but it can be difficult to factor these into the economics of the forestry wood chain. These issues are being addressed, although much remains to be done. If these undoubted benefits can be taken into account in the finances of timber production, then this will improve the economic competitiveness of wood products.

9.6.2 Standards

The standards that have been developed to date are concerned with determining the properties of conventional wood products. There is now an urgent need to develop appropriate standards and agree them at an international level. At present, claims made with respect to modified wood are based upon the current set of standards, which may not always be appropriate to the performance of these new materials in real-life situations. This is an area that has to be addressed in the very near future.

9.6.3 Future Developments

The past 50 years has seen intense research activity in wood modification research, and this will continue. Over the past decade, we have seen the first tentative steps towards the development of an entirely new industry. The next decade will see the maturing of this industry, with a large increase in the number of products available. This is a very exciting time for all of those involved in this area. This book reports upon the situation as of the beginning of 2005. In 10 years' time, it may be necessary to devote an entire book solely to commercial aspects of wood modification.

References

Aboltins, A. and Morozovs, A. (1998). Mathematical aspects of absorption of acetic anhydride in wood. *Transactions of the Estonian Agricultural University*, **196**, 167–171.

Ahmad, A.J. and Harun, H.J. (1992). Reaction of butyl isocyanate and butylene oxide with *Acacia mangium* (Willd). In: *Lignocellulosics, Science, Technology, Development and Use*, Kennedy, J.F., Phillips, G.O. and Williams, P.A. (Eds.). Ellis Horwood, Chichester, UK, pp. 779–789.

Ahmed Kabir, F.R., Nicholas, D.D., Vasishth, R.C. and Barnes, H.M. (1992). Laboratory methods to predict the weathering characteristics of wood. *Holzforschung*, **46**(5), 395–401.

Akitsu, H., Norimoto, M. and Morooka, T. (1991). Vibrational properties of chemically modified wood. *Mokuzai Gakkaishi*, **37**(7), 590–597.

Akitsu, H., Gril, J., Morooka, T. and Norimoto, M. (1992). Dynamic mechanical properties of chemically modified wood (II). In: *Pacific Rim Bio-Based Composites Symposium: Chemical Modification of Lignocellulosics*, Plackett, D.V. and Dunningham, E.A. (Eds.). FRI Bulletin, **176**, pp. 130–139.

Akitsu, H., Norimoto, M., Morooka, T. and Rowell, R.M. (1993a). Effect of humidity on vibrational properties of chemically modified wood. *Wood and Fiber Science*, **25**(3), 250–260.

Akitsu, H., Gril, J. and Norimoto, M. (1993b). Uniaxial modelling of vibrational properties of chemically modified wood. *Mokuzai Gakkaishi*, **39**(3), 258–264.

Arni, P.C., Gray, J.D. and Scougall, R.K. (1961a). Chemical modification of wood. I. Use of trifluoroacetic acid in the esterification of wood by carboxylic acids. *Journal of Applied Chemistry*, **11**, 157–163.

Arni, P.C., Gray, J.D. and Scougall, R.K. (1961b). Chemical modification of wood II. Use of trifluoroacetic acid as a catalyst for the acetylation of wood. *Journal of Applied Chemistry*, **11**, 163–170.

Arora, M., Rajawat, M.S. and Gupta, R.C. (1979). Studies on the acetylation of wood. *Holforschung und Holzverwertung*, **31**(6), 138–141.

Arora, M., Rajawat, M.S. and Gupta, R.C. (1980). Effect of acetylation time on degree of acetylation in wood. *Holforschung und Holzverwertung*, **32**(6), 138–139.

Arora, M., Rajawat, M.S. and Gupta, R.C. (1981). Effect of acetylation on properties of particleboards prepared from acetylated and normal particles of wood. *Holforschung und Holzverwertung*, **33**(1), 8–10.

Ashaari, Z., Barnes, H.M., Vasishth, R.C., Nicholas, D.D. and Lyon, D.E. (1990a). Effect of aqueous polymer treatments on wood properties. Part I: treatability and dimensional stability. International Research Group on Wood Preservation, Doc. no. IRG/WP 3610.

Ashaari, Z., Barnes, H.M., Vasishth, R.C., Nicholas, D.D. and Lyon, D.E. (1990b). Effect of aqueous polymer treatments on wood properties. Part II: Mechanical properties. International Research Group on Wood Preservation, Doc. no. IRG/WP 3611.

Ayadi, N., Lejeune, F., Charrier, F., Charrier, B. and Merlin, A. (2003). Color stability of heat-treated wood during artificial weathering. *Holz als Roh- und Werkstoff*, **61**(3), 221–226.

Ayres, R.U. and Ayres, L.W. (1996). *Industrial Ecology*. Edward Elgar, UK.

Wood Modification: Chemical, Thermal and Other Processes C. Hill

Back, E.L. and Danielsson, S. (1987). Oxidative activation of wood and paper surfaces for bonding and for paint adhesion. *Nordic Pulp and Paper Research Journal*, **2**(special issue), 53–62.
Baechler, R.H. (1959). Improving wood's durability through chemical modification. *Forest Products Journal*, **9**, 166–171.
Baird, B.R. (1969). Dimensional stabilisation of wood by vapor phase chemical treatments. *Wood and Fiber*, **1**(1), 54–63.
Balaba, W.M. and Subramanian, R.V. (1985). A viscometric investigation of the polymerisation of furfuryl alcohol catalysed by bound wood acid. *Holzforschung*, **39**(3), 143–147.
Banks, W.B. (1990). Derivatisation of wood. In: *Cellulosic Sources and Exploitation*, Kennedy, J.F., Phillips, G. and Williams, P.A. (Eds.). Ellis Horwood, Chichester, UK, pp. 221–234.
Banks, W.B. and Lawther, M.L. (1994). Derivatization of wood in composites. In: *Cellulosic Polymers, Blends and Composites*, Gilbert, R.D. (Ed.). Hanser/Gardner Cincinnati, USA, pp. 131–155.
Banks, W.B., Din, R.H. and Owen, N.L. (1995). Fibre activation and subsequent copolymerisation. *Holzforschung*, **49**(2), 104–108.
Barnes, H.M., Choong, E.T. and McIlhenny, R.C. (1969). An evaluation of several vapor phase chemical treatments for dimensional stabilisation of wood. *Forest Products Journal*, **19**(3), 35–39.
Barnett, J.R. and Jeronimidis, G. (Eds.) (2003). *Wood Quality and its Biological Basis*. Blackwell, Oxford, UK.
Barsberg, S. (2002). Modification phenomena of solid-state lignin caused by electron-abstracting oxidative systems. *Archives of Biochemistry and Biophysics*, **404**(1), 62–70.
Barsberg, S. and Hassingboe, J. (2003). Laccase catalysed enhancement of the autoadhesion of TMP fibres. In: *Proceedings of the First European Conference on Wood Modification*, Ghent, Belgium, Van Acker, J. and Hill, C.A.S. (Eds.), pp. 381–388.
Barsberg, S. and Thygesen, L.G. (1999). Spectroscopic properties of oxidation species generated in the lignin of wood fibres by a laccase catalysed treatment: electronic hole state migration and stabilization in the lignin matrix. *Biochimica et Biophysica Acta*, **1472**, 625–642.
Baysal, E., Ozaki, S.K. and Yalinkilic, M.K. (2004). Dimensional stabilisation of wood treated with furfuryl alcohol catalysed by borates. *Wood Science and Technology*, **38**(6), 405–415.
Beall, F.C. (1969). Thermogravimetric analysis of wood lignin and hemicelluloses. *Wood and Fiber*, **1**(3), 215–226.
Beall, F.C. (1971). Differential calometric analysis of wood and wood components. *Wood Science and Technology*, **5**(3), 159–175.
Beall, F.C. and Eickner, H.W. (1970). Thermal degradation of wood components: a review of the literature. USDA Forest Service Research Paper, **FPL 130**, 23 pp.
Beckers, E.P.J. and Militz, H. (1994). Acetylation of solid wood. Initial trials on lab and semi industrial scale. In: *Second Pacific Rim Bio-Based Composites Symposium*, Vancouver, Canada, pp. 125–135.
Beckers, E.P.J., Militz, H. and Stevens, M. (1994). Resistance of acetylated wood to basidiomycetes, soft rot and blue stain. International Research Group on Wood Preservation, Doc. no. IRG/WP 94-40021.
Beckers, E.P.J., Militz, H. and Stevens, M. (1995). Acetylated solid wood laboratory durability tests (part II) and field trials. International Research Group on Wood Preservation, Doc. no. IRG/WP 95-40048.
Beckers, E.P.J., Meijer, M. de, Militz, H. and Stevens, M. (1998). Performance of finishes on wood that is chemically modified by acetylation. *Journal of Coatings Technology*, **70**(878), 59–67.
Beckers, E.P.J., Bongers, H.P.M., Zee, M.E. van der and Sander, C. (2003). Acetyl content determination using different analytical techniques. In: *Proceedings of the First European Conference on Wood Modification*, Ghent, Belgium, Van Acker, J. and Hill, C.A.S. (Eds.), pp. 83–101.
Bekere, M., Shvalbe, K. and Ozolina, I. (1978). Some factors affecting the quality of boards made from wood fibers. *Latvijas Lauksaimniecibas Akadenija. Raksti.*, **163**, 31–35.
Bekhta, P. and Niemz, P. (2003). Effect of high temperature on the change in color, dimensional stability and mechanical properties of spruce wood. *Holzforschung*, **57**(5), 539–546.
Bender, M.L. (1960). Mechanisms of catalysis of nucleophilic reactions of carboxylic acid derivatives. *Chemical Reviews*, **60**, 53–113.
Bengtsson, C., Jermer, J. and Brem, F. (2002). Bending strength of heat-treated spruce and pine timber. International Research Group on Wood Preservation, Doc. No. IRG/WP 02-40242.

Bengtsson, C., Jermer, J., Clang, A. and Ek-Olausson, B. (2003). Investigation of some technical properties of heat-treated wood. International Research Group on Wood Preservation, Doc. No. IRG/WP 03-40266.

Bente, M., Avramidis, G., Förster, S., Rohwer, E.G. and Vlöl, W. (2004). Wood surface modification in dielectric barrier discharges at atmospheric pressure for creating water repellent characteristics. *Holz als Roh- und Werkstoff*, **62**(3), 157–163.

Bergervoet, T. (1983). Presteaming radiata pine heartwood to improve treatability. International Research Group on Wood Preservation, Doc. No. IRG/WP 3239.

Bhuiyan, M.T.R., Hirai, N. and Sobue, N. (2000). Changes of crystallinity of wood cellulose by heat treatment under dried and moist conditions. *Journal of Wood Science*, **46**(6), 431–436.

Bhuiyan, M.T.R., Hirai, N. and Sobue, N. (2001). Effect of intermittent heat treatment on crystallinity in wood cellulose. *Journal of Wood Science*, **47**(5), 336–341.

Bhuiyan, M.T.R., Hirai, N. and Sobue, N. (2002). Behaviour of piezoelectric. dielectric, and elastic constants of wood during about 40 repeated measurements between 100 °C and 220 °C. *Journal of Wood Science*, **48**(1), 1–7.

Birkinshaw, C. (1998). Chemically modified wood—a review with consideration of the opportunities for application to Irish timber. *Irish Forestry*, **55**(2), 21–34.

Birkinshaw, C. and Hale, M.D. (2002). Mechanical properties and fungal resistance of acetylated fast grown softwoods. I. Small specimens. *Irish Forestry*, **59**(1–2), 49–58.

Bledzki, A.K., Reihmane, S. and Gassan, J. (1998). Thermoplastics reinforced with wood fillers: a literature review. *Polymer and Plastics Technology*, **37**(4), 451–468.

Bolton, A.J., Dinwoodie, J.M. and Davies, D.A. (1988). The validity of the use of SEM/EDAX as a tool for the detection of UF resin penetration into wood cell walls in particleboard. *Wood Science and Technology*, **22**(4), 345–356.

Bongers, H.P.M. and Beckers, E.P.J. (2003). Mechanical properties of acetylated solid wood treated on pilot plant scale. In: *Proceedings of the First European Conference on Wood Modification*, Ghent, Belgium, Van Acker, J. and Hill, C.A.S. (Eds.), pp. 341–350.

Boonstra, M.G., Pizzi, A., Tekely, P. and Pendlebury, J. (1996). Chemical modification of Norway spruce and Scots pine. A ^{13}C NMR CP-MAS study of the reactivity of polymeric wood components with acetic anhydride. *Holzforschung*, **50**(3), 215–220.

Boonstra, M.G., Pizzi, A., Tekely, P. and Pendlebury, J. (1997). A study of the influence of solvents on the modification of wood with organic anhydrides. *Holzforschung*, **51**(1), 62–66.

Boonstra, M.J., Tjeerdsma, B.F. and Groeneveld, H.A.C. (1998). Thermal modification of non-durable wood species. 1. The PLATO technology; thermal modification of wood. International Research Group on Wood Preservation, Doc. No. IRG/WP/98-40123.

Booth, R. and Elliott, P. (1993). Forestry biomass energy and the carbon cycle. In: *New Crops for Temperate Regions*, Anthony, K.R.M., Meadley, J. and Robbelen, G. (Eds.). Chapman and Hall, London, UK, pp. 109–119.

Bourgois, J. and Guyonnet, R. (1988). Characterization and analysis of torrified wood. *Wood Science and Technology*, **22**(2), 143–155.

Bourgois, J., Bartholin, M.C. and Guyonnet, R. (1989). Thermal treatment of wood: analysis of the obtained product. *Wood Science and Technology*, **23**(4), 303–310.

Bourgois, J., Janin, G. and Guyonnet, R. (1991). Measuring colour: a method of studying and optimising the chemical transformations of thermally-treated wood. *Holzforschung*, **45**(5), 377–382.

Bowyer, J.L., Shmulsky, R. and Haygreen, J.G. (2003). *Forest Products and Wood Science: an Introduction*, 4th edn. Iowa State University Press, USA.

Brebner, K.I. and Schneider, M.H. (1985). Wood–polymer combinations: bonding of alkoxysilane coupling agents to wood. *Wood Science and Technology*, **19**(1), 75–81.

Brooks, D., Pajuoja, H., Peck, T.J., Solberg, B. and Wardle, P.A. (1996). Long-term trends and prospects in world supply and demand for wood. In: *Long-term Trends and Prospects in World Supply and Demand for Wood and Implications for Sustainable Forest Management*, Solberg, B. (Ed.). European Forest Institute, Research Report 6, EFI, Joensuu, Finland, pp. 75–106.

Buchanan, C.M., Gardner, R.M. and Komarek, R.J. (1993). Aerobic biodegradation of cellulose acetate. *Journal of Applied Polymer Science*, **47**(10), 1709–1719.

Burmester, A. (1967). Tests for wood treatment with monomeric gas of formaldehyde using gamma rays. *Holzforschung*, **21**(1), 13–20.

Burmester, A. (1971a). On the improvement of wood with formaldehyde—Part II: combined treatment of wood with formaldehyde. *Holz als Roh- und Werkstoff*, **29**(3), 97–102.

Burmester, A. (1971b). On the improvement of wood with formaldehyde—Part III: the treatment of reactivated heartwood. *Holz als Roh- und Werkstoff*, **29**(5), 184–188.
Burmester, A. (1973). Effects of heat-pressure treatments of semi-dry wood on its dimensional stability. *Holz als Roh- und Werkstoff*, **31**(6), 237–243.
Burmester, A. (1975). The dimensional stabilization of wood. *Holz als Roh- und Werkstoff*, **33**(9), 333–335.
Burmester, A. (1981). Dimensional stabilisation of wood. International Research Group on Wood Preservation, Doc. No. IRG/WP 3171.

Caulfield, D.F., Koutsky, J.A. and Quillen, D.J. (1993). Cellulose/polypropylene composites: the use of AKD and ASA sizes as compatibilizers. In: *Wood Fiber/Polymer Composites: Fundamental Concepts, Processes, and Material Options*, Wolcott, M.P. (Ed.). Forest Products Society, pp. 128–134.
Caldwell, E.S. and Steelink, C. (1969). Phenoxy radical intermediates in the enzymatic degradation of lignin model compounds. *Biochimica et Biophysica Acta*, **184**, 420–431.
Callow, H.J. (1947). Action of light upon jute. *Nature*, **159**, 309.
Callow, H.J. (1951). Acetylation of cellulose and lignin in jute. *Journal of the Indian Chemical Society*, **28**(11), 605–610.
Cannell, M.G.R. (2003). Carbon sequestration and biomass energy offset: theoretical, potential and achievable capacities globally, in Europe and the UK. *Biomass and Bioenergy*, **24**(2), 97–116.
Cardias Williams, F. and Hale, M.D. (1999). The resistance of wood chemically modified with isocyanates. Part 1. Brown rot, white rot and acid chlorite delignification. *Holzforschung*, **53**(3), 230–236.
Cardias Williams, F. and Hale, M.D. (2003). The resistance of wood chemically modified with isocyanates: the role of moisture content in decay suppression. *International Biodeterioration and Biodegradation*, **52**(4), 215–221.
Carrington, H. (1922). The elastic constants of spruce as influenced by moisture. *The Aeronautical Journal*, **24**, 462–471.
Çetin, N.S. (1999). Surface activation of lignocellulosics by chemical modification. PhD thesis, University of Wales Bangor.
Çetin, N.S. and Hill, C.A.S. (1999). An investigation of the reaction of epoxides with wood. *Journal of Wood Chemistry and Technology*, **19**(3), 247–264.
Çetin, N.S. and Ozmen, N. (2001). Dimensional changes in Corsican and Scots pine sapwood due to reaction with crotonic anhydride. *Wood Science and Technology*, **35**(3), 257–267.
Chadwick, D.H. and Cleveland, T.H. (1981). Isocyanates, organic. In: *Kirk–Othmer Encyclopaedia of Chemical Technology*, **13**, 3rd edn, John Wiley & Sons, Inc., New York, pp. 789–818.
Chang, S.T. and Chang, H.T. (2001). Comparisons of the photostability of esterified wood. *Polymer Degradation and Stability*, **71**(2), 261–266.
Chang, H.T. and Chang, S.T. (2002). Moisture excluding efficiency and dimensional stability of wood improved by acylation. *Bioresource Technology*, **85**(2), 201–204.
Chang, H.T. and Chang, S.T. (2003). Improvements in dimensional stability and lightfastness of wood by butyrylation using microwave heating. *Journal of Wood Science*, **49**(5), 455–460.
Chang, C.I. and Keith, C.T. (1978). Properties of heat-darkened wood. II. Mechanical properties and gluability. Report, Eastern Forest Products Laboratory, Canada, No. OPX214E.
Chang, S.T., Chang, H.T., Huang, Y.S. and Hsu, F.L. (2000). Effects of chemical modification reagents on acoustic properties of wood. *Holzforschung*, **54**(6), 669–675.
Chanzy, H. (1990). Aspects of cellulose structure. In: *Cellulose Sources and Exploitation: Industrial Utilisation, Biotechnology and Physico-chemical Properties*, Kennedy, J.F., Phillips, G.O. and Williams, P.A. (Eds.). Ellis Horwood, Chichestir, UK, pp. 3–12.
Chauhan, S.S., Aggrawal, P., Karmarkar, A. and Pandey, K.K. (2001). Moisture adsorption behaviour of esterified rubber wood (*Hevea brasiliensis*). *Holz als Roh- und Werkstoff*, **59**(4), 250–253.
Chen, G.C. (1992a). Fungal resistance of loblolly pine reacted with para-toluene, sulfonyl chloride or isocyanate. *Wood and Fiber Science*, **24**(2), 161–167.
Chen, G.C. (1992b). Fungal resistance of loblolly pine reacted with methyl or phenyl isocyanate. *Holzforschung*, **46**(1), 77–80.
Chen, G.C. (1992c). Fungal decay resistance and dimensional stability of loblolly pine reacted with 1,6-diisocyanatohexane. *Wood and Fiber Science*, **24**(3), 307–314.

Chen, G.C. (1994). Fungal decay resistance of wood reacted with chlorosulfonyl isocyanate or epichlorohydrin. *Holzforschung*, **48**(3), 181–185.

Chen, R. and Kokta, B.V. (1982). Graft copolymerisation of lignosulphonate with methacrylic acid and acrylate monomers. In: *Graft Polymerisation of Lignocellulosic Fibers*, Hon, D.N.S. (Ed.). ACS Symposium Series, **187**, pp. 285–299.

Chen, G.C., Rowell, R.M. and Ellis, W.D. (1990). Fungal resistance of southern pine impregnated with methyl fluorophenyl carbamates or reacted with fluorophenyl isocyanates. *Wood and Fiber Science*, **22**(2), 165–172.

Chen, M.-J., Meister, J.J., Gunnells, D.W. and Gardner, D.J. (1995). Alteration of the surface energy of wood using lignin-(-1-phenylethene) graft copolymers. *Journal of Wood Chemistry and Technology*, **15**(2), 287–302.

Chow, P., Harp, T., Meimban, R., Youngquist, J.A. and Rowell, R.M. (1994). Biodegradation of acetylated southern pine and aspen composition board. International Research Group on Wood Preservation, Doc. No. IRG/WP 94-40020.

Chow, P., Bao, Z., Youngquist, J.A., Rowell, R.M., Muehl, J.H. and Krzysik, A.M. (1996a). Effects of two fiber treatments on properties of hemlock hardboard. *Forest Products Journal*, **46**(7/8), 62–66.

Chow, P., Bao, Z., Youngquist, J.A., Rowell, R.M., Muehl, J.H. and Krzysik, A.M. (1996b). Properties of hardboards made from acetylated aspen and southern pine. *Wood and Fiber Science*, **28**(2), 252–258.

Chow, S.-Z. (1971). Infrared spectral characteristics and surface inactivation of wood at high temperatures. *Wood Science and Technology*, **5**(1), 27–39.

Chow, S.-Z. (1972). Hydroxyl accessibility, moisture content, and biochemical activity in cell walls of Douglas-fir trees. *Tappi*, **55**(4), 539–544.

Chow, S.-Z. and Mukai, H.N. (1972). Effect of thermal degradation of cellulose on wood–polymer bonding. *Wood Science*, **4**(4), 202–208.

Chowdhury, M.J.A. and Humphrey, P.E. (1999). The effect of acetylation on the shear strength development kinetics of phenolic resin-to-wood bonds. *Wood and Fiber Science*, **31**(3), 293–299.

Chtourou, H., Riedl, B. and Ait-Kadi, A. (1992). Reinforcement of recycled polyolefins with wood fibres. *Journal of Reinforced Plastic and Composites*, **11**, 372–394.

Clausen, C.A. (2004a). Inproving the two-step remediation process for CCA-treated wood: part I. Evaluating oxalic acid extraction. *Waste Management*, **24**(4), 401–405.

Clausen, C.A. (2004b). Improving the two-step remediation process for CCA-treated wood: part II. Evaluating bacterial nutrient sources. *Waste Management*, **24**(4), 407–411.

Clemons, C., Young, R.A. and Rowell, R.M. (1992). Moisture sorption properties of composite boards from esterified aspen fiber. *Wood and Fiber Science*, **24**(3), 353–363.

Clermont, L.P. and Bender, F. (1957). The effect of swelling agents and catalysts on acetylation of wood. *Forest Products Journal*, **7**(5), 167–170.

Codd, P., Banks, W.B., Cornfield, J.A. and Williams, G.R. (1992). The biological effectiveness of wood modified with heptadecylsuccinic anhydride against two brown rot fungi: *Coniophora puteana* and *Gloeophyllum trabeum*. International Research Group on Wood Preservation, Doc. No. IRG/WP 3705-92.

Cox, T.R.G. and Richardson, B.A. (1978). Chromium in wood preservation: health and environmental aspects. International Research Group on Wood Preservation, Doc. No. IRG/WP 3120.

Dawson, B. and Torr, K. (1992). Spectroscopic and colour studies on acetylated radiata pine exposed to UV and visible light. In: *Pacific Rim Bio-Based Composites Symposium: Chemical Modification of Lignocellulosics*, Plackett, D.V. and Dunningham, E.A. (Eds.). FRI Bulletin, **176**, pp. 41–51.

Dawson, B.S.W., Cowan, F.J., Uprichard, J.M., Cronshaw, D.R. and Coombridge, B.A. (1995). Chemical treatment and coating of *Pinus radiata* to inhibit photoyellowing. *Surface Coatings Australia*, **32**(10), 16–19, 22–25.

Dawson, S.W., Franich, R.A., Kroese, H.W. and Steward, D. (1999). Reactivity of radiata pine sapwood towards carboxylic acid anhydrides. *Holzforschung*, **53**(2), 195–198.

Deka, M. and Saikia, C.N. (2000). Chemical modification of wood with thermosetting resin: effect on dimensional stability and strength property. *Bioresource Technology*, **73**(2), 179–181.

Denes, A.R. and Young, R.A. (1999). Reduction of weathering degradation of wood through plasma-polymer coating. *Holzforschung*, **53**(6), 632–640.

Denes, A.R., Tshabalala, M.A., Rowell, R., Denes, F. and Young, R.A. (1999). Hexamethyldisiloxane-plasma coating of wood surfaces for creating water repellent characteristics. *Holzforschung*, **53**(3), 318–326.

Desch, H.E. and Dinwoodie, J.M. (1996). *Timber: Structure, Properties, Conversion and Use*, 7th edn. Macmillan Press, Basingstoke, Hampshire, UK.

Dewispelaere, W., Raemdonck, J. and Stevens, M. (1977). Decay resistance of wood treated for dimensional stabilisation with monomers and formaldehyde. *Material und Organismen*, **12**(3), 211–222.

Dinwoodie, J.M. (2000). *Timber: its Nature and Behaviour*, 2nd edn. E. and F.N. Spon, London, UK.

Dirol, D. and Guyonnet, R. (1993). The improvement of wood durability by retification process. International Research Group on Wood Preservation, Doc. No. IRG/WP 93-40015.

Dobbs, A.J. and Grant, C. (1978). The volatilisation of arsenic on burning copper–chrome–arsenic (CCA) treated wood. *Holzforschung*, **32**(1), 32–35.

Doi, S., Kurimoto, Y., Takahashi, M. and Yoshimura, T. (1997). Effects of steaming heat treatment on the stimulation of termite feeding. International Research Group on Wood Preservation, Doc. No. IRG/WP 97-10212.

Doi, S. Kurimot, Y., Ohmura, W., Ohara, S. Aoyama, M. and Yoshimura, T. (1999). Effects of heat treatments on the feeding behaviour of two subterranean termites. *Holzforschung*, **53**(3), 225–229.

Donath, S., Militz, H. and Mai, C. (2004). Wood modification with alkoxysilanes. *Wood Science and Technology*, **38**(7), 555–566.

Dreher, W.A., Goldstein, I.S. and Cramer, G.R. (1964). Mechanical properties of acetylated wood. *Forest Products Journal*, **14**(2), 66–68.

Dunningham, E.A., Plackett, D.V. and Singh, A.P. (1992). Weathering of chemically modified wood. Natural weathering of acetylated radiata pine: preliminary results. *Holz als Roh- und Werkstoff*, **50**(11), 429–432.

Dwianto, W., Tanaka, F., Inoue, M. and Norimoto, M. (1996). Crystallinity changes of wood by heat or steam treatment. *Wood Research*, **83**, 47–49.

Eaton, R.A. and Hale, M.D.C. (1993). *Wood: Decay, Pests and Protection*. Chapman and Hall, London, UK.

Ebrahimzadeh, P.R. (1998). Effect of impregnation on mechanosorption in wood and paper studied by dynamic mechanical analysis. *Wood Science and Technology*, **32**(2), 101–118.

Efanov, M.V. (2001). Transformations of aspen wood and its principal components by *o*-acylation. *Chemistry of Natural Compounds*, **37**(5), 482–485.

Elder, T. (1991). The pyrolysis of wood. In: *Wood and Cellulosic Chemistry*, Hon, D.N.S. and Shiraishi, N. (Eds.). Marcel Dekker, New York, USA, pp. 665–699.

Ellis, W.D. and Rowell, R.M. (1984). Reaction of isocyanates with southern pine to improve dimensional stability and decay resistance. *Wood and Fiber Science*, **16**(3), 349–356.

Engonga, P.E., Schneider, R., Gérardin, P. and Loubinoux, B. (1999). Chemical modification of wood with perfluoroalkyl ethanol and 4,4′-diphenylmethane diisocyanate. *Holzforschung*, **53**(3), 272–276.

Evans, P. (2003). Emerging technologies in wood protection. *Forest Products Journal*, **53**(1), 14–22.

Evans, P.D. (1998). Weather resistance of wood esterified with dicarboxylic acid anhydrides. *Holz als Roh- und Werkstoff*, **56**(5), 294.

Evans, P.D., Wallis, A.F.A. and Owen, N.L. (2000). Weathering of chemically modified wood surfaces. Natural weathering of Scots pine acetylated to different weight gains. *Wood Science and Technology*, **34**(2), 151–165.

Faison, B.D. and Kirk, T.K. (1983). Relationship between lignin degradation and production of reduced oxygen species by *Phanerochaete chrysosporium*. *Applied and Environmental Microbiology*, **46**(5), 1140–1145.

FAO (2001). *State of the World's Forests*. Food and Agriculture Organisation of the United Nations.

Feist, W.C. and Rowell, R.M. (1982). Ultraviolet degradation and accelerated weathering of chemically modified wood. In: *Graft Copolymerization of Lignocellulosic Fibers*, Hon, D.N.S. (Ed.). ACS Symposium Series, **187**, pp. 349–370.

Feist, W.C. and Sell, J. (1987). Weathering behaviour of dimensionally stabilized wood by heating under pressure of nitrogen gas. *Wood and Fiber Science*, **19**(2), 183–195.

Feist, W.C., Rowell, R.M. and Ellis, W.D. (1991a). Moisture sorption and accelerated weathering of acetylated and methacrylated aspen. *Wood and Fiber Science*, **23**(1), 128–136.

Feist, W.C., Rowell, R.M. and Youngquist, J.A. (1991b). Weathering and finish performance of acetylated aspen fibreboard. *Wood and Fiber Science*, **23**(2), 260–272.

Felby, C., Hassingboe, J. and Lund, M. (2002). Pilot-scale production of fibreboards made by laccase oxidized wood fibers: board properties and evidence for cross-linking of lignin. *Enzyme and Microbial Technology*, **31**(6), 736–741.

Felby, C., Pedersen, L.S. and Nielsen, B.R. (1997a). Enhanced auto adhesion of wood fibers using phenol oxidases. *Holzforschung*, **51**(3), 281–286.

Felby, C., Nielsen, B.R., Olesen, P.O. and Skibsted, L.H. (1997b). Identification and quantification of radical reaction intermediates by electron spin resonance spectrometry of laccase-catalyzed oxidation of wood fibres from beech (*Fagus sylvatica*). *Applied Microbiology and Biotechnology*, **48**(4), 459–464.

Felix, J.M. and Gatenholm, P. (1991). The nature of adhesion in composites of modified cellulose fibers and polypropylene. *Journal of Applied Polymer Science*, **42**(3), 609–620.

Fengel, D. and Wegener, G. (1989). *Wood: Chemistry, Ultrastructure, Reactions*. Walter De Gruyter, Berlin, Germany.

Finlay, M.R. (2004). Old efforts at new uses: a brief history of chemurgy and the American search for biobased materials. *Journal of Industrial Ecology*, **7**(3–4), 33–46.

Flournoy, D.S. (1991). Changes in pore structure and cell wall volume in wood decayed by brown- and white-rot fungi. International Research Group on Wood Preservation, Doc. no. IRG/WP 1501.

Forster, S.C., Hale, M.D. and Williams, G.R. (1997). Efficacy of anhydrides as wood protection chemicals. International Research Group on Wood Preservation, Doc. no. IRG/WP 97-30162.

Forster, S.C., Hale, M.D. and Williams, G.R. (1998). Efficacy of anhydrides as wood protection chemicals. II. Performance against soft rot fungi. International Research Group on Wood Preservation, Doc. no. IRG/WP 98-30174

Frosch, R.A. and Gallopoulos, N.E. (1989). Strategies for manufacturing. *Scientific American*, **261**(3), 144–153.

Fujimoto, H. (1992). Weathering behaviour of chemically modified wood with a maleic acid – glycerol (MG) mixture. In: *Pacific Rim Bio-Based Composites Symposium: Chemical Modification of Lignocellulosics*, Plackett, D.V. and Dunningham, E.A. (Eds.). FRI Bulletin **176**, pp. 87–96.

Fujimoto, H. and Yamagishi, K. (1991). Chemical treatment of wood with a maleic acid and glycerol mixture. In: *Proceedings of the International Symposium on Chemical Modification of Wood (Kyoto)*, Takahashi, M. (Ed.), pp. 83–91.

Fujimoto, H., Anazawa, T. and Yamagishi, K. (1987). The production of antiswelling particleboard. Treatment with maleic acid/glycerol mixture. *Mokuzai Gakkaishi*, **33**(7), 610–612.

Fujimoto, H., Anazawa, T. and Yamagishi, K. (1988). The production of antiswelling particleboard I. Treatment with a maleic acid and glycerol mixture. *Mokuzai Gakkaishi*, **34**(11), 904–909.

Fujimoto, H., Anazawa, T., Ohmiya, Y. and Yamagishi, K. (1991). Dimensional stability of a maleic acid – glycerol (MG) treated particleboard. Effects of MG contents and hot-press temperature. *Mokuzai Gakkaishi*, **37**(5), 456–461.

Fujimura, T. and Inoue, M. (1991). Improvement of the durability of wood with acryl-high-polymer III. Dimensional stability of wood with crosslinked epoxy-copolymer. *Mokuzai Gakkaishi*, **37**(8), 719–726.

Fujimura, T. and Inoue, M. (1992). Improvement of the durability of wood with acrylic high polymers IV. Effects of bulking on the dimensional stability of composites. *Mokuzai Gakkaishi*, **38**(3), 264–269.

Fujimura, T., Inoue, M. and Uemura, I. (1990). Durability of wood with acryl-high-polymer II. Dimensional stability with crosslinked acryl polymer in wood. *Mokuzai Gakkaishi*, **36**(10), 851–859.

Fujimura, T., Ryu, J.Y., Imamura, Y., Furuno, T. and Jodai, S. (1993a). Improvement of the durability of wood with acryl-high-polymer VII. Biological resistance of acrylic-copolymer treated wood. *Mokuzai Gakkaishi*, **39**(9), 1042–1048.

Fujimura, T., Inoue, M., Furuno, T., Imamura, Y. and Jodai, S. (1993b). Improvement of the durability of wood with acryl-high-polymer V. Adsorption of hydrophilic acrylic polymer onto wood swollen with acetone. *Mokuzai Gakkaishi*, **39**(3), 315–321.

Funakoshi, H., Shiraishi, N., Norimoto, M., Aoki, T., Hayashi, H. and Yokota, T. (1979). Studies on the thermoplasticization of wood. *Holzforschung*, **33**(5), 159–166.

Fung, D.P.C., Stevenson, J.A. and Shields, J.K. (1974). The effect of heat and $NH_4H_2PO_4$ on the dimensional and anatomical properties of Douglas-fir. *Wood Science*, **7**(1), 13–20.

Furuno, T. and Goto, T. (1979). Structure of the interface between wood and synthetic polymer. XII. Distribution of styrene polymer in the cell wall of wood–polymer composite (WPC) and dimensional stability. *Mokuzai Gakkaishi*, **25**(7), 488–495.

Furuno, T., Uehara, T. and Jodai, S. (1991). Combinations of wood and silicate I. Impregnation by water glass and applications of aluminium sulfate and calcium chloride as reactants. *Mokuzai Gakkaishi*, **37**(5), 462–472.

Furuno, T., Uehara, T. and Jodai, S. (1993). Combinations of wood and silicate III. Some properties of wood–mineral composites using water–glass–boron compound system. *Mokuzai Gakkaishi*, **39**(5), 561–570.

Furuno, T., Shimada, K., Uehara, T. and Jodai, S. (1992). Combinations of wood and silicate II. Wood–mineral composites using water glass and reactants of barium chloride, boric acid, borax, and their properties. *Mokuzai Gakkaishi*, **38**(5), 448–457.

Fuse, G. and Nishimoto, K. (1961). Preservation of wood by chemical modification. I. Preservative properties of cyanoethylated wood. *Mokuzai Gakkaishi*, **7**(4), 157–161.

Fuwape, J.A. and Oyagade, A.O. (1999). Strength and dimensional stability of acetylated *Gmelina* and Sitka spruce particle board. *Journal of the Timber Development Association of India*, **45**(1/2), 5–10.

Fuwape, J.A. and Oyagade, A.O. (2000). Strength and dimensional stability of acetylated *Gmelina* and Sitka spruce particle board. *Journal of Tropical Forest Products*, **6**(2), 184–189.

Galperin, A.S., Kuleshov, G.G., Tarashkevich, V.I. and Shutov, G.M. (1995). Manufacturing and properties of modified wood: A review of 25 years work. *Holzforschung*, **49**(1), 45–50.

Garrote, G. Domínguez, H. and Parajó, J.C. (1999). Hydrothermal processing of lignocellulosic materials. *Holz als Roh- und Werkstoff*, **57**(3), 191–202.

Geiser, K. (2001). *Materials Matter: Towards a Sustainable Materials Policy*. MIT Press, Cambridge, Massachusetts, USA.

Gérardin, P., Maurin, E. and Loubinoux, B. (1995). Reaction of wood with isocyanates generated *in situ* from acyl azides. *Holzforschung*, **49**(4), 379–381.

Gierer, J., Yang, E. and Reitberger, T. (1992). The reactions of hydroxyl radicals with aromatic rings in lignins, studied with creosol and 4-methylveratrol. *Holzforschung*, **46**(6), 495–504.

Gierlinger, N., Hansmann, C., Roder, T., Sixta, H., Gindl, W. and Wimmer, R. (2005). Comparison of UV and confocal Raman spectroscopy to measure the melamine – formaldehyde resin content within cell walls of impregnated spruce wood. *Holzforschung*, **59**(2), 210–213.

Gjovik, L.R. and Davidson, H.L. (1973). Comparison of wood preservatives in stake tests. Research Note, Forest Products Laboratory, USDA Forest Service, No. FPL-02.

Goethals, P. and Stevens, M. (1994). Dimensional stability and decay resistance of wood upon modification with some new type chemical reactants. International Research Group on Wood Preservation, Doc. No. IRG/WP 94-40028.

Goldstein, I.S. (1955). The impregnation of wood to impart resistance to alkali and acid. *Forest Products Journal*, **5**, 265–267.

Goldstein, I.S. and Dreher, W.A. (1960). Stable furfuryl alcohol impregnating solutions. *Industrial and Engineering Chemistry*, **52**(1), 57–58.

Goldstein, I.S., Dreher, W.A. and Jeroski, E.B. (1959). Wood processing inhibition against swelling and decay. *Industrial and Engineering Chemistry*, **51**(10), 1313–1317.

Goldstein, I.S., Jeroski, E.B., Lund, A.E., Nielson, J.F. and Weaver J.W. (1961). Acetylation of wood in lumber thickness. *Forest Products Journal*, **11**(8), 363–370.

Gomez-Bueso, J., Westin, M., Torgilsson, R., Olesen, P.O. and Simonson, R. (1999a). Composites made from acetylated lignocellulosic fibers of different origin. Part I. Properties of dry-formed fiber boards. *Holz als Roh- und Werkstoff*, **57**(6), 433–438.

Gomez-Bueso, J., Westin, M., Torgilsson, R., Olesen, P.O. and Simonson, R. (1999b). Composites made from acetylated lignocellulosic fibers of different origin. Part II. The effect of nonwoven fiber mat composition upon molding ability. *Holz als Roh- und Werkstoff*, **57**(3), 178–184.

Gomez-Bueso, J., Westin, M., Torgilsson, R., Olesen, P.O. and Simonson, R. (2000). Composites made from acetylated lignocellulosic fibers of different origin. Part I. Properties of dry-formed fiber boards. *Holz als Roh- und Werkstoff*, **58**(1–2), 9–14.

Goodell, B., Nicholas, D.D. and Schultz, T.P. (2003). *Wood Deterioration and Preservation: Advances in our Changing World.* ACS Symposium Series, **845**. American Chemical Society.

Goroyias, G.J. and Hale, M.D. (2002). Heat treatment of wood strands for OSB production: effect on the mechanical properties, water absorption and dimensional stability. International Research Group on Wood Preservation, Doc. No. IRG/WP 02-40238.

Grelier, S., Castellan, A., Desrousseaux, S., Nourmamode, A. and Podgorski, L. (1997). Attempt to protect wood colour against UV/visible light by using antioxidants bearing isocyanate groups and grafted to the material with microwave. *Holzforschung*, **51**(6), 511–518.

Grönqvist, S., Suurnäkki, A., Niku-Paavola, M.L., Kruus, K., Buchert, J. and Viikari, L. (2003). Lignocellulosic processing with oxidative enzymes. In: *Applications of Enzymes to Lignocellulosics*, Mansfield, D. and Saddler, J.N. (Eds.). ACS Symposium Series, **855**, pp. 46–65.

Guevera, R., and Moslemi, A.A. (1983). Hygroexpansive and sorptive behaviour of wood modified with propylene oxide and oligomeric diisocyanate. *Journal of Wood Chemistry and Technology*, **3**(1), 95–114.

Guevera, R. and Moslemi, A.A. (1984). Effect of alkylene oxides, furan resin and vinylpyrrolidinone on wood dimensional stability. *Wood Science and Technology*, **18**(3), 225–240.

Hadi, Y.S. (1992). Acetylated flakeboard properties In: *Pacific Rim Bio-Based Composites Symposium: Chemical Modification of Lignocellulosics*, Plackett, D.V. and Dunningham, E.A. (Eds.). FRI Bulletin, **176**, pp. 9–15.

Hadi, Y.S., Darma, I.G.K.T., Febrianto, F. and Herliyana, E.N. (1995). Acetylated rubberwood flakeboard resistance to bio-deterioration. *Forest Products Journal*, **45**(10), 64–66.

Hager, R. (1995). Waterborne silicones as wood preservatives. International Research Group on Wood Preservation, Doc. No. IRG/WP 95-30062.

Han, G.S., Mori, A. Yoshioka, M. and Shiraishi, N. (1991). Composites of wood and unsaturated polyesters. *Mokuzai Gakkaishi*, **37**(1), 44–49.

Hanger, J., Huber, H., Lackner, R., Wimmer, R. and Fellner, J. (2002). Improving the natural durability of heat-treated spruce, pine and beech. *Holzforschung und Holzverwertung*, **54**(5), 92–93.

Hansmann, C., Schwanninger, M., Stefke, B., Hinterstoisser, B. and Gindl, W. (2004). UV-microscopic analysis of acetylated spruce and birch cell walls. *Holzforschung*, **58**(5), 483–488.

Haque, M.N. and Hill, C.A.S. (1998). Chemical modification of wood flour and fibre with acetic anhydride. *Journal of the Timber Development Association of India*, **54**(3), 25–33.

Haque, M.N. and Hill, C.A.S. (2000). Chemical modification, wood flour and fibre with acetic anhydride. *Journal of the Institute of Wood Science*, **15**(3), 109–115.

Haw, J.F. and Schultz, T.P. (1985). Carbon-13 CP/MAS NMR and FT–IR study of low-temperature lignin pyrolysis. *Holzforschung*, **39**(5), 289–296.

Helsen, L., Van de Bulck, E. and Hery, J.S. (1998). Total recycling of CCA treated wood waste by low-temperature pyrolysis. *Waste Management*, **18**(6–8), 571–578.

Henningsson, B. and Carlsson, B. (1984). Leaching of copper, chrome and srsenic from preservative-treated timber in playground equipment. International Research Group on Wood Preservation, Doc. No. IRG/WP 3149.

Hietala, S., Maunu, S.L., Sundholm, F., Jämsä, S. and Viitaniemi, P. (2002). Structure of thermally modified wood studied by liquid state NMR measurements. *Holzforschung*, **56**(5), 522–528.

Hill, C.A.S. (2000). Wood–plastic composites: strategies for compatibilising the phases. *Journal of the Institute of Wood Science*, **15**(3), 140–146.

Hill, C.A.S. (2003). The kinetics of acetylation reactions. In: *Proceedings of the First European Conference on Wood Modification*, Ghent, Belgium, Van Acker, J. and Hill, C.A.S. (Eds.), pp. 103–111.

Hill, C.A.S. and Çetin, N.S. (2000). Surface activation of wood for graft polymerisation. *International Journal of Adhesion and Adhesives*, **20**, 71–76.

Hill, C.A.S. and Hillier, J. (1998). The kinetics of anhydride modification reactions of wood. Experimental results and theoretical modelling. International Research Group on Wood Preservation, Doc. No. IRG/WP/98-40125.

Hill, C.A.S. and Hillier, J. (1999). Kinetic studies of the reaction of carboxylic acid anhydrides with wood. Experimental determination and modelling of kinetic profiles. *Physical Chemistry and Chemical Physics*, **1**, 1569–1576.

Hill, C.A.S. and Jones, D. (1996a). The dimensional stabilisation of Corsican pine sapwood by reaction with carboxylic acid anhydrides. *Holzforschung*, **50**(5), 457–462.

Hill, C.A.S. and Jones, D. (1996b). A chemical kinetics study of the propionic anhydride modification of Corsican pine. Determination of activation energies. *Journal of Wood Chemistry and Technology*, **16**(3), 235–247.

Hill, C.A.S. and Jones, D. (1996c). Chemical modification of wood new results and potentials. In: *British Wood Preserving and Damp-proofing Association Annual Convention*, Stratford-upon-Avon, UK, pp. 42–46.

Hill, C.A.S. and Jones, D. (1999). Dimensional changes in Corsican pine sapwood due to chemical modification with linear chain anhydrides. *Holzforschung*, **53**(3), 267–271.

Hill, C.A.S. and Mallon, S. (1998). The chemical modification of Scots pine with succinic anhydride or octenyl succinic anhydride. I. Dimensional stabilisation. *Holzforschung*, **52**(4), 427–433.

Hill, C.A.S. and Ormondroyd, G.A. (2004). Dimensional changes in Corsican pine (*Pinus nigra* Arnold) modified with acetic anhydride measured using a helium pycnometer. *Holzforschung*, **58**(5), 544–547.

Hill, C.A.S. and Papadopoulos, A.N. (2001). A review of methods used to determine the size of cell wall microvoids of wood. *Journal of the Institute of Wood Science*, **15**(6), 337–345.

Hill, C.A.S. and Papadopoulos, A.N. (2002). The pyridine-catalysed acylation of pine sapwood and phenolic model compounds with carboxylic acid anhydrides. Determination of activation energies and entropy of activation. *Holzforschung*, **56**(2), 150–156.

Hill, C.A.S., Çetin, N.S. and Ozmen, N. (2000). Potential catalysts for the acetylation of wood. *Holzforschung*, **54**(3), 269–272.

Hill, C.A.S., Farahani, M.R.M. and Hale, M.D.C. (2004a). The use of organo alkoxysilane coupling agents for wood preservation. *Holzforschung*, **58**(3), 316–325.

Hill, C.A.S., Papadopoulos, A.N. and Payne, D. (2004b). Chemical modification employed as a means of probing the cell-wall micropore of pine sapwood. *Wood Science and Technology*, **37**(6), 475–488.

Hill, C.A.S., Jones, D., Strickland, G. and Çetin, N.S. (1998). Kinetic and mechanistic aspects of the acetylation of wood with acetic anhydride. *Holzforschung*, **52**(6), 623–629.

Hill, C.A.S., Çetin, N.S., Quinney, R.F., Derbyshire, H. and Ewen, R.J. (2001). An investigation of the potential for chemical modification and subsequent polymeric grafting as a means of protecting wood against photodegradation. *Polymer Degradation and Stability*, **72**(1), 133–139.

Hill, C.A.S., Forster, S., Farahani, M.R.M., Hale, M.D.C. and Williams, G. (2005). An investigation of cell wall micropore blocking as a possible mechanism for the decay resistance of anhydride modified wood. *International Biodeterioration and Biodegradation*, **55**(1), 69–76.

Hill, C.A.S., Hale, M.D., Farahani, M.R., Forster, S., Suttie, E.D., Jones, D. and Papadopoulos, A.N. (2003). Decay of anhydride modified wood. In: *Proceedings of the First European Conference on Wood Modification*, Ghent, Belgium, Van Acker, J. and Hill, C.A.S. (Eds.), pp. 143–152.

Hillier, W. and Murphy, R. (2000). Life cycle assessment of wood products—a good story to tell. *Journal of the Institute of Wood Science*, **15**(4), 221–232.

Hillier, W., Murphy, R.J., Dickinson, D.J. and Bell, J.N.B. (1996). Life-cycle assessment of treated wood: a view from the road. International Research Group on Wood Preservation, Doc. No. IRG/WP 96-50078.

Hirai, N., Sobue, N. and Asano, I. (1972). Studies on piezoelectric effect of wood. IV. Effects of heat treatment on cellulose crystallites and piezoelectric effect of wood. *Mokuzai Gakkaishi*, **18**(11), 535–542.

Hon, D.N.S. (1995). Stabilization of wood color: is acetylation blocking effective? *Wood and Fiber Science*, **27**(4), 360–367.

Hon, D.N.S. (Ed.) (1996). *Chemical Modification of Lignocellulosic Materials*. Marcel Dekker, New York, USA.

Hon, D.N.S. and Bangi, A.P. (1996). Chemical modification of juvenile wood. Part 1. Juvenility and response of southern pine OSB flakes to acetylation. *Forest Products Journal*, **46**(7/8), 73–78.

Hon, D.N.S. and Ou, N.-H. (1989). Thermoplasticization of wood. I. Benzylation of wood. *Journal of Polymer Science, Part A: Polymer Chemistry*, **27**, 2457–2482.

Hon, D.N.S. and San Luis, J.M. (1989). Thermoplasticization of wood. II. Cyanoethylation. *Journal of Polymer Science, Part A: Polymer Chemistry*, **27**, 4143–4160.

Hon, D.N.S. and Shiraishi, N. (2001). *Wood and Cellulosic Chemistry*. Marcel Dekker, New York, USA.

Honma, S. and Nakano, T. (1991). Carboxymethylation of wood in organic solvent – water reaction media. *Mokuzai Gakkaishi*, **37**(7), 625–630.

Houts, J.H. van, Winistorfer, P.M. and Wang, S.Q. (2003). Improving dimensional stability by acetylation of discrete layers within flakeboard. *Forest Products Journal*, **53**(1), 82–88.

Hsu, W.E., Schwald, W. and Shields, J.A. (1989). Chemical and physical changes required for producing dimensionally stable wood-based composites. Part 2: heat post-treatment. *Wood Science and Technology*, **23**(3), 281–288.

Ibach, R.E. and Lee, B.-G. (2002). The effect on biological and moisture resistance of epichlorohydrin chemically modified wood. International Research Group on Wood Preservation, Doc. No. IRG/WP 02-40224.

Ibach, R.E. and Rowell, R.M. (2000). Improvements in decay resistance based on moisture exclusion. *Molecular Crystals and Liquid Crystals*, **353**, 23–33.

Imamura, Y. (1991). Morphological characters of acetylated wood exposed to weathering. In: *Proceedings of the International Symposium on Chemical Modification of Wood (Kyoto)*, Takahashi, M. (Ed.), pp. 130–135.

Imamura, Y. (1993). Morphological changes in acetylated wood exposed to weathering. *Wood Research Kyoto*, **79**, 54–61.

Imamura, Y. and Nishimoto, K. (1986). Resistance of acetylated wood to attack by subterranean termites. *Wood Research Kyoto*, **72**, 37–44.

Imamura, Y. and Nishimoto, K. (1987a). Some aspects of the resistance of acetylated wood against biodeterioration. *Wood Research Kyoto*, **74**, 33–44.

Imamura, Y. and Nishimoto, K. (1987b). Bending properties of acetylated particleboards exposed to decay fungi. *Mokuzai Gakkaishi*, **33**(1), 25–30.

Imamura, Y. Nishimoto, K., Yoshida, Y., Kawai, S., Sato, T. and Nakaji, M. (1986). Production technology for acetylated low-density particleboard II. Decay and termite resistance. *Wood Research Kyoto*, **73**, 35–43.

Imamura, Y., Nishimoto, K. and Rowell, R.M. (1987). Internal bond strength of acetylated flakeboard exposed to decay hazards. *Mokuzai Gakkaishi*, **33**(12), 986–991.

Imamura, Y., Rowell, R.M., Simonson, R. and Tillman A.-M. (1988). Bending creep tests on acetylated pine and birch particleboards during white- and brown-rot fungal attack. *Paperi ja Puu*, **9**, 816–820.

Imamura, Y., Subiyanto, B., Rowell, R.M. and Nilsson, T. (1989). Dimensional stability and biological resistance of particleboard from acetylated albizzia wood particles. *Wood Research Kyoto*, **76**, 49–58.

Inoue, M. and Norimoto, M. (1991). Permanent fixation of compressive deformation of wood. In: *Proceedings of the International Symposium on Chemical Modification of Wood (Kyoto)*, Takahashi, M. (Ed.), pp. 45–50.

Inoue, M., Norimoto, M., Tanahashi, M. and Rowell, R.M. (1993a). Steam or heat fixation of compressed wood. *Wood and Fiber Science*, **25**(3), 224–235.

Inoue, M., Marooka, T., Norimoto, M., Rowell, R.M. and Egawa, G. (1992). Permanent fixation of compressive deformation of wood. (II) Mechanisms of permanent fixation. In: *Pacific Rim Bio-Based Composites Symposium: Chemical Modification of Lignocellulosics*, Plackett, D.V. and Dunningham, E.A. (Eds.). FRI Bulletin, **176**, pp. 181–189.

Inoue, M., Ogata, S., Kawai, S., Rowell, R.M. and Norimoto, M. (1993b). Fixation of compressed wood using melamine – formaldehyde resin. *Wood and Fiber Science*, **25**(4), 404–410.

Ito, Y., Tanahashi, M., Shigematsu, M. and Shinoda, Y. (1998a). Compressive-molding of wood by high-pressure steam-treatment. Part 2. Mechanism of permanent fixation. *Holzforschung*, **52**(2), 217–221.

Ito, Y., Tanahashi, M., Shigematsu, M., Shinoda, Y. and Ohta, C. (1998b). Compressive-molding of wood by high-pressure steam-treatment. Part 1. Development of compressively molded squares from thinnings. *Holzforschung*, **52**(2), 211–216.

Jaksland, C., Rasmussen, E. and Rohde, T. (2000). A new technology for the treatment of PVC waste. *Waste Management*, **20**(5–6), 463–467.

Jermer, J. (1990). Production of preservative treated wood in some countries, (1990). International Research Group on Wood Preservation, Doc. No. IRG/WP 3598.

Johnson, B.R. and Rowell, R.M. (1988). Resistance of chemically modified wood to marine borers. *Material und Organismen*, **23**(2), 147–156.

Kajita, H. and Imamura, Y. (1993). Chemically modified particleboards. In: *Recent Research on Wood and Wood-based Materials*, Shiraishi, N., Kajita, H. and Norimoto, M. (Eds.). Elsevier, Barking, UK, pp. 67–74.

Kalnins, M.A. (1982). Chemical modification of wood for improved decay resistance. *Wood Science*, **15**(2), 81–89.

Kalnins, M.A. (1984). Photochemical degradation of acetylated, methylated, phenylhydrazine-modified and ACC-treated wood. *Journal of Applied Polymer Science*, **29**(1), 105–115.

Kamden, D.P., Pizzi, A. and Triboulot, M.C. (2000). Heat-treated timber: potentially toxic byproducts presence and extent of cell wall degradation. *Holz als Roh- und Werkstoff*, **58**(4), 253–257.

Kamden, D.P., Pizzi, A. and Jermannaud, A. (2002). Durability of heat-treated wood. *Holz als Roh- und Werkstoff*, **60**(1), 1–6.

Kamden, D.P., Pizzi, A. Guyonnet, R. and Jermannaud, A. (1999). Durability of heat-treated wood. International Research Group on Wood Preservation, Doc. No. IRG/WP 99-40145.

Kameyama, N., Yano, H., Minato, K. and Norimoto, M. (1994). Relationships between dielectric properties and dimensional stabilities of formaldehyde-treated wood. *Mokuzai Gakkaishi*, **40**(4), 399–406.

Kawai, S., Wang, Q., Sasaki, H. and Tanahashi, M. (1992). Production of compressed laminated veneer lumber by steam pressing. In: *Pacific Rim Bio-Based Composites Symposium: Chemical Modification of Lignocellulosics*, Plackett, D.V. and Dunningham, E.A. (Eds.). FRI Bulletin, **176**, pp. 121–128.

Kazi, F.K.M. and Cooper, P.A. (2002). Rapid extraction oxidation process to recover and reuse copper chromium and arsenic from industrial wood preservative sludge. *Waste Management*, **22**(3), 293–301.

Keith, C.T. and Chang, C.I. (1978). Properties of heat-darkened wood. I. Hygroscopic properties. Report, Eastern Forest Products Laboratory, Canada, No. OPX213E.

Kenaga, D.L. and Sproull, R.C. (1951). Further experiments on dimensional stabilization of wood by allylation. *Journal of the Forest Products Research Society*, **1**, 28–32.

Kenaga, D.L., Sproull, R.C. and Esslinger, J. (1950). Preliminary experiments on dimensional stabilization of wood by allylation. *Southern Lumberman*, **180**, 45, 48.

Kharazipour, A., Hüttermann, A. and Luedemann, H.D. (1997). Enzymatic activation of wood fibres as a means of the production of wood composites. *Journal of Adhesion Science and Technology*, **11**(3), 419–427.

Kharazipour, A., Bergmann, K., Nonninger, K. and Hüttermann, A. (1998). Properties of fibre boards obtained by activation of the middle lamella of wood fibres with peroxidase and H_2O_2 before conventional pressing. *Journal of Adhesion Science and Technology*, **12**(10), 1045–1053.

Kiguchi, M. (1990a). Chemical modification of wood surfaces by etherification. I. Manufacture of surface hot-melted wood by etherification. *Mokuzai Gakkaishi*, **36**(8), 651–658.

Kiguchi, M. (1990b). Chemical modification of wood surfaces by etherification. II. Weathering ability of hot-melted wood surfaces and manufacture of self hot-melt bonded particleboard. *Mokuzai Gakkaishi*, **36**(10), 867–875.

Kiguchi, M. (1993). Chemical modification of wood surfaces by etherification. IV. Benzylation with solvent-dilution and vapor-phase methods. *Mokuzai Gakkaishi*, **39**(1), 80–85.

Kiguchi, M. (1995). Chemical modification of wood in Japan: research and commercial applications. *Wood Science Digest*, **55**(March), 3–4.

Kiguchi, M. (1997). Photo-deterioration of chemically modified wood surfaces: acetylated wood and alkylated wood. *Japan Agricultural Research Quarterly*, **31**(2), 147–154.

Kiguchi, M. and Evans, P.D. (1998). Photostabilisation of wood surfaces using a grafted benzophenone UV absorber. *Polymer Degradation and Stability*, **61**(1), 33–45.

Kiguchi, M. and Suzuki, M. (1985). Physical properties of particleboard with acetylated chips and binder. *Mokuzai Gakkaishi*, **31**(3), 200–208.

Kiguchi, M. and Yamamoto, K. (1991). Dimensional stability and decay resistance of hot-melt self bonded particleboard by surface benzylated pine chips. International Research Group on Wood Preservation, Doc. No. IRG/WP 3652.

Kiguchi, M. and Yamamoto, K. (1992). Chemical modification of wood surfaces by esterification III. Some properties of self-bonded benzylated particleboard. *Mokuzai Gakkaishi*, **38**(2), 150–158.

Kiguchi, M., Karaoke, Y., Evans, P.D., Kadegaru, Y. and Imamura, Y. (2001). Pretreatments of wood surfaces for improving weatherability of clear finishing. International Research Group on Wood Preservation, Doc. No. IRG/WP 01-40196.

Kim, D.-Y., Nishiyama, Y., Wada, M., Kuga, S. and Okano, T. (2001). Thermal decomposition of cellulose crystallites in wood. *Holzforschung*, **55**(5), 521–524.

Kim, G.-H., Yun, K.-E. and Kim, J.-J. (1998). Effect of heat treatment on the decay resistance and bending properties of radiata pine sapwood. *Material und Organismen*, **32**(2), 101–108.

Kirk, D.N. (1973). Selectivity in reactions of epoxides. *Chemistry and Industry*, no. 3, 109–110.

Kishi, H. and Shiraishi, N. (1986). Wood–phenol adhesives prepared from carboxymethylated wood II. *Mokuzai Gakkaishi*, **32**(7), 520–526.

Kishi, H., Yoshioka, M., Yamanoi, A. and Shiraishi, N. (1988). Composites of wood and polypropylenes I. *Mokuzai Gakkaishi*, **34**(2), 133–139.

Klinga, L.O. and Tarkow, H. (1966). Dimensional stabilization of hardboard by acetylation. *Tappi*, **49**(1), 23–27.

Kokta, B.V., Lo, R.C. and Daneault, C. (1982). The xanthate method of grafting IX. Grafting of acrylonitrile onto high yield hardwood pulp. In: *Graft Copolymerization of Lignocellulosic Fibers*, Hon, D.N.S. (Ed.). ACS Symposium Series, **187**, pp. 269–283.

Komarek, R.J., Gardner, R.M., Buchanan, C.M. and Gedon, S. (1993). Biodegradation of radiolabelled cellulose acetate and cellulose propionate. *Journal of Applied Polymer Science*, **50**(10), 1739–1746.

Korai, H. (2001). Effects of low bondability of acetylated fibers on mechanical properties and dimensional stability of fiberboard. *Journal of Wood Science*, **47**(6), 430–436.

Korai, H. and Suzuki, M. (1995). Dimensional stability and dynamic viscoelasticity of double-chemically treated wood. *Mokuzai Gakkaishi*, **41**(1), 51–62.

Korai, H., Kiguchi, M., Hosoya, S., Suzuki, S., Hirano, Y. and Iwata, R. (2001). Improvements of fiberboard made from acetylated fibers by ozonation I: effect of ozonation on mechanical properties. *Journal of Wood Science*, **47**(1), 24–29.

Koshijima, T. and Watanabe, T. (2003). *Association Between Lignin and Carbohydrates in Wood and Other Plant Tissues*. Springer Series in Wood Science, Springer-Verlag, Berlin, Germany.

Krause, A., Hof, C. and Militz, H. (2004). Novel wood modification processes for window and cladding products. International Research Group on Wood Preservation, Doc. No. IRG/WP 04-40285.

Kringstad, K. (1969). Degradation of wood and high-yield pulps by light—a survey of the present state of knowledge. *Tappi*, **52**(6), 1070.

Kubojima, Y., Okano, T. and Ohta, M. (1998). Vibrational properties of Sitka spruce heat-treated in nitrogen gas. *Journal of Wood Science*, **44**(1), 73–77.

Kubojima, Y. Okano, T. and Ohta, M. (2000a). Bending strength and toughness of heat-treated wood. *Journal of Wood Science*, **46**(1), 8–15.

Kubojima, Y. Okano, T. and Ohta, M. (2000b). Vibrational properties of heat-treated green wood. *Journal of Wood Science*, **46**(1), 63–67.

Kumar, S. (1994). Chemical modification of wood. *Wood and Fiber Science*, **26**(2), 270–280.

Kumar, S. and Agarwal, S.C. (1983). Biological degradation resistance of wood acetylated with thioacetic acid. International Research Group on Wood Preservation, Doc. No. IRG/WP 3223.

Kumar, S., Dev, I. and Singh, S.P. (1991). Hygroscopicity and dimensional stability of wood acetylated with thioacetic acid and acetyl chloride. *Journal of the Timber Development Association of India*, **37**(1), 25–32.

Kumar, S., Singh, S.P. and Sharma, M. (1979). Effect of acetylation on the permeability of wood. *Journal of the Timber Development Association of India*, **25**(3), 5–9.

Lande, S., Westin, M. and Schneider, M.H. (2004). Environmental friendly wood protection—furfurylated wood as an alternative to wood preservation. In: *International Conference on Environmentally-Compatible Forest Products*, Opporto, Portugal, Jorge, F.C. (Ed.), pp. 163–175.

Larsson, P. and Simonson, R. (1994). A study of the strength, hardness and deformation of acetylated Scandinavian softwoods. *Holz als Roh- und Werkstoff*, **52**(2), 83–86.

Larsson, P. and Tillman, A.-M. (1989). Acetylation of lignocellulosic materials. International Research Group on Wood Preservation, Doc. No. IRG/WP 3516.

Larsson, P., Mahlberg, R., Vick, C., Simonson, R. and Rowell, R.M. (1992). Adhesive bonding of acetylated pine and spruce. In: *Pacific Rim Bio-Based Composites Symposium: Chemical Modification of Lignocellulosics*, Plackett, D.V. and Dunningham, E.A. (Eds.). FRI Bulletin, **176**, pp. 16–24.

Larsson Brelid, P. (2002). The influence of post-treatments on acetyl content for removal of chemicals after acetylation. *Holz als Roh- und Werkstoff*, **60**(2), 92–95.

Larsson Brelid, P. and Simonson, R. (1999). Acetylation of solid wood using microwave heating. Part 2. Experiments in laboratory scale. *Holz als Roh- und Werkstoff*, **57**(5), 383–389.

Larsson Brelid, P., Simonson, R. and Bergman, Ö. (1997). Resistance of acetylated wood of biological degradation. Evaluation of field test. International Research Group on Wood Preservation, Doc. No. IRG/WP 97-30139.

Larsson Brelid, P., Simonson, R. and Risman, P.O. (1999). Acetylation of wood using microwave heating. Part 1. studies of dielectric properties. *Holz als Roh- und Werkstoff*, **57**(4), 259–263.

Larsson Brelid, P., Simonson, R., Bergman, Ö. and Nilsson, T. (2000). Resistance of acetylated wood to biological degradation. *Holz als Roh- und Werkstoff*, **58**(5), 331–337.

Leary, G.J. (1968). The yellowing of wood by light: Part II. *Tappi*, **51**(6), 257–260.

Leithoff, H., Peek, R.D., Borck, V., Gottsche, R., Kirk, H. and Grinda, M. (1999). Toxic values derived from EN113 tests—are they determined by the virulence of the test fungus?—Results from a round robin test. International Research Group on Wood Preservation, Doc. No. IRG/WP 99-20176.

Li, J.Z., Furuno, T. and Katoh, S. (2000a). Dimensional stability and flame resistance of silicate-acetylated and propionylated wood composites. *Journal of Wood Chemistry and Technology*, **20**(4), 441–453.

Li, J.Z., Furuno, T. and Katoh, S. (2001a). Wood propionylation in the presence of catalysts. *Wood and Fiber Science*, **33**(2), 255–263.

Li, J.Z., Furuno, T. and Katoh, S. (2001b). Preparation and properties of acetylated and propionylated wood-silicate composites. *Holzforschung*, **55**(1), 93–96.

Li, J.Z., Furuno, T., Katoh, S. and Uehara, T. (2000b). Chemical modification of wood without solvents or catalysts. *Journal of Wood Science*, **46**(3), 215–221.

Lin, R.T. (1969). Effect of heat treatment on some properties of Douglas-fir. *Wood Science*, **2**(2), 112–119.

Lindroos, L. (2002). Balance of arsenic and recycling. International Research Group on Wood Preservation, Doc. No. IRG/WP 02-50189.

Lippke, B., Wilson, J., Perez-Garcia, J., Bowyer, J. and Meil, J. (2004). CORRIM: life-cycle environmental performance of renewable building materials. *Forest Products Journal*, **54**(6), 8–19.

Liski, J., Karjalainen, T., Pussinen, A., Nabuurs, G.J. and Kauppi, P. (2000). Trees as carbon sinks and sources in the European Union. *Environmental Science and Policy*, **3**(2–3), 91–97.

Liu, F.P., Wolcott, M.P., Gardner, D.J. and Rials, T.G. (1994). Characterisation of the interface between cellulosic fibres and a thermoplastic matrix. *Composite Interfaces*, **2**(6), 419–432.

Loras, V. (1968). Bleaching and stabilizing of flash-dried mechanical pulp. *Pulp and Paper Magazine of Canada*, **69**(2), 57–63.

Loubinoux, B., Gérardin, P., Kilbertus, G. and Miazimbakana, J. (1992). Reaction of wood and cellulose with quinone methides prepared *in situ* from phenolic precursors. *Holzforschung*, **46**(2), 175–176.

Lu, J.Z., Wu, Q. and McNabb, H.S. (2000). Chemical coupling in wood fiber and polymer composites: a review of coupling agents and treatments. *Wood and Fiber Science*, **32**(1), 88–104.

Lukowsky, D. (2002). Influence of the formaldehyde content of waterbased melamine formaldehyde resins on physical properties of Scots pine impregnated therewith. *Holz als Roh- und Werkstoff*, **60**(5), 349–355.

Lukowsky, D. and Peek, R.D. (1998). Time dependent over-uptake of etherificated melamine resins. International Research Group on Wood Preservation, Doc. No. IRG/WP 98-40109.

Lukowsky, D., Buschelberger, F. and Schmidt, O. (1999). *In situ* testing of the influence of melamine resins on the enzymatic activity of basidiomycetes. International Research Group on Wood Preservation, Doc. No. IRG/WP 99–30194.

Lukowsky, D., Peek, R.D. and Rapp, A.O. (1997). Water-based silicones on wood. International Research Group on Wood Preservation, Doc. No. IRG/WP 97-30144.

Lukowsky, D., Peek, R.D. and Rapp, A.O. (1998). Curing conditions for a low formaldehyde etherificated melamine resin. International Research Group on Wood Preservation, Doc. No. IRG/WP 98-40108.

Lutomski, K. (1975). Resistance of beechwood modified with styrene, methyl methacrylate, and diisocyanate against the action of fungi. *Material und Organismen*, **10**(4), 255–262.

MacLean, J.D. (1951). Rate of disintegration of wood under different heating conditions. *Proceedings of the American Wood Preservers Association*, **47**, 155–169.

MacLean, J.D. (1953). Effect of steaming on the strength properties of wood. *Proceedings of the American Wood Preservers Association*, **49**, 88–112.

MacLean, J.D. (1954). Effect of heating in water on the strength properties of wood. *Proceedings of the American Wood Preservers Association*, **50**, 253–281.

Mahlberg, R., Niemi, H.E.M., Denes, F. and Rowell, R.M. (1998). Effect of oxygen and hexamethyldisiloxane plasma on morphology, wettability and adhesion properties of polypropylene and lignocellulosics. *International Journal of Adhesion and Adhesives*, **18**(4), 283–297.

Mahlberg, R., Paajanen, L., Nurmi, A., Kivisto, A., Koskela, K. and Rowell, R.M. (2001). Effect of chemical modification of wood on the mechanical and adhesion properties of wood fiber/polypropylene fiber and polypropylene/veneer composites. *Holz als Roh- und Werkstoff*, **59**(5), 319–326.

Mai, C. and Militz, H. (2004a). Modification of wood with silicon compounds. Inorganic silicon compounds and sol-gel systems: a review. *Wood Science and Technology*, **37**(5), 339–348.

Mai, C. and Militz, H. (2004b). Modification of wood with silicon compounds. Treatment systems based on organic silicon compounds—a review. *Wood Science and Technology*, **37**(6), 453–461.

Makinaga, M., Norimoto, M. and Inoue, M. (1997). Permanent fixation of bending deformation of wood by stem treatment. *Wood Research*, **84**, 39–41.

Mallari, V.C., Fukuda, K., Morohoshi, N. and Haraguchi, T. (1988). Biodegradation of particleboard. I. Decay resistance of chemically modified wood and qualities of particleboard. *Mokuzai Gakkaishi*, **35**(9), 832–838.

Mallari, V.C., Fukuda, K., Morohoshi, N. and Haraguchi, T. (1990). Biodegradation of particleboard. II. Decay resistance of chemically modified particleboard. *Mokuzai Gakkaishi*, **36**(2), 139–146.

Mallon, S. and Hill, C.A.S. (2002). Covalent bonding of wood through chemical activation. *International Journal of Adhesion and Adhesives*, **22**(6), 465–469.

Manninen, A.M., Pasanen, P. and Holopainen, J.K. (2002). Comparing the atmospheric emissions between air-dried and heat-treated Scots pine wood. *Atmospheric Environment*, **36**(11), 1763–1768.

Mansour, O.Y., Nagaty, A. and Beshay, A-F. (1982). Grafting of some vinyl monomers onto lignocellulose and cellulose in the presence of lignin. In: *Graft Polymerisation of Lignocellulosic Fibers*, Hon, D.N.S. (Ed.). ACS Symposium, **187**, pp. 253–268.

Mantanis, G.I., Young, R.A. and Rowell, R.M. (1994). Swelling of wood. Part II. Swelling in organic liquids. *Holzforschung*, **48**(6), 480–490.

Marcea, R.L. and Lau, K.K. (1992). Carbon dioxide implications of building materials. *Journal of Forest Engineering*, **3**(2), 37–43.

Marchetti, V., Gérardin, P., Tekely, P. and Loubinoux, B. (1998). Graft copolymerisation of acrylic acid onto sawdust using $KMnO_4$ as initiator. *Holzforschung*, **52**(6), 654–660.

Marcovich, N.E., Reboredo, M.M. and Aranguren, M.I. (1996). Sawdust modification, maleic anhydride treatment. *Holz als Roh- und Werkstoff*, **54**(3), 189–193.

Martins, V.A. and Banks, W.C.B. (1991). Sorptive properties of chemically modified wood. *Wood Protection*, **1**(2), 69–75.

Mateus, E.P., Ribeiro, A.B. and Ottosen, L. (2002). Electrodialytic Remediation of creosote and CCA treated timber wastes. International Research Group on Wood Preservation, Doc. No. IRG/WP 02–50190.

Matsuda, H. (1987). Preparation and utilisation of esterified woods bearing carboxyl groups. *Wood Science and Technology*, **21**(1), 75–88.

Matsuda, H. (1993). Preparation and properties of oligoesterified wood blocks based on anhydride and epoxide. *Wood Science and Technology*, **27**(1), 23–34.

Matsuda, H. and Ueda, M. (1985a). Preparation and utilisation of esterified woods bearing carboxyl groups. III. Heat and moisture stability of esterified woods. *Mokuzai Gakkaishi*, **31**(2), 103–108.

Matsuda, H. and Ueda, M. (1985b). Preparation and utilisation of esterified woods bearing carboxyl groups. IV. Plasticization of esterified woods. *Mokuzai Gakkaishi*, **31**(3), 215–221.

Matsuda, H. and Ueda, M. (1985c). Preparation and utilisation of esterified woods bearing carboxyl groups. V. Addition reaction of esterified woods with epoxide and oligoesterification of the adducts obtained. *Mokuzai Gakkaishi*, **31**(4), 267–273.

Matsuda, H. and Ueda, M. (1985d). Preparation and utilisation of esterified woods bearing carboxyl groups. VI. Stepwise alternately adding esterification reaction of esterified woods with epoxides and dicarboxylic acid anhydrides. *Mokuzai Gakkaishi*, **31**(6), 468–474.

Matsuda, H. and Ueda, M. (1985e). Preparation and utilisation of esterified woods bearing carboxyl groups. VII. Plasticity of epoxide-adducted esterified woods and oligoesterified woods derived from esterified woods. *Mokuzai Gakkaishi*, **31**(7), 579–586.

Matsuda, H. and Ueda, M. (1985f). Preparation and utilisation of esterified woods bearing carboxyl groups. VIII. Crosslinking of esterifed woods with bisepoxide and properties of crosslinked woods. *Mokuzai Gakkaishi*, **31**(11), 903–911.

Matsuda, H., Dohi, H. and Ueda, M. (1987). Catalysts for the etherification reaction of wood with an epoxy compound. *Mokuzai Gakkaishi*, **33**(11), 884–891.

Matsuda, H., Murakami, K. and Ueda, M. (1988a). Oligoesterifed woods based upon anhydride and epoxide III. Preparation and dimensional stability of oligoesterified woods by heating wood impregnated with anhydride–epoxide solution. *Mokuzai Gakkaishi*, **34**(10), 844–850.

Matsuda, H., Murakami, K. and Ueda, M. (1988b). Oligoesterifed woods based upon anhydride and epoxide IV. Preparation and dimensional stability of oligoesterified woods by heating–suction of wood impregnated with reactant solution. *Mokuzai Gakkaishi*, **34**(12), 1004–1011.

Matsuda, H., Ueda, M. and Hara, M. (1984a). Preparation and utilisation of esterified woods bearing carboxyl groups. I. Esterification of wood with dicarboxylic acid anhydries in the presence of a solvent. *Mokuzai Gakkaishi*, **30**(9), 735–741.

Matsuda, H., Ueda, M. and Mori, H. (1988c). Preparation and crosslinking of oligoesterified woods based on maleic anhydride and allyl glycidyl ether. *Wood Science and Technology*, **22**(1), 21–32.

Matsuda, H., Ueda, M. and Mori, H. (1988d). Preparation and crosslinking of oligoesterified woods based on phthalic anhydride and glycidyl methacrylate. *Wood Science and Technology*, **22**(4), 335–344.

Matsuda, H., Ueda, M. and Murakami, K. (1984b). Preparation and utilisation of esterified woods bearing carboxyl groups. II. Esterification of wood with dicarboxylic acid anhydries in the absence of solvent. *Mokuzai Gakkaishi*, **30**(12), 1003–1010.

Matsuda, H., Ueda, M. and Murakami, K. (1988e). Oligoesterifed woods based upon anhydride and epoxide I. Preparation and dimensional stability of oligoesterified woods by stepwise addition reactions. *Mokuzai Gakkaishi*, **34**(2), 140–148.

Matsuda, H., Ueda, M. and Murakami, K. (1988f). Oligoesterifed woods based upon anhydride and epoxide II. Preparation and dimensional stability of oligoesterified woods by heating wood immersed in anhydride–epoxide solution. *Mokuzai Gakkaishi*, **34**(7), 597–603.

Mayes, D. and Oksanen, O. (2002). *ThermoWood Handbook*. Finnforest, Finland.

McKelvey, J.B., Webre, B.G. and Klein, E. (1959). Reaction of epoxides with cotton cellulose in the presence of sodium hydroxide. *Textile Research Journal*, **29**, 918–925.

McMillan, C.W. (1963). Chemical treatment of wood. Dimensional stabilization with polymerisable vapor of ethylene oxide. *Forest Products Journal*, **13**(2), 56–61.

Messner, K., Bruce, A. and Bongers, H.P.M. (2003). Treatability of refractory wood species after fungal pre-treatment. In: *Proceedings of the First European Conference on Wood Modification*, Ghent, Belgium, Van Acker, J. and Hill, C.A.S. (Eds.), pp. 389–401.

Militz, H. (1991a). The improvement of dimensional stability and durability of wood through treatment with non-catalysed acetic acid anhydride. *Holz als Roh- und Werkstoff*, **49**(4), 147–152.

Militz, H. (1991b). Improvements of stability and durability of beechwood (*Fagus sylvatica*) by means of treatment with acetic anhydride. International Research Group on Wood Preservation, Doc. No. IRG/WP 3645.

Militz, H. (1993). Treatment of timber with water soluble dimethylol resins to improve their dimensional stability and durability. *Wood Science and Technology*, 27(5), 347–355.

Militz, H. (2002). Thermal treatment of wood: European processes and their background. International Research Group on Wood Preservation, Doc. No. IRG/WP 02–40241.

Militz, H. and Peek, R.D. (1993). Possibilities of improving some characteristics of poplar wood by impregnation with water-soluble resins. *Material und Organismen*, **28**(1), 55–73.

Militz, H., Beckers, E.P.J. and Homan, W.J. (1997). Modification of solid wood: research and practical potential. International Research Group on Wood Preservation, Doc. No. IRG/WP 97–40098.

Militz, H., Son, D.W., Gómez-Hernández, L. and Sierra-Alvarez, R. (2003). Effect of fungal degradation on the chemical composition of acetylated beech wood. International Research Group on Wood Preservation, Doc. No. IRG/WP 03–40267.

Millett, M.A. and Gerhards, G.C. (1972). Accelerated aging: residual weight and flexural properties of wood heated in air at 115 °C to 175 °C. *Wood Science*, **4**(4), 193–201.

Minato, K. (1993). Moisture adsorption characteristics of medium-density fibreboard and its raw wood filler treated with formaldehyde. *Mokuzai Gakkaishi*, **39**(10), 1162–1168.

Minato, K. and Mizukami, F. (1982). A kinetic study of the reaction between wood and vaporous formaldehyde. *Mokuzai Gakkaishi*, **28**(6), 346–354.

Minato, K. and Norimoto, M. (1985). Moisture absorption characteristics and the dimensional stabilisation mechanism of paper and wood cross-linked by formaldehyde. *Mokuzai Gakkaishi*, **31**(3), 209–214.

Minato, K. and Yano, H. (1990). Improvement of dimensional stability and acoustic properties of wood for musical instruments by sulfur dioxide catalysed formalization. *Mokuzai Gakkaishi*, **36**(5), 362–367.

Minato, K. and Yasuda, R. (1992). Improvement of acoustic and hygrocopic properties of wood by some non-formaldehyde cross-linking agents. In: *Pacific Rim Bio-Based Composites Symposium: Chemical Modification of Lignocellulosics*, Plackett, D.V. and Dunningham, E.A. (Eds.). FRI Bulletin, **176**, pp. 97–106.

Minato, K., Yasuda, R. and Yano, H. (1990a). Improvement of dimensional stability and acoustic properties of wood for musical instruments with cyclic oxymethylenes. I. Formalization with trioxane. *Mokuzai Gakkaishi*, **36**(10), 860–866.

Minato, K., Yasuda, R. and Yano, H. (1990b). Improvement of dimensional stability and acoustic properties of wood for musical instruments with cyclic oxymethylenes. II. Formalization with tetraoxane. *Mokuzai Gakkaishi*, **36**(11), 990–996.

Minato, K., Kubo, N., Norimoto, M., Sasaki, H., Sawada, M. and Yamamoto, T. (1992). Dimensional stabilisation of medium-density fibreboard by formaldehyde treatment. *Mokuzai Gakkaishi*, **38**(1), 67–72.

Mitchell, P.H. (1988). Irreversible property changes of small loblolly pine specimens heated in air, nitrogen, or oxygen. *Wood and Fiber Science*, **20**(3), 320–355.

Mitchell, R.L., Seborg, R.M. and Millett, M.A. (1953). Effect of heat on the properties and chemical composition of Douglas-fir wood and its major components. *Journal of the Forest Products Research Society*, **3**(4), 38–42, 72–73.

Mitsui, K., Murata, A., Kohara, M. and Tsuchikawa, S. (2003). Colour modification of wood by light-irradiation and heat treatment. In: *Proceedings of the First European Conference on Wood Modification*, Ghent, Belgium, Van Acker, J. and Hill, C.A.S. (Eds.), pp. 43–52.

Miyafuji, H. and Saka, S. (1996). Wood-inorganic composites prepared by the sol-gel process V. Fire-resisting properties of the SiO_2–P_2O_5–B_2O_3 wood–inorganic composites. *Mokuzai Gakkaishi*, **42**(1), 74–80.

Miyafuji, H. and Saka, S. (1997). Fire-resisting properties in several TiO_2 wood–inorganic composites and their topochemistry. *Wood Science and Technology*, **31**(6), 449–455.

Miyafuji, H. and Saka, S. (2001). Na_2O–SiO_2 wood–inorganic composites prepared by the sol-gel process and their fire-resistance properties. *Journal of Wood Science*, **47**(6), 483–489.

Miyafuji, H., Kokaji, H. and Saka, S. (2004). Photostable wood–inorganic composites prepared by the sol-gel process with UV absorbent. *Journal of Wood Science*, **50**(2), 130–135.

Miyafuji, H., Saka, S. and Yamamoto, A. (1998). SiO_2–P_2O_5–B_2O_3 wood–inorganic composites prepared by metal alkoxides oligomers and their fire-resisting properties. *Holzforschung*, **52**(4), 410–416.

Mohebby, B. and Militz, H. (2002). Soft rot decay in acetylated wood. Chemical and anatomical changes in decayed wood. International Research Group on Wood Preservation, Doc. No. IRG/WP 02-40231.

Mohebby, B., Mai, C. and Militz, H. (2003). Soft rot decay in acetylated wood: microcalorimetry and ergosterol assay in decayed wood. In: *Proceedings of the First European Conference on Wood Modification*, Ghent, Belgium, Van Acker, J. and Hill, C.A.S. (Eds.), pp. 197–202.

Morita, M. and Sakata, I. (1988). Development of the thermoplasticity and solubility of cyanoethylated woods and barks of various species. *Mokuzai Gakkaishi*, **34**(11), 917–922.

Morita, M., Shigematsu, M. and Sakata, I. (1987). An improvement of the thermoplasticity and the solvent-solubility of cyanoethylated wood by halogen treatment. *Cellulose Chemistry and Technology*, **21**(3), 255–265.

Morozovs, A. and Aboltins, A. (1999). Acetylation of solid wood and modelling of acetic anhydride diffusion. In: *Proceedings of the International Scientific Colloquium, Modelling of Material Processing*, Riga, Latvia, pp. 124–129.

Morozovs, A., Aboltins, A., Zoldners, J. and Akerfelds, I. (2003). Wood modification in Latvia. In: *Proceedings of the First European Conference on Wood Modification*, Ghent, Belgium, Van Acker, J. and Hill, C.A.S. (Eds.), pp. 351–362.

Murakami, K. and Matsuda, H. (1989a). Oligoesterified woods based on anhydride and epoxide V. Properties of oligoesterified woods obtained by heating–suction of wood impregnated with reactants solution. *Mokuzai Gakkaishi*, **35**(4), 328–335.

Murakami, K. and Matsuda, H. (1989b). Oligoesterified woods based on anhydride and epoxide VI. Effects of size of wood block on preparation and properties of oligoesterified wood. *Mokuzai Gakkaishi*, **35**(10), 924–932.

Murakami, K. and Matsuda, H. (1990a). Oligoesterified woods based on anhydride and epoxide VII. Dynamic mechanical properties of oligoesterified woods. *Mokuzai Gakkaishi*, **36**(1), 49–56.

Murakami, K. and Matsuda, H. (1990b). Oligoesterified woods based on anhydride and epoxide VIII. Resistances of oligoesterified woods against weathering and biodeterioration. *Mokuzai Gakkaishi*, **36**(7), 538–544.

Nakagami, T. (1978). Esterification of wood with unsaturated carboxylic acids. V. Effect of delignification treatments on dissolution of wood esterified by the TFAA method. *Mokuzai Gakkaishi*, **24**(5), 318–323.

Nakagami, T. and Yokota, T. (1978a). Esterification of wood with unsaturated carboxylic acids IV: Chemical composition and molecular weight of the acetone-soluble fraction of β–methylcrotonylated woods. *Mokuzai Gakkaishi*, **24**(5), 311–317.

Nakagami, T. and Yokota, T. (1978b). Esterification of wood with unsaturated carboxylic acids V: Effect of delignification treatments on dissolution of wood esterified by the TFAA method. *Mokuzai Gakkaishi*, **24**(5), 318–323.

Nakagami, T. and Yokota, T. (1981a). A possible crosslinking of the methacrylated wood by the covalent bond. *Mokuzai Gakkaishi*, **27**(2), 87–97.

Nakagami, T. and Yokota, T. (1981b). The characteristics of copolymers isolated from reaction products between methacrylated wood and styrene. *Mokuzai Gakkaishi*, **27**(1), 32–38.

Nakagami, T. and Yokota, T. (1983). Estimation of crosslink-formation among wood components by dimensional stability measurements. *Mokuzai Gakkaishi*, **29**(3), 240–247.

Nakagami, T., Oishi, T. and Yokota, T. (1983). An attempt at bonding wood cross-sections and veneers through covalent crosslinks. *Mokuzai Gakkaishi*, **29**(3), 248–257.

Nakano, T. (1993a). Relaxation properties of carboxymethylated wood plus metallic salts in aqueous solution. *Holzforschung*, **47**(3), 202–206.

Nakano, T. (1993b). Dependence of relaxation property of succinylated wood on side-chain ionisation. *Holzforschung*, **47**(4), 278–282.

Nakano, T. (1993c). Reaction of glyoxal and glyoxal/glycol with wood. *Wood Science and Technology*, **28**(1), 23–33.

Nakano, T. (1994). Mechanism of thermoplasticity for chemically modified wood. *Holzforschung*, **48**(4), 318–324.

Nakano, T. (1996). Characterization of chemically modified wood. In: *Chemical Modification of Lignocellulosic Materials*, Hon, D.N.S. (Ed.). Marcel Dekker, New York, USA, pp. 247–275.

Nakano, T. and Nakamura, H. (1986a). Viscoelasticity of esterified wood specimens I. Stress relaxation behaviour of caprated wood and the effect of ester content. *Mokuzai Gakkaishi*, **32**(3), 176–183.

Nakano, T. and Nakamura, H. (1986b). Viscoelasticity of esterified wood specimens II. The effect of the introduced acyl groups on stress relaxation. *Mokuzai Gakkaishi*, **32**(5), 337–343.

Nakano, T. and Nakamura, H. (1986c). Viscoelasticity of esterified wood specimens III. The effect of ester content on dynamic mechanical properties. *Mokuzai Gakkaishi*, **32**(10), 820–826.

Nakano, T. and Nakamura, H. (1987). Viscoelasticity of esterified wood specimens IV. The effect of introduced acyl groups on dynamic mechanical properties. *Mokuzai Gakkaishi*, **33**(6), 472–477.

Nakano, T., Honma, S. and Matsumoto, A. (1991). Physical properties of chemically modified wood containing metal. II. Interaction between modified wood and metal. *Mokuzai Gakkaishi*, **37**(10), 924–929.

Nakano, T., Honma, S., Ehata, S. and Matsumoto, A. (1990a). Carboxymethylation of wood by ethanol–water reaction medium. *Mokuzai Gakkaishi*, **36**(3), 193–199.

Nakano, T., Honma, S., Ehata, S. and Matsumoto, A. (1990b). Introduction of metals to chemically-modified woods. Effects of conformation and location of side chain on bonding. *Mokuzai Gakkaishi*, **36**(8), 644–650.

Nakao, T., Okano, T. and Asano, I. (1983). Effects of heat treatment on the loss tangent of wood. *Mokuzai Gakkaishi*, **29**(10), 657–662.

Nami Kartal, S. (2003). Removal of copper, chromium, and arsenic from CCA-C treated wood by EDTA extraction. *Waste Management*, **23**(6), 537–546.

Narayan, R. and Shay, M. (1986). Graft polymerization onto cellulose acetate and wood using anionic polymerisation. In: *Renewable Resource Materials*, Carraher, E. and Springer, L.H. (Eds.). Plenum Publishing Corporation, pp. 137–146.

Narayanamurti, D. and Handa, B.K. (1953). Acetylated wood. *Das Papier*, **7**(5–6), 87–92.

Nassar, M.M. and MacKay, G.D.M. (1984). Mechanism of thermal decomposition of lignin. *Wood and Fiber Science*, **16**(3), 441–453.

Navi, P. and Girardet, F. (2000). Effects of thermo-hydro-mechanical treatment on the structure and properties of wood. *Holzforschung*, **54**(3), 287–293.

Nearn, W.T. (1974). Application of the ultrastructure concept in industrial wood products research. *Wood Science*, **6**(3), 285–293.

Nicholas, D.D. and Williams, A.D. (1987). Dimensional stabilisation of wood with dimethylol compounds. International Research Group on Wood Preservation, Doc. No. IRG/WP 3412.

Nilsson, T. and Rowell, R.M. (1982). Decay patterns observed in butylene oxide modified ponderosa pine after exposure in unsterile soil. International Research Group on Wood Preservation, Doc. No. IRG/WP/3211.

Nilsson, T. and Rowell, R.M. (1983). Decay patterns observed in butylene oxide modified ponderosa pine attacked by *Fomitopsis pinicola*. International Research Group on Wood Preservation, Doc. No. IRG/WP/1183.

Nilsson, T., Rowell, R.M., Simonson, R. and Tillman, A.-M. (1988). Fungal resistance of pine particle boards made from various types of acetylated chips. *Holzforschung*, **42**(2), 123–126.

Nishino, Y. (1991). Simplified vapor phase acetylation of small specimens of hinoki (*Chamaecyparis obtusa*) wood with acetic anhydride. *Mokuzai Gakkaishi*, **37**(4), 370–374.

Nishino, Y., Nakao, T., Tanaka, C. and Takahashi, A. (1991). Acylation of hinoki wood by maleic anhydride and the dynamic properties. *Bulletin of the Faculty of Agriculture Shimane University*, **25**, 161–164.

Norimoto, M. (1988). Acetylated Wood. *Wood Research and Technical Notes*, **24**, 13–30.

Norimoto, M. and Gril, J. (1993). Structure and properties of chemically treated woods. In: *Recent Research on Wood and Wood-based Materials*, Shiraishi, N., Kajita, H. and Norimoto, M. (Eds.). Elsevier, Barking, UK, pp. 135–154.

Norimoto, M., Gril, J. and Rowell, R.M. (1992). Rheological creep properties of chemically modified wood. Relationship between dimensional and creep stability. *Wood and Fiber Science*, **24**(10), 25–35.

Norimoto, M., Ota, C. Akitsu, H. and Yamada, T. (1993). Permanent fixation of bending deformation in wood by heat treatment. *Wood Research*, **79**, 23–33.

Nuopponen, M., Vuorinen, T., Jämsä, S. and Viitaniemi, P. (2003). The effects of a heat treatment on the behaviour of extractives in softwood studies by FTIR spectroscopic methods. *Wood Science and Technology*, **37**(2), 109–115.

Nuopponen, M., Vuorinen, T., Jämsä, S. and Viitaniemi, P. (2004). Thermal modifications in softwood studied by FT–IR and UV resonance Raman spectroscopies. *Journal of Wood Chemistry and Technology*, **24**(1), 13–26.

Obataya, E. (1999). Suitability of acetylated woods for clarinet reed. *Journal of Wood Science*, **45**(2), 106–112.

Ogawa, T. and Ohkoshi, M. (1997). Properties of medium density fibreboards produced from thermoplasticized wood fibers by allylation without adhesives. *Mokuzai Gakkaishi*, **43**(1), 61–67.

Ogiso, K. and Saka, S. (1993). Wood-inorganic composites prepared by sol-gel process II. Effects of ultrasonic treatments on preparation of wood–inorganic composites. *Mokuzai Gakkasihi*, **39**(3), 301–307.

Ogiso, K. and Saka, S. (1994). Wood–inorganic composites prepared by sol-gel processes IV. Effects of chemical bonds between wood and inorganic substances on property enhancement. *Mokuzai Gakkaishi*, **40**(10), 1100–1106.

Ohkoshi, M. (1990). Bonding of wood by thermoplasticising the surfaces I. Effects of allylation and hot-press conditions. *Mokuzai Gakkaishi*, **36**(1), 57–63.

Ohkoshi, M. (1991). Bonding of wood by thermoplasticizing the surfaces II. Possible crosslinking of wood by the graft-copolymerizing of styrene onto allylated surfaces. *Mokuzai Gakkaishi*, **37**(10), 917–923.

Ohkoshi, M. (2002). FTIR–PAS study of light-induced changes in the surface of acetylated or polyethylene glycol-impregnated wood. *Journal of Wood Science*, **48**(5), 394–401.

Ohkoshi, M. and Kato, A. (1992). Distribution of substituents in acetylated wood as determined by nuclear magnetic resonance analysis. In: *Pacific Rim Bio-Based Composites Symposium: Chemical Modification of Lignocellulosics*, Plackett, D.V. and Dunningham, E.A. (Eds.). FRI Bulletin, **176**, pp. 25–32.

Ohkoshi, M. and Kato, A. (1993). Determination of substituent distribution of DMSO-soluble portion of acetylated wood meal by ^{13}C-NMR spectroscopy. *Mokuzai Gakkaishi*, **39**(7), 849–854.

Ohkoshi, M. and Kato, A. (1997a). ^{13}C-NMR analysis of acetyl groups in acetylated wood I. Acetyl groups in cellulose and hemicellulose. *Mokuzai Gakkaishi*, **43**(4), 327–336.

Ohkoshi, M. and Kato, A. (1997b). ^{13}C-NMR analysis of acetyl groups in acetylated wood II. Acetyl groups in lignin. *Mokuzai Gakkaishi*, **43**(4), 364–369.

Ohkoshi, M., Kato, A., Suzuki, K., Hayashi, N. and Ishihara, M. (1999). Characterisation of acetylated wood decayed by brown-rot and white-rot fungi. *Journal of Wood Science*, **45**(1), 69–75.

Ohmae, K., Minato, K. and Norimoto, M. (2002). The analysis of dimensional changes due to chemical treatments and water soaking for hinoki (*Chamaecyparis obtusa*) wood. *Holzforschung*, **56**(1), 98–102.

Okino, E.Y.A., Santata, M.A.E., Souza, M.R. de and Sousa, M.E. de (2001). Dimensional stability and mechanical properties of acetylated *Hevea brasiliensis* flakeboards. *Journal of Tropical Forest Products*, **7**(1), 96–102.

Okino, E.Y.A., Alves, M.V. da S., Santana, M.A.E., Souza, M.R. de and Sousa, M.E. de (2003). Accelerated laboratory test of acetylated *Hevea brasiliensis* flakeboards. *Journal of Tropical Forest Products*, **9**(1–2), 54–58.

Okino, E.Y.A., Souza, M.R. de, Santana, M.A.E., Alves, M.V. da S., Sousa, M.E. de and Teixeira, D.E. (2004). Evaluation of the physical and biological properties of particleboard and flakeboard made from *Cupressus* spp. *International Biodeterioration and Biodegradation*, **53**(1), 1–5.

Oniki, T. and Takahama, U. (1997). Free radicals produced by the oxidation of compounds containing syringyl and guaiacyl groups. *Mokuzai Gakkaishi*, **43**(6), 493–498.

Ono, T. and Norimoto, M. (1984). On physical criteria for the selection of wood for soundboards of musical instruments. *Rheologica Acta*, **23**, 652–656.

Ota, M., Abe, K. and Sekiguchi, T. (1996). Light-induced color change of acetylated wood meal of kiri (*Paulownia tomentosa* Steud.). *Mokuzai Gakkaishi*, **42**(2), 216–221.

Owens, C.W., Shortle, W.C. and Shigo, A.L. (1980a). Preliminary evaluation of silicon tetrachloride as a wood preservative. *Holzforschung*, **34**(6), 223–225.

Owens, C.W., Shortle, W.C. and Shigo, A.L. (1980b). Silicon tetrachloride: a potential wood preservative. International Research Group on Wood Preservation, Doc. No. IRG/WP 3133.

Papadopoulos, A.N. and Hill, C.A.S. (2002). The biological effectiveness of wood modified with linear chain carboxylic acid anhydrides against *Coniophora puteana. Holz als Roh- und Werkstoff*, **60**(5), 329–332.

Papadopoulos, A.N. and Hill, C.A.S. (2003). The sorption of water vapour by anhydride modified softwood. *Wood Science and Technology*, **37**(3–4), 221–231.

Papadopoulos, A.N. and Traboulay, E. (2002). Dimensional stability of OSB made from acetylated fir strands. *Holz als Roh- und Werkstoff*, **60**(2), 92–95.

Patscheke, G. and Dilg, K. (1958). Investigations on preparation and reactivity of inclusion woods. *Archiv für Forstwesen*, **7**(3), 203–219.

Peill, P.L.D. (1946). Permanent bleaching of ligno-cellulosic materials. *Nature*, **158**, 554.

Peterson, M.D. and Thomas, R.J. (1978). Protection of wood from decay fungi by acetylation—an ultrastructural and chemical study. *Wood and Fiber*, **10**(3), 149–163.

Pétrissans, M., Géradin, P., El-Bakali, I. and Seraj, M. (2003). Wettability of heat-treated wood. *Holzforschung*, **57**(3), 301–307.

Philippou, J.L. (1982). Applicability of oxidative systems to initiate grafting on and bonding of wood. *Journal of Wood Chemistry and Technology*, **1**(2), 199–221.

Philippou, J.L., Johns, W.E. and Nguyen, T. (1982). Bonding wood by graft polymerisation. The effect of hydrogen peroxide concentration on the bonding and properties of particleboard. *Holzforschung*, **36**(1), 37–42.

Pizzi, A. (1990). Extended durability by the chemical fixation of unsaturated alkyd finishes to wood. *Holzforschung und Holzverwertung*, **42**(6), 107–109.

Pizzi, A., Stephanou, A., Boonstra, M.J. and Pendlebury, A.J. (1994). A new concept on the chemical modification of wood by organic anhydrides. *Holzforschung*, **48**(suppl.), 91–94.

Plackett, D.V. and Dunningham, E.A. (1991). Ultraviolet resistance and colour stability of esterified wood. In: *Proceedings of the International Symposium on Chemical Modification of Wood*, Kyoto, Japan, Takahashi, M. (Ed.), pp. 136–141.

Plackett, D.V., Dunningham, E.A. and Singh, A.P. (1992). Weathering of chemically modified wood. Accelerated weathering of acetylated radiata pine. *Holz als Roh- und Werkstoff*, **50**(4), 135–140.

Plackett, D.V., Dunningham, E.A. and Singh, A.P. (1996). Weathering of chemically modified wood. In: *Chemical Modification of Lignocellulosic Materials*, Hon, D.N.S. (Ed.). Marcel Dekker, New York, USA, pp. 277–294.

Podgorski, L. and Roux, M. (1999). Wood modification to improve the durability of coatings. *Surface Coatings*, **82**(12), 590–596.

Podgorski, L., Chevet, B., Onic, L. and Merlin, A. (2000). Modification of wood wettability by plasma and corona treatments. *International Journal of Adhesion and Adhesives*, **20**(2), 103–111.

Popper, R. and Bariska, M. (1972). Acylation of wood. Part I: The sorption behaviour of water vapour. *Holz als Roh- und Werkstoff*, **30**(8), 289–294.

Popper, R. and Bariska, M. (1973). Acylation of wood. Part II: Thermodynamics of water vapour sorption. *Holz als Roh- und Werkstoff*, **31**(2), 65–70.

Popper, R. and Bariska, M. (1975). Acylation of wood. Part III: Swelling and shrinking behaviour. *Holz als Roh- und Werkstoff*, **33**(11), 415–419.

Powell, M.R. (2003). Treatment of wood with Royale stabilising oil. In: *Proceedings of the First European Conference on Wood Modification*, Gent, Belgium, Van Acker, J. and Hill, C.A.S. (Eds.), pp. 15–22.

Ramiah, M.V. (1970). Thermogravimetric and differential thermal analysis of cellulose, hemicellulose, and lignin. *Journal of Applied Polymer Science*, **14**(5), 1323–1337.

Ramiah, M.V. and Goring, D.A.I. (1967). Some dilatomeric measurements of the thermal decomposition of cellulose, hemicellulose and lignin. *Cellulose Chemistry and Technology*, **1**(3), 277–285.

Ramsden, M.J., Blake, F.S.R. and Fey, N.J. (1997). The effect of acetylation on the mechanical properties, hydrophobicity and dimensional stability of *Pinus sylvestris*. *Wood Science and Technology*, **31**(2), 97–104.

Rapp, A.O. and Peek, R.D. (1995). New principles for the protection of wood: impregnation with water-borne resins. International Research Group on Wood Preservation, Doc. No. IRG/WP 95-40047.

Rapp, A.O. and Peek, R.D. (1996). Melamine resins as wood preservatives—results of biological testing. International Research Group on Wood Preservation, Doc. No. IRG/WP 96-40061.

Rapp, A.O., Bestgen, H., Adam, W. and Peek, R.D. (1999). EELS (Electron Energy Loss Spectroscopy) a technique for the quantification of nitrogen and other light elements in the cell wall. International Research Group on Wood Preservation, Doc. No. IRG/WP 99-20163.

Razzaque, M.A. and Banks, W.B. (1982). Dimensional stability and sorption characteristics of wood treated with alkylene oxide. *Bano Biggyan Patrika*, **11**(1/2), 17–23.

Rehn, P. and Vlöl, W. (2003). Dielectric barrier discharge treatments at atmospheric pressure for wood surface modification. *Holz. als Roh- und Werkstoff*, **61**(2), 145–150.

Reiterer, A. and Sinn, G. (2002). Fracture behaviour of modified spruce wood: a study using linear and non-linear fracture mechanics. *Holzforschung*, **56**(2), 191–198.

Rep, G., Pohleven, F. and Bucar, B. (2004). Characteristics of thermally modified wood in vacuum. International Research Group on Wood Preservation, Doc. No. IRG/WP 04-40287.

Repellin, V. and Guyonnet, R. (2005). Evaluation of heat-treated wood swelling by differential scanning calorimetry in relation to chemical composition. *Holzforschung*, **59**(1), 28–34.

Richardson, B.A. and Cox, T.R.G. (1985). The future of chromium in wood preservation. International Research Group on Wood Preservation, Doc. No. IRG/WP 3332.

Rijckaert, V., De Geyter, S., Van Acker, J. and Stevens, M. (2000). Screening of the technical performance and aquatic toxicity of *N*-methylolacrylamide treated wood. International Research Group on Wood Preservation, Doc. No. IRG/WP 00-40166.

Risi, J. and Arseneau, D.F. (1957a). Dimensional stabilization of wood. *Forest Products Journal*, **7**(6), 210–213.

Risi, J. and Arseneau, D.F. (1957b). Dimensional stabilization of wood. Part II: Crotonylation and crotylation. *Forest Products Journal*, **7**(7), 245–246.

Risi, J. and Arseneau, D.F. (1957c). Dimensional stabilization of wood. Part III: Butylation. *Forest Products Journal*, **7**(8), 261–265.

Risi, J. and Arseneau, D.F. (1957d). Dimensional stabilization of wood. Part IV: Allylation. *Forest Products Journal*, **7**(9), 293–295.

Risi, J. and Arseneau, D.F. (1958). Dimensional stabilization of wood. Part V: Phthalolylation. *Forest Products Journal*, **8**(9), 252–255.

Ritschkoff, A.C., Rättö, M., Nurmi, A., Kokko, H., Rapp, A. and Militz, H. (1999). Effect of some resin treatments on fungal degradation reactions. International Research Group on Wood Preservation, Doc. No. IRG/WP 99-10318.

Rosenqvist, M. (2000). The distribution of introduced acetyl groups and a linseed oil model substance in wood examined by microradiography and ESEM. International Research Group on Wood Preservation, Doc. No. IRG/WP 00-40169.

Rosenqvist, M. (2002). Acetyl group distribution in acetylated wood investigated by microautoradiography. *Holzforschung*, **55**(3), 270–275.

Roussel, C., Marchetti, V., Lemor, A., Wozniak, E., Loubinoux, B. and Gérardin, P. (2001). Chemical modification of wood by polyglycerol/maleic anhydride treatment. *Holzforschung*, **55**(1), 57–62.

Rowell, R.M. (1977a). Nonconventional wood preservation methods. In: *Wood Technology: Chemical Aspects*, Goldstein, I.S. (Ed.). ACS Symposium Series, **43**, pp. 47–56.

Rowell, R.M. (1977b). Effects of drying method and penetration of chemicals on chemically modified southern pine. *Wood Science*, **9**(3), 144–148.

Rowell, R.M. (1978). Distribution of reacted chemicals in southern pine treated with alkylene oxides. *Wood Science*, **10**(4), 193–197.

Rowell, R.M. (1980). Distribution of reacted chemicals in southern pine modified with methyl isocyanate. *Wood Science*, **13**(2), 102–110.

Rowell, R.M. (1982). Distribution of acetyl groups in southern pine reacted with acetic anhydride. *Wood Science*, **15**(2), 172–182.

Rowell, R.M. (1983a). Chemical modification of wood. *Forest Products Abstracts*, **6**(12), 363–382.

Rowell, R.M. (1983b). Chemical modification of wood. In: *Wood and Cellulosic Chemistry*, Hon, D.N.S., and Shiraishi, N. (Eds.). Marcel Dekker, New York, USA, pp. 703–756.

Rowell, R.M. (1984a). Penetration and reactivity of cell wall components. In: *The Chemistry of Solid Wood*, Rowell, R.M. (Ed.). Advances in Chemistry Series, **207**, American Chemical Society, pp. 175–210.

Rowell, R.M. (1984b). Bonding of toxic chemicals to wood. *Applied Biochemistry and Biotechnology*, **9**, 447–453.

Rowell, R.M. (1996). Physical and mechanical properties of chemically modified wood. In: *Chemical Modification of Lignocellulosic Materials*, Hon, D.N.S. (Ed.). Marcel Dekker, New York, USA, pp. 295–310.

Rowell, R.M. and Banks, W.B. (1987). Tensile strength and toughness of acetylated pine and lime flakes. *British Polymer Journal*, **19**(5), 479–482.

Rowell, R.M. and Ellis, W.D. (1978). Determination of the dimensional stabilization of wood using the water-soak method. *Wood and Fiber*, **10**(4), 104–111.

Rowell, R.M. and Ellis, W.D. (1979). Chemical modification of wood: reaction of methyl isocyanate with southern pine. *Wood Science*, **12**(1), 52–58.

Rowell, R.M. and Ellis, W.D. (1984). Effects of moisture on the chemical modification of wood with epoxides and isocyanates. *Wood and Fiber Science*, **16**(2), 257–267.

Rowell, R.M. and Gutzmer, D.I. (1975). Chemical modification of wood: reactions of alkylene oxides with southern yellow pine. *Wood Science*, **7**(3), 240–246.

Rowell, R.M. and Plackett, D.V. (1988). Dimensional stability of flakeboards made from acetylated *Pinus radiata* heartwood or sapwood flakes. *New Zealand Journal of Forestry Science*, **18**(1), 124–131.

Rowell, R.M. and Rowell, J.S. (1989). Moisture sorption properties of acetylated lignocellulosic fibers. In: *Cellulose and Wood: Chemistry and Technology, Proceedings of the 10th Cellulose Conference*, Schuerch, C. (Ed.). John Wiley & Sons, Inc., New York, pp. 343–355.

Rowell, R.M., Feist, W.C. and Ellis, W.D. (1981). Weathering of chemically modified southern pine. *Wood Science*, **13**(4), 202–208.

Rowell, R.M., Hart, S.V. and Esenther, G.R. (1979). Resistance of alkylene-oxide-modified southern pine to attack by subterranean termites. *Wood Science*, **11**(4), 271–274.

Rowell, R.M., Lichtenberg, R.S. and Larsson, P. (1992). Stability of acetyl groups in acetylated wood to changes in pH, temperature, and moisture In: *Pacific Rim Bio-Based Composites Symposium: Chemical Modification of Lignocellulosics*, Plackett, D.V. and Dunningham, E.A. (Eds.). FRI Bulletin, **176**, pp. 33–40.

Rowell, R.M., Lichtenberg, R.S. and Larsson, P. (1993a). Stability of acetylated wood to environmental changes. *Wood and Fiber Science*, **25**(4), 359–364.

Rowell, R.M., Kawai, S. and Inoue, M. (1995). Dimensionally stabilized, very low density fibreboard. *Wood and Fiber Science*, **27**(4), 428–436.

Rowell, R.M., Moisuk, R. and Meyer, J.A. (1982). Wood–polymer composites: cell wall grafting with alkylene oxides and lumen treatments with methyl methacrylate. *Wood Science*, **15**(2), 90–96.

Rowell, R.M., Simonson, R. and Tillman, A.-M. (1986a). A simplified procedure for the acetylation of chips for dimensionally stabilized particleboard products. *Paperi ja Puu*, **68**(10), 740–744.

Rowell, R.M., Simonson, R. and Tillman, A.-M. (1990). Acetyl balance for the acetylation of wood particles by a simplified procedure. *Holzforschung*, **44**(4), 263–269.

Rowell, R.M., Simonson, R. and Tillman, A.-M. (1986b). Dimensional stability of particleboard made from vapor-phase acetylated pine wood chips. *Nordic Pulp and Paper Research Journal*, **1**, 11–17.

Rowell, R.M., Tillman A.-M. and Simonson, R. (1986c). A simplified procedure for the acetylation of hardwood and softwood flakes for flakeboard production. *Journal of Wood Chemistry and Technology*, **6**(3), 427–448.

Rowell, R.M., Tillman, A.-M. and Simonson, R. (1986d). Vapor phase acetylation of southern pine, Douglas-fir, and aspen wood flakes. *Journal of Wood Chemistry and Technology*, **6**(3), 449–471.

Rowell, R.M., Tillman, A.-M. and Zhengtian, L. (1986e). Dimensional stabilization of flakeboard by chemical modification. *Wood Science and Technology*, **20**(1), 83–95.

Rowell, R.M., Wang, R.H.S. and Hyatt, J.A. (1986f). Flakeboards made from aspen and southern pine wood flakes reacted with gaseous ketene. *Journal of Wood Chemistry and Technology*, **6**(3), 449–471.

Rowell, R.M., Youngquist, J.A. and Imamura, Y. (1988). Strength tests on acetylated aspen flake boards exposed to a brown-rot fungus. *Wood and Fiber Science*, **20**(2), 266–271.

Rowell, R.M., Youngquist, J.A. and Sachs, I.B. (1987a). Adhesive bonding of acetylated aspen flakes, part 1. Surface changes, hydrophobicity, adhesive penetration and strength. *International Journal of Adhesion and Adhesives*, **7**(4), 183–188.

Rowell, R.M., Esenther, G.R., Nicholas, D.D. and Nilsson, T. (1987b). Biological resistance of southern pine and aspen flakeboards made from acetylated flakes. *Journal of Wood Chemistry and Technology*, **7**(3), 427–440.

Rowell, R.M., Gutzmer, D.I., Sachs, I.B. and Kinney, R.E. (1976). Effects of alkylene oxide treatments on dimensional stability of wood. *Wood Science*, **9**(1), 51–54.

Rowell, R.M., Imamura, Y., Kawai, S. and Norimoto, M. (1989). Dimensional stability, decay resistance, and mechanical properties of veneer-faced low-density particleboards made from acetylated wood. *Wood and Fiber Science*, **21**(1), 67–79.

Rowell, R.M., Youngquist, J.A., Rowell, J.S. and Hyatt, J.A. (1991). Dimensional stability of aspen fiberboard made from acetylated fiber. *Wood and Fiber Science*, **23**(4), 558–566.

Rowell, R.M., Cleary, B.A., Rowell, J.S., Clemons, C. and Young, R.A. (1993b). Results of chemical modification of lignocellulosic fibers for use in composites. In: *Wood Fiber/Polymer Composites: Fundamental Concepts, Processes, and Material Options*, Wolcott, M.P. (Ed.). Forest Products Society, Madison, Wiseconsin, USA, pp. 121–127.

Rowell, R.M., Simonsen, R., Hess, S., Plackett, D.V., Cronshaw, D. and Dunningham, E. (1994). Acetyl distribution in acetylated whole wood and reactivity of isolated cell wall components to acetic anhydride. *Wood and Fiber Science*, **26**(1), 11–18.

Rowell, R.M., Dawson, B.S., Hadi, Y.S., Nicholas, D.D., Nilsson, T., Plackett, D.V., Simonson, R. and Westin, M. (1997). Worldwide in-ground stake test of acetylated composite boards. International Research Group on Wood Preservation, Doc. No. IRG/WP 97–40088.

Rozman, H.D., Banks, W.B. and Lawther, M.L. (1994a). Improvements of fibreboard properties through fiber activation and subsequent copolymerisation with vinyl monomer. *Journal of Applied Polymer Science*, **54**(2), 191–200.

Rozman, H.D., Banks, W.B. and Lawther, M.L. (1994b). FTIR characterization of chemically modified fibreboard. *International Journal of Polymer Materials*, **26**(1–2), 19–24.

Rozman, H.D., Abusamah, A., Kumar, R.N. and Abdul Khalil, H.P.S. (1997a). Rubberwood–polymer composites based on methacrylate silane and methyl methacrylate. *Journal of Tropical Forest Products*, **2**(2), 227–237.

Rozman, H.D., Kumar, R.N., Abdul Khalil, H.P.S., Abusamah, A. and Abu, R. (1997b). Chemical modification of wood with maleic anhydride and subsequent copolymerisation with diallyl phthalate. *Journal of Wood Chemistry and Technology*, **17**(4), 419–433.

Rudkin, A.W. (1950). The role of hydroxyl group in the gluing of wood. *Australian Journal of Applied Science*, **1**, 270–283.

Rusche, H. (1973a). Thermal degradation of wood at temperatures up to 200 deg C. I. Strength properties of wood after heat treatment. *Holz als Roh- und Werkstoff*, **31**(7), 273–281.

Rusche, H. (1973b). Thermal degradation of wood at temperatures up to 200 deg C. Part II. Reaction kinetics of loss of mass during heat treatment of wood. *Holz als Roh- und Werkstoff*, **31**(8), 307–312.

Ryu, J.Y., Takahashi, M., Imamura, Y. and Sato, T. (1991). Biological resistance of phenol-resin treated wood. *Mokuzai Gakkaishi*, **37**(9), 852–858.

Ryu, J.Y., Imamura, Y. and Takahashi, M. (1992). Biological resistance of furfuryl alcohol-treated wood. International Research Group on Wood Preservation, Doc. No. IRG/WP 3703-92.

Ryu, J.Y., Imamura, Y., Takahashi, M. and Kajita, H. (1993). Effects of molecular weight and some other properties of resins on the biological resistance of phenolic resin treated wood. *Mokuzai Gakkaishi*, **39**(4), 486–492.

Sailer, M. and Rapp, A.O. (1997). Influence of different resin treatments on the dimensional stability of wood. International Research Group on Wood Preservation, Doc. No. IRG/WP 97-40090.

Sailer, M., Rapp, A.O. and Peek, R.D. (1998). Biological resistance of wood treated with water-based resins and drying oils in a mini-block test. International Research Group on Wood Preservation, Doc. No. IRG/WP 98-40107.

Sailer, M., Rapp, A.O. and Leithoff, H. (2000a). Improved resistance of Scots pine by application of an oil-heat treatment. International Research Group on Wood Preservation, Doc. No. IRG/WP 00-40162.

Sailer, M., Rapp, A.O., Leithoiff, H. and Peek, R.D. (2000b). Upgrading of wood by application of an oil-heat treatment. *Holz als Roh- und Werkstoff*, **58**(1/2), 15–22.

Saka, S. and Tanno, F. (1996). Wood–inorganic composites prepared by the sol-gel process VI. Effects of a property-enhancer on fire-resistance in SiO_2–P_2O_5 and SiO_2–B_2O_3 wood–inorganic composites. *Mokuzai Gakkaishi*, **42**(1), 81–86.

Saka, S. and Ueno, T. (1997). Several SiO_2 wood–inorganic composites and their fire-resisting properties. *Wood Science and Technology*, **31**(6), 457–466.

Saka, S. and Yakake, Y. (1993). Wood–inorganic composites prepared by sol-gel process III. Chemically modified wood–inorganic composites. *Mokuzai Gakkaishi*, **39**(3), 308–314.

Saka, S., Sasaki, M. and Tanahashi, M. (1992). Wood–inorganic composites prepared by the sol-gel process I. Wood–inorganic composites with porous structure. *Mokuzai Gakkaishi*, **38**(11), 1043–1049.

Sakuragawa, S. (1996). Dyeing of wood and protection of discoloration. *Mokuzai Kogyo*, **51**(3), 102–106.

Sander, C. and Koch, G. (2001). Effects of acetylation and hydrothermal treatment on lignin as revealed by cellular UV-spectroscopy in Norway spruce (*Picea abies* [L.] Karst.) *Holzforschung*, **55**(2), 193–198.

Sander, C., Beckers, E.P.J., Militz, H. and Van Veenendaal, W. (2003). Analysis of acetylated wood by electron microscopy. *Wood Science and Technology*, **37**(1), 39–46.

Sanderman, W. and Augustin, H. (1964). Chemical investigations on the thermal decomposition of wood—Part III: chemical investigation on the course of decomposition. *Holz als Roh- und Werkstoff*, **22**(10), 377–386.

Sarni, F., Moutounet, M., Puech, J.-L. and Rabier, P. (1990). Effect of heat treatment on oak wood extractable compounds. *Holzforschung*, **44**(6), 461–466.

Sasaki, H. and Kawai, S. (1994). Recent research and development work on composites in Japan. *Wood Science and Technology*, **28**(4), 241–248.

Sasaki, T., Norimoto, M., Yamada, T. and Rowell, R.M. (1988). Effect of moisture on the acoustical properties of wood. *Mokuzai Gakkaishi*, **34**(10), 794–803.

Savage, R.A. (2003). Tree growth and wood quality. In: *Wood Quality and its Biological Basis*, Barnett, J.R. and Jeronimidis, G. (Eds.). Blackwell, Oxford, UK, pp. 1–29.

Schmidt, E.L. (1983). Effects of a formaldehyde and sulphur dioxide treatment on decay and mechanical properties of aspen waferboard. International Research Group on Wood Preservation, Doc. No. IRG/WP 3242.

Schneider, M.H. (1995). New cell wall and cell lumen wood polymer composites. *Wood Science and Technology*, **29**(4), 121–127.

Schneider, M.H. and Brebner, K.I. (1985). Wood–polymer combinations: the chemical modification of wood by alkoxysilane coupling agents. *Wood Science and Technology*, **19**(1), 67–73.

Schneider, A. and Rusche, H. (1973). Sorption behaviour of beech and spruce wood after heat treatment in air and in vacuum. *Holz als Roh- und Werkstoff*, **31**(8), 313–319.

Schneider, M.H. and Witt, A.E. (2004). History of wood polymer composite commercialization. *Forest Products Journal*, **54**(2), 19–24.

Schneider, M.H., Brebner, K.I. and Hartley, I.D. (1991). Swelling of a cell lumen filled and a cell-wall bulked wood polymer composite in water. *Wood and Fiber Science*, **23**(2), 165–172.

Sèbe, G. and Brook, M. (2001). Hydrophobization of wood surfaces: covalent grafting of silicone polymers. *Wood Science and Technology*, **35**(3), 269–282.

Sèbe, G. and De Jéso, B. (2000). The dimensional stabilisation of maritime pine sapwood (*Pinus pinaster*) by chemical reaction with organosilicon compounds. *Holzforschung*, **54**(5), 474–480.

Sèbe, G., Tingaut, P., Safou-Tchiama, R., Pétraud, M., Grelier, S. and De Jeso, B. (2004). Chemical reaction of maritime pine sapwood (*Pinus pinaster* Soland) with alkoxysilane molecules: a study of chemical pathways. *Holzforschung*, **58**(5), 511–518.

Seborg, R.M., Millett, M.A. and Stamm, A.J. (1945). Heat-stabilized compressed wood (Staypak). *Mechanical Engineering*, **67**(1), 25–31.

Seborg, R.M., Tarkow, H. and Stamm, A.J. (1953). Effect of heat upon the dimensional stabilization of wood. *Journal of the Forest Products Research Society*, **3**(3), 59–67.

Sepall, O. and Mason, S.G. (1961). Hydrogen exchange between cellulose and water. I. Measurement of accessibility. *Canadian Journal of Chemistry*, **39**, 1934–1943.

Shafizadeh, F. (1984). The chemistry of pyrolysis and combustion. In: *The Chemistry of Solid Wood*, Rowell, R.M. (Ed.). ACS Symposium Series, **207**, pp. 489–529.

Shafizadeh, F. and Chin, P.P.S. (1977). Thermal deterioration of wood. In: *Wood Technology: Chemical Aspects*, Goldstein, I.S. (Ed.). ACS Symposium Series, **43**, pp. 57–81.

Sheen, A.D. (1992). The preparation of acetylated wood fibre on a commercial scale. In: *Pacific Rim Bio-Based Composites Symposium: Chemical Modification of Lignocellulosics*, Plackett, D.V. and Dunningham, E.A. (Eds.). FRI Bulletin, **176**, pp. 1–8.

Sherman, S., Gannon, J., Buchi, G. and Howell, W.R. (1980). Epoxy resins. In: *Kirk–Othmer Encyclopaedia of Chemical Technology*, **9**, 3rd edn, John Wiley and Sons, Inc., New York, USA, pp. 267–290.

Shimizu, K., Sudo, H. and Fijii, T. (1989). Total utilisation of wood components by steam explosion pre-treatment. In: *Wood Processing and Utilization*, Kennedy, J.F., Phillips, G.O. and Williams, P.A. (Eds.). Ellis Harwood, Chichester, UK, pp. 407–412.

Shimizu, K., Teratini, F., Hashi, M. and Miyazaki, K. (1972). Effect of the thermal treatment on wood hemicelluloses. VI. Studies on the thermal analysis of arabinogalactan, and O-acetyl and deacetylated-galactoglucomannans. *Mokuzai Gakkaishi*, **18**(2), 79–84.

Shiraishi, N. and Kishi, H. (1986). Wood–phenol adhesives prepared from carboxymethylated wood. I. *Journal of Applied Polymer Science*, **32**(1), 3189–3209.

Shiraishi, N., Matsunaga, T. and Yokota, T. (1979). Thermal softening and melting of esterified wood prepared in an N_2O_4–DMF cellulose solvent medium. *Journal of Applied Polymer Science*, **24**(12), 2361–2368.

Shiraishi, N., Murata, M. and Yokota, T. (1972). Polymerization of vinyl monomer within the cell wall of wood. II. Polymerization of methyl methacrylate in the presence of wood, water and carbon tetrachloride. *Mokuzai Gakkaishi*, **18**(6), 299–306.

Shiraishi, N., Sumizawa, K. and Yokota, T. (1973). The interaction of wood with organic solvents. IV. Some additional characterizations of woods acetylated after pretreatments with organic solvents. *Mokuzai Gakkaishi*, **19**(5), 241–249.

Shiraishi, N., Aoki, T., Norimoto, M. and Okumara, M. (1983). Make cellulosics thermoplastic. *Chemtech*, **13**(6), 366–373.

Siau, J.F. (1984). *Transport Processes in Wood*. Springer-Verlag, Berlin.

Simonsen, J. (1998). Lack of dimensional stability in cross-linked wood-polymer composites. *Holzforschung*, **52**(1), 102–104.

Simonson, R. and Rowell, R.M. (2000). A new process for the continuous acetylation of lignocellulosic fiber. In: *Proceedings of the 5th Pacific Rim Bio-Based Composites Symposium*, Canberra, Australia, pp. 190–196.

Singh, S.P., Dev, I. and Kumar, S. (1979). Chemical modification of wood. Vapor phase acetylation with thioacetic acid. *Wood Science*, **11**(4), 268–270.

Singh, S.P., Dev, I. and Kumar, S. (1981). Chemical modification of wood. II. Vapor phase acetylation with acetyl chloride. *International Journal of Wood Preservation*, **1**(4), 169–171.

Singh, D, Dev, I. and Kumar, S. (1992). Chemical modification of wood with acetic anhydride. *Journal of the Timber Development Association of India,* **38**(1), 5–8.

Sivonen, H., Maunu, S.L., Sundholm, F., Jämsä, S. and Viitaniemi, P. (2002). Magnetic resonance studies of thermally modified wood. *Holzforschung*, **56**(6), 648–654.

Sjöström, E. (1981). *Wood Chemistry: Fundamentals and Applications*. Academic Press, London, UK.

Skaar, C. (1972). *Water in Wood*. Syracuse University Press, New York.

Smith, L.A. and Côté, W.A. (1971). Studies of the penetration of phenol-formaldehyde resin into wood cell walls with the SEM and energy dispersive X-ray analyser. *Wood and Fiber*, **3**(1), 56–57.

Smith, L.A. and Côté, W.A. (1972). Resin penetration into wood cell walls. *Journal of Paint Technology*, **44**(564), 71.

Soclow, R., Andrews, C. Berkhout, F. and Thomas, V. (Eds.) (1994). *Industrial Ecology and Global Change*. Cambridge University Press, UK.

Soulounganga, P., Loubinoux, B., Wozniak, E., Lemor, A. and Gérardin, P. (2004). Improvement of wood properties by impregnation with polyglycerol methacrylate. *Holz als Roh- und Werkstoff*, **62**(4), 281–285.

Spalt, H.A. (1958). The fundamentals of water vapor sorption by wood. *Forest Products Journal*, **8**(10), 288–295.

Srebotnik, E. and Messner, K. (1990). Enzymatic attack of wood is limited by the inaccessibility of the substrate. In: *Biotechnology in Pulp and Paper Manufacture*, Kirk, T.K. and Chang, H. (Eds.). Butterworth-Heinemann, London, UK, pp. 111–122.

Srebotnik, E. and Messner, K. (1991). Immunoelectron microscopical study of the porosity of brown-rot degraded wood. *Holzforschung*, **45**(2), 95–101.

Stamm, A.J. (1956). Thermal degradation of wood and cellulose. *Industrial and Engineering Chemistry*, **48**(3), 413–417.

Stamm, A.J. (1959a). Dimensional stabilisation of wood by thermal reactions and formaldehyde cross-linking. *Tappi*, **42**(1), 39–44.

Stamm, A.J. (1959b). Dimensional stabilisation of wood by catalysed heat treatment and cross-linking with formaldehyde. *Tappi*, **42**(1), 44–50.

Stamm, A.J. (1959c). Dimensional stabilisation of wood by thermal reactions and formaldehyde cross-linking. *Tappi*, **42**(1), 39–44.

Stamm, A.J. (1964). *Wood and Cellulose Science*. Ronald Press, New York, USA.

Stamm, A.J. (1977). Dimensional stabilization of wood with furfuryl alcohol. In: *Wood Technology: Chemical Aspects*, Goldstein, I.S. (Ed.). ACS Symposium Series, **43**, American Chemical Society, Washington, DC, pp. 141–149.

Stamm, A.J. and Baechler, R.H. (1960). Decay resistance and dimensional stability of five modified woods. *Forest Products Journal*, **10**(1), 22–26.

Stamm, A.J. and Hansen, L.A. (1937). Minimizing wood shrinkage and swelling. Effect of heating in various gases. *Industrial and Engineering Chemistry*, **29**(7), 831–833.

Stamm, A.J. and Seborg, R.M. (1936). Minimizing wood shrinkage and swelling, treatment with synthetic resin forming materials. *Industrial and Engineering Chemistry*, **28**(10), 1164–1169.

Stamm, A.J. and Seborg, R.M. (1939). Resin-treated plywood. *Industrial and Engineering Chemistry*, **31**(7), 897–902.

Stamm, A.J. and Seborg, R.M. (1941). Resin-treated, compressed wood. *Transactions of the American Institute of Chemical Engineers*, **37**, 385–397.

Stamm, A.J., and Tarkow, H. (1947). Dimensional stabilization of wood. *Journal of Physical and Colloid Chemistry*, **51**(2), 493–505.

Stamm, A.J., Burr, H.K. and Kline, A.A. (1946). Staybwood. Heat stabilized wood. *Industrial and Engineering Chemistry*, **38**(6), 630–634.

Stevens, M. (1981). Preliminary results of the treatment of wood with chlorosilanes. International Research Group on Wood Preservation, Doc. no. IRG/WP3172.

Stevens, M. (1985). Restrictions on the use of chlorosilanes as potential wood preservatives. International Research Group on Wood Preservation, Doc. no. IRG/WP3345.

Stevens, M. and Parameswaran, N. (1981a). Effect of formaldehyde-acid catalysed reactions on wood ultrastructure. *Wood Science and Technology*, **15**(4), 287–300.

Stevens, M. and Parameswaran, N. (1981b). Microscopical analysis of formaldehyde-acid modified wood. International Research Group on Wood Preservation, Doc. No. IRG/WP 3182.

Stevens, M. and Schalck, J. (1978). Combined effects of the treatment of wood with formaldehyde. International Research Group on Wood Preservation, Doc. No. IRG/WP 3117.

Stevens, M., Schalck, J. and Raemdonck, J.V. (1979). Chemical modification of wood by vapour-phase treatment with formaldehyde and sulfur dioxide. *International Journal of Wood Preservation*, **1**(2), 57–68.

Subiyanto, B., Yusuf, S., Kawai, S. and Imamura, Y. (1989). Particleboard from acetylated albizzia particles. I. The effect of acetyl weight gain on mechanical properties and dimensional stability. *Mokuzai Gakkaishi*, **35**(5), 412–418.

Sudo, K., Shimizu, K. and Sakurai, K. (1985). Characterization of steamed wood lignin from beech wood. *Holzforschung*, **39**(5), 281–288.

Sugiyama, M. and Norimoto, M. (1996). Temperature dependence of dynamic viscoelasticities of chemically treated woods. *Mokuzai Gakkaishi*, **42**(11), 1049–1056.

Sumi, Y., Hale, R.D., Meyer, J.A., Leopold, B. and Ranby, B.G. (1964). Accessibility of wood and wood carbohydrates measured with tritiated water. *Tappi*, **47**(10), 621–624.

Suttie, E.D. (1997). Novel wood preservatives. *Chemistry and Industry*, **18**, 720–724.

Suttie, E.D., Hill, C.A.S., Jones, D. and Orsler, R.J. (1997). Assessing the bioresistance conferred to solid wood by chemical modification. International Research Group on Wood Preservation, Doc. No. IRG/WP 97-40099.

Suttie, E.D., Hill, C.A.S., Jones, D. and Orsler, R.J. (1998). Chemically modified solid wood. I. Resistance to fungal attack. *Material und Organismen*, **32**(3), 159–182.

Suttie, E.D., Hill, C.A.S., Jones, D. and Orsler, R.J. (1999). Chemically modified solid wood. II. Resistance to *Hylotrupes bajulus* attack. *Material und Organismen*, **33**(2), 81–90.

Suzuki, M., Araki, M. and Goto, T. (1963a). Water sorption and dynamic viscoelastic properties of untreated, HCl-treated and HCHO-treated wood. *Mokuzai Gakkaishi*, **9**(1), 11–17.

Suzuki, M., Araki, M. and Goto, T. (1963b). Studies on dynamic behaviour of untreated, HCl-treated and HCHO-treated wood. *Mokuzai Gakkaishi*, **9**(2), 43–48.

Syrjänen, T. and Kangas, E. (2000). Heat treated timber in Finland. International Research Group on Wood Preservation, Doc. No. IRG/WP 00-40158.

Takahashi, M. (1981). Bending creep behaviour of plywoods under long term exposure to fungal attack. International Research Group on Wood Preservation, Doc. No. IRG/WP 81-2163.

Takahashi, M. (1993). Wood preserving techniques to prevent biodeterioration. In: *Recent Research on Wood and Wood-based Materials*, Shiraishi, N., Kajita, H. and Norimoto, M. (Eds.). Elsevier, Barking, UK, pp. 223–240.

Takahashi, M. (1996). Biological properties of chemically modified wood. In: *Chemical Modification of Lignocellulosic Materials*, Hon, D.N.S. (Ed.). Marcel Dekker, New York, USA, pp. 331–361.

Takahashi, M. and Imamura, Y. (1990). Biological resistance of phenol-resin treated wood. International Research Group on Wood Preservation, Doc. No. IRG/WP 3602.

Takahashi, M., Imamura, Y. and Tanahashi, M. (1989). Effect of acetylation on decay resistance of wood against brown rot, white rot and soft rot fungi. International Research Group on Wood Preservation, Doc. No. IRG/WP 3540.

Takino, S., Norimoto, M., Kawai, S. and Sasaki, H. (1989). Particleboard from acetylated albizzia particles II. Creep under humidity changes. *Mokuzai Gakkaishi*, **35**(7), 625–632.

Taniguchi, T., Harada, H. and Nakato, K. (1978). Determination of water adsorption sites in wood by a hydrogen–deuterium exchange. *Nature*, **272**, 230–231.

Tanno, F., Saka, S. and Takabe, K. (1997). Antimicrobial TMSAC-added wood–inorganic composites prepared by the sol-gel process. *Materials Science Research International*, **3**(3), 137–142.

Tanno, F., Saka, S. Yamamoto, A. and Takabe, K. (1998). Antimicrobial TMSAH-added wood–inorganic composites prepared by the sol-gel process. *Holzforschung*, **52**(4), 365–370.

Tarkow, H. and Stamm, A.J. (1953). Effect of formaldehyde treatments upon the dimensional stailization of wood. *Journal of the Forest Products Research Society*, **3**(2), 33–37.

Tarkow, H., Stamm, A.J. and Erickson, E.C.O. (1946). Acetylated wood. Report, Forest Products Laboratory, USDA Forest Service, **1593**.

Thiebaud, S. and Borredon, M.E. (1995). Solvent-free wood esterification with fatty acid chlorides. *Bioresource Technology*, **52**(2), 169–173.

Thiebaud, S., Borredon, M.E., Baziard, G. and Senocq, F. (1997). Properties of wood esterified by fatty acid chlorides. *Bioresource Technology*, **59**(2–3), 103–107.

Tiemann, H.D. (1915). The effect of different methods of drying on the strength of wood. *Lumber World Review*, **28**(7), 19–20.

Tillman, A.-M. (1988). Chemical modification of wood products with gaseous alkylene oxides. *Journal of Wood Chemistry and Technology*, **8**(2), 235–259.

Timar, M.C., Mihai, M.D. and Baciu, G. (1997a). About the water and biological resistance of some new chemically modified wood composites. International Research Group on Wood Preservation, Doc. No. IRG/WP 97-40077.

Timar, M.C., Pitman, A. and Mihai, M.D. (1998). The decay resistance of chemically modified aspen composites to the white rot fungus *Coriolus versicolor* (L.) Quelet. International Research Group on Wood Preservation, Doc. No. IRG/WP 98-40122.

Timar, M.C., Pitman, A. and Mihai, M.D. (1999a). Biological resistance of chemically modified aspen composites. *International Biodeterioration and Biodegradation*, **43**(4), 181–187.

Timar, M.C., Maher, K., Irle, M. and Mihai, M.D. (1999b). High performance wood-fibre/plastic composites by chemical modification. In: *Proceedings of the Third European Panel Products Symposium*, Hague, J., Griffiths, S., Loxton, C., McLauchlin, A., Mayhead, G. and Tomkinson, J. (Eds.), pp. 352–358.

Timar, M.C., Maher, K., Irle, M. and Mihai, M.D. (2000a). Preparation of wood with thermoplastic properties. Part 2. Simplified technologies. *Holzforschung*, **54**(1), 77–82.

Timar, M.C., Maher, K., Irle, M. and Mihai, M.D. (2004). Thermal forming of chemically modified wood to make high performance plastic-like wood composites. *Holzforschung*, **58**(5), 519–528.

Timar, M.C., Mihai, M.D., Irle, M. and Maher, K. (1997b). Chemically modified wood for injection moulding. *Proceedings of the First European Panel Products Symposium*, Hague, J., Loxton, C., Bolton, J. and Mott, L. (Eds.), pp. 156–164.

Timar, M.C., Mihai, M.D., Maher, K. and Irle, M. (2000b). Preparation of wood with thermoplastic properties. Part 1. Classical synthesis. *Holzforschung*, **54**(1), 71–76.

Timmons, T.K., Meyer, J.A. and Cote, W.A. (1971). Polymer location in the wood polymer composite. *Wood Science*, **4**(1), 13–24.

Tjeerdsma, B.F., Boonstra, M. and Militz, H. (1998a). Thermal modification of non-durable wood species 2. Improved wood properties of thermal treated wood. International Research Group on Wood Preservation, Doc. No. IRG/WP/98-40124.

Tjeerdsma, B.F., Stevens, M. and Militz, H. (2000). Durability aspects of (hydro)thermal treated wood. International Research Group on Wood Preservation, Doc. No. IRG/WP 00-40160.

Tjeerdsma, B.F., Stevens, M., Militz, H. and Van Acker, J. (2002). Effect of process conditions on moisture content and decay resistance of hydro-thermally treated wood. *Holforschung und Holzverwertung*, **54**(5), 94–96, 98–99.

Tjeerdsma, B.F., Boonstra, M., Pizzi, A., Tekely, P. and Militz, H. (1998b). Characterisation of thermally treated wood: molecular reasons for wood performance improvement. *Holz als Roh- und Werkstoff*, **56**(3), 149–153.

Torgovnikov, G. and Vinden, P. (2000). Microwave modification of wood properties: improvements in wood permeability. International Research Group on Wood Preservation, Doc. No. IRG/WP 00-40181.

Torr, K.M., Dawson, B.S.W., Ede, R.M. and Singh, J. (1996). Surface changes on acetylation and exposure to ultraviolet radiation of *Pinus radiata* using X-ray photoelectron spectroscopy. *Holzforschung*, **50**(5), 449–456.

Treu, A., Habicht, J., Klaucke, R. and Militz, H. (2003). Improvement of wood properties by a combined impregnation process—the Royal process. In: *Proceedings of the First European Conference on Wood Modification*, Ghent, Belgium, Van Acker, J. and Hill, C.A.S. (Eds.), pp. 3–14.

Troya, M.T. and Navarette, A. (1994). Study of the degradation of retified wood through ultrasonic and gravimetric techniques. International Research Group on Wood Preservation, Doc. No. IRG/WP 94-40030.

Tsujiyama, S. and Nakano, N. (1996). Distribution of acetyl esterase in wood-rotting fungi. *Mycoscience*, **37**(3), 289–294.

Turner, G.P.A. (1967). *Introduction to Paint Chemistry*. Chapman and Hall, London.

Ueda, M., Matsuda, H. and Matsumoto, Y. (1994). Chemical modification of wood by simultaneous oligoesterification and vinyl polymerisation under hot pressing. *Mokuzai Gakkaishi*, **40**(7), 725–732.

Ueyama, A., Araki, M. and Goto, T. (1961). Dimensional stability of wood. X. Decay resistance of formaldehyde treated wood. *Wood Research*, **26**, 67–73.

Van Acker, J. (2003). Assessing performance potential of modified wood focussing on dimensional stability and biological durability. In: *Proceedings of the First European Conference on Wood Modification*, Ghent, Belgium, Van Acker, J. and Hill, C.A.S. (Eds.), pp. 153–168.

Van Acker, J. and Jones, D. (2002). Test methods for modified wood: The EU-thematic network approach. International Research Group on Wood Preservation, Doc. No. IRG/WP 02-20255.

Van Acker, J. and Stevens, M. (2000). Increased biological durability differs for traditional wood preservation and new non-biocidal systems (NBS). International Research Group on Wood Preservation, Doc. No. IRG/WP 00-20212.

Van Acker, J., Nurmi, A., Gray, S., Militz, H., Hill, C., Kokko, H. and Rapp, A. (1999). Decay resistance of resin treated wood. International Research Group on Wood Preservation, Doc. No. IRG/WP 99-30206.

Vasishth, R.C. (1983). Importance of depositing polymers in wood-cell walls for wood treatment. *Proceedings of the American Wood Preservers Association*, **79**, 129–135.

Vick, C.B. and Rowell, R.M. (1990). Adhesive bonding of acetylated wood. *International Journal of Adhesion and Adhesives*, **10**(4), 263–272.

Vick, C.B., Krzysik, A. and Wood, J.E. (1991). Acetylated isocyanate-bonded flakeboards after accelerated aging. Dimensional stability and mechanical properties. *Holz als Roh- und Werkstoff*, **49**(6), 221–228.

Videlov, C.L. (1989). Biological degradation resistance of pine wood treated with dimethylol compounds. International Research Group on Wood Preservation, Doc. No. IRG/WP 3528.

Vihavainen, T., Piispanen, K. and Mansikkmäki, P. (1980). Treatment of wood with formaldehyde. Acid catalysis of the reaction between formaldehyde and wood. International Research Group on Wood Preservation, Doc. No. IRG/WP 3146.

Viitanen, H. and Paajanen, L. (1988). The critical moisture and temperature conditions for the growth of some mould fungi and the brown rot fungus *Coniophora puteana* on wood. International Research Group on Wood Preservation, Doc. No. IRG/WP 1369.

Viitanen, H., Jämsä, S., Paajanen, L., Nurmi, A. and Viitaniemi, P. (1994). The effect of heat treatment on the properties of spruce. A preliminary report. International Research Group on Wood Preservation, Doc. No. IRG/WP 94-40032.

Wakeling, A.N., Plackett, D.V. and Cronshaw, D.R. (1991). The susceptibility of acetylated *Pinus radiata* to mould and stain fungi. In: *Proceedings of the International Symposium on Chemical Modification of Wood*, Kyoto, Japan, Takahashi, M. (Ed.), pp. 142–147.

Wakeling, A.N., Plackett, D.V. and Cronshaw, D.R. (1992). The susceptibility of acetylated *Pinus radiata* to mould and stain fungi. International Research Group on Wood Preservation, Doc. No. IRG/WP 1548-92.

Wakita, H., Onishi, H., Jodai, S. and Goto, T. (1977). Studies on the improvement of wood materials. XVIII. Vapor phase reaction of 2,4-tolylene diisocyanate in wood. *Zairo*, **26**(284), 460–464.

Walker, J.C. (1994). *Primary Wood Processing: Principles and Practice*. Chapman and Hall, USA.

Weaver, J.W., Nielson, J.F. and Goldstein, I.S. (1960). Dimensional stabilization of wood with aldehydes and related compounds. *Forest Products Journal*, **10**(6), 306–310.

Weiland, J.J. and Guyonnet, R. (2003). Study of chemical modifications and fungi degradation of thermally modified wood using DRIFT spectroscopy. *Holz als Roh- and Werkstoff*, **61**(3), 216–220.

Wellons, J.D. (1977). Adhesion to wood substrates. In: *Wood Technology: Chemical Aspects*, Goldstein, I.S. (Ed.). ACS Symposium Series, **43**, pp. 150–168.

Welzbacher, C.R. and Rapp, A.O. (2002). Comparison of thermally treated wood originating from four industrial scale processes-durability. International Research Group on Wood Preservation, Doc. No. IRG/WP 02-40229.

West, H. and Banks W.B. (1986). Topochemistry of the wood isocyanate reaction. An analysis of reaction profiles. *Journal of Wood Chemistry and Technology*, **6**(3), 411–425.

West, H. and Banks, W.B. (1987). Preparation of butyl carbamates of wood. In: *Wood and Cellulosics, Industrial Utilisation, Biotechnology, Structure and Properties*, Kennedy, J.F., Phillips, G.O. and Williams, P.A. (Eds.). Ellis Horwood, Chichester, UK, pp. 135–142.

Westin, M., Lande, S. and Schneider, M. (2003). Furfurylation of wood—process, properties and commercial production. In: *Proceedings of the First European Conference on Wood Modification*, Ghent, Belgium, Van Acker, J. and Hill, C.A.S. (Eds.), pp. 289–306.

Westin, M., Lande, S. and Schneider, M. (2004a). Wood furfurylation process and properties of furfurylated wood. International Research Group on Wood Preservation, Doc. No. IRG/WP 04-40289.

Westin, M., Rapp, A.O. and Nilsson, T. (2004b). Durability of pine modified by 9 different methods. International Research Group on Wood Preservation, Doc. No. IRG/WP 04-40288.

Widsten, P., Qvintus-Leino, P., Tuominen, S. and Laine, J.L. (2003). Manufacture of fibreboard from wood fibers activated with Fenton's reagent ($H_2O_2/FeSO_4$). *Holzforschung*, **57**(4), 447–452.

Williams, R.S. (1983). Effect of grafted UV stabilizers on wood surface erosion and clear coating performance. *Journal of Applied Polymer Science*, **28**(6), 2093–2103.

Wilson, K. and White, D.J.B. (1986). *The Anatomy of Wood: its Diversity and Variability*. Stobart and Son, London, UK.

Yamaguchi, H. (1994a). Properties of silicic acid as chemical agents for impregnation and fixation of wood. *Mokuzai Gakkaishi*, **40**(8), 830–837.

Yamaguchi, H. (1994b). Preparation and physical properties of wood fixed with silicic acid compounds. *Mokuzai Gakkaishi*, **40**(8), 838–845.

Yamaguchi, H. (2002). Low molecular weight silicic acid – inorganic compound complex as wood preservative. *Wood Science and Technology*, **36**(5), 399–417.

Yamawaki, Y., Morita, M. and Sakata, I. (1990). Mechanical and dielectric properties of cyanoethylated wood. *Journal of Applied Polymer Science*, **40**(9–10), 1757–1769.

Yano, H. and Minato, K. (1992). Improvement of the acoustic and hygroscopic properties of wood by a chemical treatment and application to the violin parts. *Journal of the Acoustical Society of America*, **92**(3), 1222–1227.

Yano, H. and Minato, K. (1993). Controlling the timbre of wooden musical instruments by chemical modification. *Wood Science and Technology*, **27**(4), 287–293.

Yano, H. Mukudai, J. and Norimoto, M. (1988). Improvements in the piano pin block. *Mokuzai Gakkaishi*, **34**(2), 94–99.

Yano, H., Norimoto, M. and Rowell, R.M. (1993). Stabilization of acoustical properties of wooden musical instruments by acetylation. *Wood and Fiber Science*, **25**(4), 395–403.

Yano, H., Norimoto, M. and Yamada, T. (1986a). Changes in acoustical properties of Sitka spruce due to acetylation. *Mokuzai Gakkaishi*, **32**(12), 990–995.

Yano, H., Yamada, T. and Minato, K. (1986b). Changes in acoustical properties of Sitka spruce due to reaction with formaldehyde. *Mokuzai Gakkaishi*, **32**(12), 984–989.

Yasuda, R. and Minato, K. (1994). Chemical modification of wood by non-formaldehyde cross-linking agents. Part 1. Improvement of dimensional stability and acoustic properties. *Wood Science and Technology*, **28**(2), 101–110.

Yasuda, R. and Minato, K. (1995). Chemical modification of wood by non-formaldehyde cross-linking agents. Part 3. Mechanism of dimensional stabilization by glyoxal treatment and effect of the addition of glycol. *Wood Science and Technology*, **29**(4), 243–251.

Yasuda, R., Minato, K. and Norimoto, M. (1994). Chemical modification of wood by non-formaldehyde cross-linking agents. Part 2. Moisture sorption and creep properties. *Wood Science and Technology*, **28**(3), 209–218.

Yasuda, R., Minato, K. and Norimoto, M. (1995). Moisture sorption thermodynamics of chemically modified wood. *Holzforschung*, **49**(6), 548–554.

Yasuda, R., Minato, K. and Yano, H. (1993). Use of trioxane for improvement of hygroscopic and acoustic properties of wood for musical instruments. *Wood Science and Technology*, **27**(2), 151–160.

Yelle, D., Goodell, B., Gardner, D.J., Amirbahman, A., Winistorfer, P. and Shaler, S. (2004). Bonding of wood fiber composites using a synthetic chelator–lignin activation system. *Forest Products Journal*, **54**(4), 73–78.

Yoshida, Y., Kawai, S., Imamura, Y., Nishimoto, K. Sato, T. and Nakaji, M. (1986). Production technology for acetylated low-density particleboard. I. Mechanical properties and dimensional stability. *Mokuzai Gakkaishi*, **32**(12), 965–971.

Young, R.A. (1977). Modification of high-yield pulps by the xanthate method. *Journal of Agricultural and Food Chemistry*, **25**(4), 738–742.

Youngquist, J.A. and Rowell, R.M. (1990). Adhesive bonding of acetylated aspen flakes. Part 3. Adhesion with isocyanates. *International Journal of Adhesion and Adhesives*, **10**(4), 273–276.

Youngquist, J.A., Krzysik, A. and Rowell, R.M. (1986a). Dimensional stability of acetylated aspen flakeboard. *Wood and Fiber Science*, **18**(1), 90–98.

Youngquist, J.A., Rowell, R.M. and Krzysik, A. (1986b). Mechanical properties and dimensional stability of acetylated aspen flakeboard. *Holz als Roh- und Werkstoff*, **44**(12), 453–457.

Youngquist, J.A., Sachs, I.B. and Rowell, R.M. (1988). Adhesive bonding of acetylated aspen flakes. Part 2. Effects of emulsifiers on phenolic resin bonding. *International Journal of Adhesion and Adhesives*, **8**(4), 197–200.

Yun, K.-E., Kim, G.-H. and Kim, J.-J. (1999). Effect of heat treatment on the dimensional stability and bending properties of radiata pine sapwood. *Journal of Korean Wood Science and Technology*, **27**(4), 30–37.

Yusuf, S., Takahashi, M. and Imamura, Y. (1989). Particleboard from acetylated albizzia particles. III. Enhancement of decay resistance, termite resistance, and weathering properties through acetylation. *Mokuzai Gakkaishi*, **35**(7), 633–639.

Yusuf, S., Imamura, Y., Takahashi, M. and Minato, K. (1994). Biological resistance of aldehyde-treated wood. (1994). International Research Group on Wood Preservation, Doc. No. IRG/WP 94-40018.

Yusuf, S., Imamura, Y., Takahashi, M. and Minato, K. (1995a). Biological resistance of wood chemically modified with non-formaldehyde cross-linking agents. *Mokuzai Gakkaishi*, **41**(2), 163–169.

Yusuf, S., Imamura, Y., Takahashi, M. and Minato, K. (1995b). Physical and biological properties of albizzia waferboard modified with cross-linking agents. International Research Group on Wood Preservation, Doc. No. IRG/WP 95-40043.

Yusuf, S., Imamura, Y., Takahashi, M. and Minato, K. (1995c). Property enhancement of albizzia waferboard by formaldehyde treatment. *Mokuzai Gakkaishi*, **41**(2), 223–228.

Yusuf, S., Imamura, Y., Takahashi, M. and Minato, K. (1995d). Property enhancement of wood chemically modified with non-formaldehyde cross-linking agents. *Mokuzai Gakkaishi*, **41**(2), 163–169.

Zabel, R.A. and Morrell, J.J. (1992). *Wood Microbiology: Decay and its Prevention*. Academic Press, San Diego, USA.

Zaman, A., Alén, R. and Kotilainen, R. (2000) Thermal behaviour of Scots pine (*Pinus sylvestris*) and silver birch (*Betula pubescens*) at 200–230 deg C. *Wood and Fiber Science*, **32**(2), 138–143.

Zee, M.E. van der, Beckers, E.P.J. and Militz, H. (1998a). Influence of concentration, catalyst, and temperature on dimensional stability of DMDHEU modified Scots pine. International Research Group on Wood Preservation, Doc. No. IRG/WP 98-40119.

Zee, M.E. van der, Stoutjesdijk, J.H., Feil, H. and Feijen, J. (1998b). Relevance of aquatic biodegradation tests for predicting degradation of polymeric materials during biological waste treatment. *Chemosphere*, **36**(3), 461–473.

Zhao, G., Nishino, Y., Nakao, T., Tanaka, C. and Takahashi, A. (1994). Dielectric relaxation of water adsorbed on acetylated wood. *Mokuzai Gakkaishi*, **40**(6), 571–576.

Zhao, G., Norimoto, M., Tanaka, F., Yamada, T. and Rowell, R.M. (1987). Structure and properties of acetylated wood. I. Changes in the degree of crystallinity and dielectric properties by acetylation. *Mokuzai Gakkaishi*, **33**(2), 136–142.

Zollfrank, C. (2001). Silylation of solid beech wood. *Wood Science and Technology*, **35**(1–2), 183–189.

Zollfrank, C. and Wegener, G. (2002). FTIR microscopy and ultrastructural investigation of silylated solid wood. *Holzforschung*, **56**(1), 39–42.

Index

Note: Page numbers given in italics refer to figures and numbers in bold refer to tables.

Wood Modification: Chemical, Thermal and Other Processes C. Hill

Printed and bound by CPI Group (UK) Ltd, Croydon, CR0 4YY
16/09/2026
14963387-0001